The Institution of Civil Engineers

Pumped storage

Proceedings of the conference organized by the Institution of Civil Engineers at Imperial College of Science, Technology and Medicine, London on 2-4 April 1990

Thomas Telford, London

Conference organized by the Institution of Civil Engineers
Organizing Committee: T. H. Douglas (Chairman),
F. J. L. Bindon, A. G. Cook, B. Hadley, F. G. Johnson,
C. Strongman, R. J. S. Ward, W. S. Williams

British Library Cataloguing in Publication Data
Pumped Storage.
1. Electricity supply
621.31
ISBN 0-7277-1586-0

Published for the Institution of Civil Engineers by Thomas Telford Ltd, Telford House, 1 Heron Quay, London E14 9XF.

Printed in Great Britain by Mackays

Contents

Linings workshop

Seals and bearings workshop

Keynote address

D. G. JEFFERIES, National Grid Company

The generation of electricity using water dates back many years, but the concept of pumped storage is a child of the twentieth century.

The oldest pumped storage plant on record, according to International Water Power and Dam construction, went into operation at Schaffhausen, in Switzerland, in 1909 and is still in operation. The oldest in the UK was at Walkerburn, in Scotland, and was commissioned in 1920.

There are over 300 pumped storage schemes around the world, either in operation or under construction, an indication that pumped storage is of continuing interest to electricity utilities.

The biggest investment in pumped storage is in the USA which uses more pumped storage than any other country in the world. Figures for 1988 show that total production was 22 900 GW hours.

Smaller pumped storage plants can still do a useful job, however. One of the smallest stations of this type exists in West Germany with a capacity of around 0.6 MW.

The majority of schemes exist in the highly industrialised nations. Japan has more than 40 schemes with a total output of more than 16 GW.

More than 50 GW of pumped storage capacity is planned in the near future. The countries developing the most plant are Japan, USA and USSR. Japan has identified over 440 potential schemes, which represent a possible capacity of 329 GW. USA and USSR have firm plans for 17 GW and 11 GW respectively.

Pumped storage is not only an established and proven technology, but, as contributions to this volume demonstrate, it is one that is developing. The electricity business is always looking for new ways to store electricity efficiently, economically and safely.

The Electric Power Research Institute (EPRI) in the USA, is funding a large research and development programme on energy storage, looking at several options, including the use of compressed air energy and batteries.

The most advanced of these technologies is compressed air. In compressed air energy storage, off-peak electricity is used to pump air into an underground cavern which may be either natural or excavated from a rock or salt formation. When electricity is needed the air is withdrawn, heated with gas or oil and run through expansion turbines to general electricity.

In Europe there is a pioneering 290 MW compressed air energy storage plant at Huntorf in West Germany. In 1991 the first American plant, constructed in a 500,000 m^3 cavern mined from a salt dome will begin operation. It will be run by the Alabama Electric Co-operative and will be capable of generating 110 MW for up to 26 hours.

Large cave plants are also being planned in the Soviet Union and Israel. According to EPRI, the total cost of air storage, including engineering, could be about half that of a pumped hydroelectric scheme, and much faster to build.

In the UK, these developments are being monitored closely. The policy of providing system reserve capacity on the British power system evolved from the electrical isolation of this system from the rest of Europe.

The new Pumped Storage Business, part of the National Grid Company, will be the fourth largest generator in England and Wales, with a staff of about 180 people.

It will have to compete in a new environment in the UK electricity industry where electricity will be bought and sold in a pool, with a market for both electricity supply and reserve provision.

As well as its daily, if not minute by minute, function of helping to stabilise the electricity supply system in England and Wales, the pumped storage business intends to play an international role where possible in this competitive energy market.

Ffestiniog and Dinorwig

Ffestiniog was completed after six years construction in 1963.

The station is capable of reaching its maximum output of 360 MW within one minute and can be operated completely under remote automatic control if necessary.

The experience of running Ffestiniog paved the way for the design and construction of Dinorwig, one of the great engineering feats of the twentieth century.

When constructed, Dinorwig was the largest hydro-electric pumped storage scheme of its type in Western Europe. It was also the largest civil engineering contract ever announced in the United Kingdom, with 16 km of tunnels excavated into the heart of the Elidir mountain.

Indeed, the 1800 MW power station at Dinorwig is one of the largest in the world and is capable, at its fastest rate, of supplying up to 1320 MW in around twelve seconds.

It has been designed for a daily pump generation cycle with a target efficiency of 78 per cent.

To achieve refill of the upper reservoir from empty takes about six hours of pumping at full load.

The building, construction and equipping of Dinorwig involved co-operation on an international scale. More than six years after its completion, the Dinorwig station frequently plays host to visitors from overseas utilities who are contemplating their own pumped storage schemes.

By careful planning at the outset and continued monitoring when building a pumped storage plant, the environment can not only be preserved but enhanced.

Both Ffestiniog and Dinorwig's massive generating equipment, and the associated 400 KV substation, were sited inside a mountain of slate. The 11 km of cable needed to connect it to the main national grid system also runs underground.

Altogether, some two million cubic meters of rock were excavated and 16 km of tunnelling carried out. It took the equivalent of 6 000 man years to drill and blast them.

Dinorwig was built on the site of an old slate quarry and the results of more than 200 years of taking slate from the hillside had caused immense environmental effects. The quarrymen had blasted into the mountain, leaving holes 600 feet in depth. During the construction of the plant, the lower lake was restored to its former size, some of the larger holes were filled in and the quarry terraces made safe.

About nine million tonnes of slate were disposed of within the old quarry workings.

After construction, great attention was paid to restoring the natural ground cover. Surveys were made of the various types of top soils and ground cover vegetation. Heather seeds were collected from the hillside and taken to the University of Wales' laboratories at Bangor for propagation. Later these were planted along the upper face of the dam.

Today it is hard to distinguish Dinorwig from its environment. Feral goats wander freely along the terraces of the Elidir mountain which contains Dinorwig and Peregrine falcons nest on the mountain.

Water quality has been maintained and long after construction, the electricity industry continues to carry out regular monitoring of the fish population and to test for water purity.

Pumped storage in the UK has shown itself to be an economic and efficient way of providing electricity supply and system stability and it will continue to do so well into the next century.

Summit hydroelectric pumped storage project

D. C. WILLETT, BSc(Eng), PEng, Acres International Corporation

SYNOPSIS. The Summit Pumped Storage Project will make use of an existing limestone mine at a depth of 670 m below ground level to provide 15000 MWh of energy storage (1500 MW x 10 hr). An environmental assessment has been completed and an application for a license to construct the project filed with the U.S. Federal Energy Regulatory Commission.

INTRODUCTION

1. The Summit hydroelectric pumped storage project which will be constructed near Akron, Ohio is believed to be the first project of its type to utilize a fully underground lower reservoir and associated power facilities. This, coupled with the relatively high head (670 m/2200 ft) to be developed, provides this project with significant benefits from both the environmental and cost standpoints when compared with more conventional surface-located pumped storage facilities. This paper briefly describes the development of the project to this point in time, and outlines the technical and environmental studies leading to the submission of an application for a construction license which was accepted for filing by the Federal Energy Regulatory Authority in April 1989.

BACKGROUND

2. In the late 1960s and early 1970s, a number of papers (Refs. 1-3) were published describing the potential benefits that might be obtained by locating the lower reservoir of a hydroelectric pumped storage facility underground. These proposals were largely in response to the increasing environmental pressures to which conventional surface-located pumped storage projects were being subjected, particularly in the U.S.; the Cornwall project of Consolidated Edison, the Blue Ridge project of American Electric Power and the Davis Mountain project of Allegheny Power, to name but a few, were abandoned in the face of intense environmental lobbying concerned at the potential effects of these projects on fish populations, wetlands, aesthetics and other matters.

3. By locating the lower reservoir underground in an excavated cavern, the underground pumped storage (UPS) concept offered a number of benefits which included:

Pumped storage. Thomas Telford, London, 1990.

- Increased freedom in the selection of sites: no longer tied to topography, the plant could be sited where the power system and other considerations such as proximity to load center and transmission facilities could best be satisfied.

- Only one reservoir to impact on the environment: if the lower reservoir could be located at sufficient depth, the size of the reservoirs for a given energy storage could be kept small with a further reduction of impact on the environment.

- Removal of the need to construct the project in an area of significant topographic relief reduces potential concerns about aesthetics both of the project itself and the associated transmission lines.

4. A number of quite detailed studies of the UPH concept were commissioned in the U.S. in the late 1970s, culminating in the EPRI/DOE study (Ref. 4) which looked at UPH in parallel with a number of compressed air energy storage concepts. The general conclusion of these studies was, however, that if UPH was to be economically competitive with other forms of energy storage or peak energy generation, a minimum capacity in excess of 1000 to 1500 MW was required and a total operating head greater than 1200 m would be needed in order to keep to a minimum the size and hence the cost of excavation of the lower reservoir.

5. These conclusions, arriving at the same time that U.S. power systems were experiencing minimal or even negative load growth, coupled with considerable surplus capacity, effectively sounded the depth knell of UPH at that time. However, the resurgence of the U.S. economy in the 1980s, accompanied by a significant increase in load growth which, under even relatively conservative assumptions, pointed to a potential shortfall of capacity in many power systems in the mid- to late-1990s, reactivated interest in energy storage both to meet peak loads and to provide assistance in system regulation (the "dynamic benefits").

6. It is within this context that the Summit pumped storage project has been initiated. Rather than excavating a completely new cavern for the lower reservoir, however, the project concept is based on the utilization of an existing cavern, previously excavated for the production of limestone for use in cement manufacture and various other chemical industry processes.

GENERAL PROJECT ARRANGEMENT

7. The proposed general arrangement of the project is illustrated in Figures 1 and 2. The mine which will form the lower reservoir is located at a depth of approximately 670 m below ground level in a thick stratum of massive limestone. It is

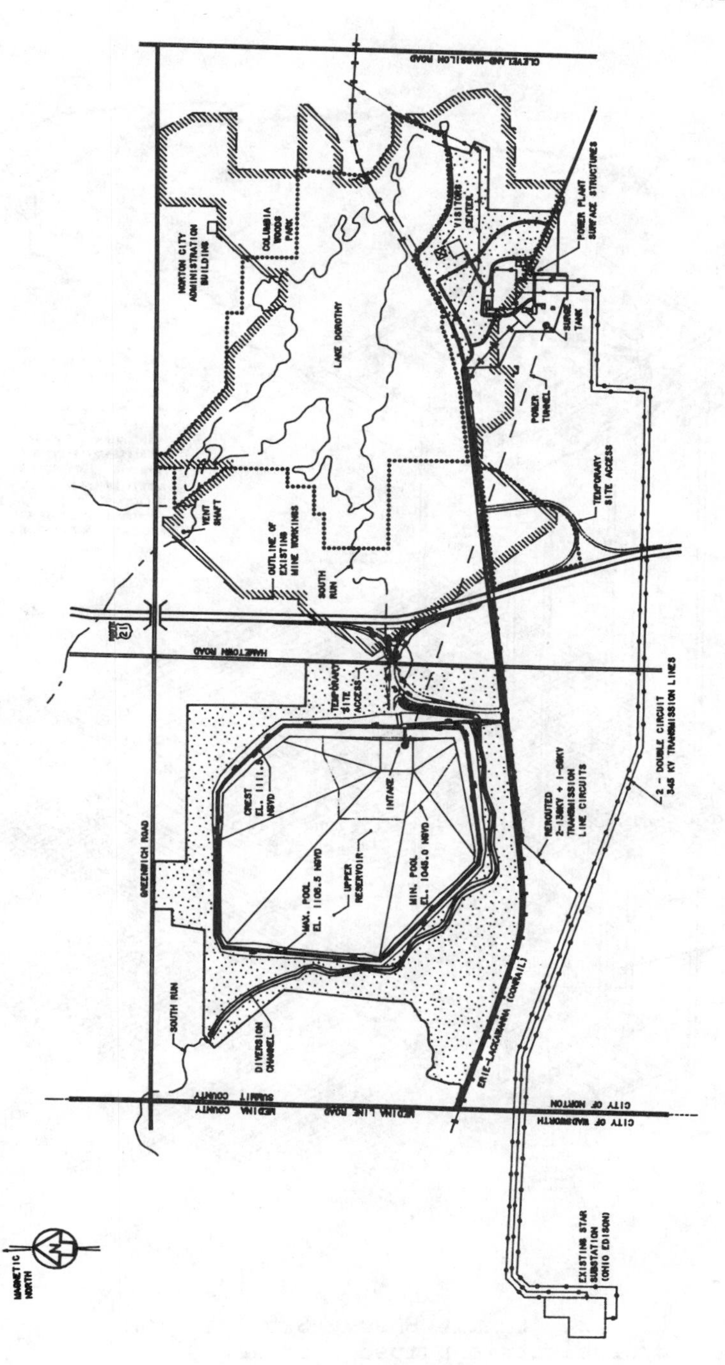

Fig. 1. Proposed plan

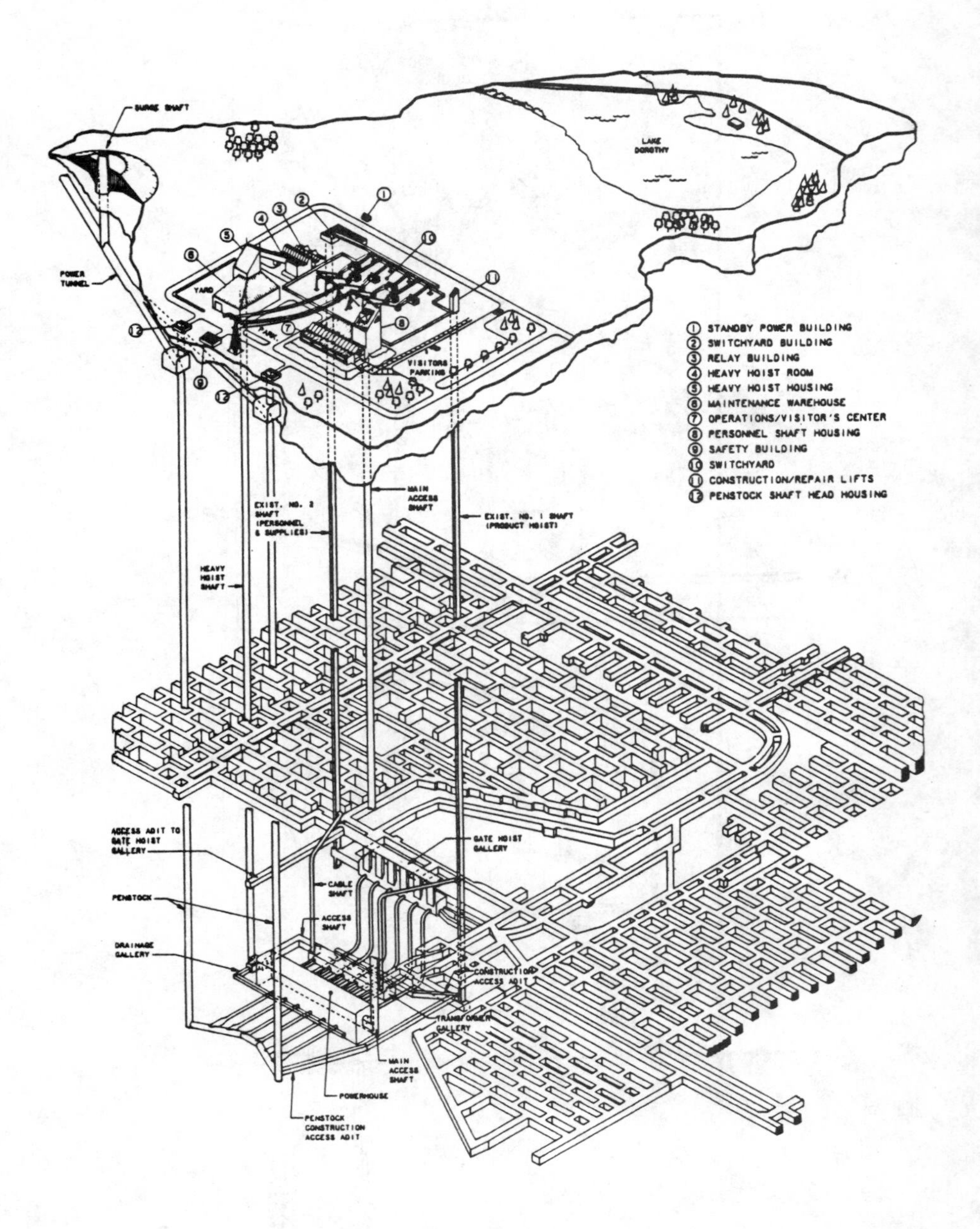

Fig. 2. Summit Energy Storage, Inc.
Hydroelectric pumped storage

currently accessed by two vertical concrete-lined shafts located about 170 m apart at the southeast side of the mine.

8. The power plant, which will have six 250-MW reversible pump-turbine, motor-generator units will be located adjacent to the existing shafts at a depth of approximately 75 m below the mine in order to provide the requisite submergence on the pump-turbines. Initial plans call for permanent access to the power plant to be provided by two vertical 7.6-m diameter shafts; one shaft would be equipped with a personnel and small materials elevator, the other with intermediate (20-ton) and heavy (150-ton) hoists. Alternative access by means of an inclined road tunnel is also being considered.

9. The upper reservoir will be located approximately 2 km west of the mine in an area currently primarily agricultural land. The reservoir will occupy an area of about 80 ha to provide a total volume of 9.2 million cubic metres, approximately equal to that of the lower reservoir. The small stream (South Run) which currently crosses the upper reservoir site will be diverted around the impoundment structure so that the reservoir will be hydraulically isolated from the environment. To minimize seepage losses and to further enhance the isolation of the upper reservoir, the impoundment will be asphalt-lined.

10. Interconnection between the upper reservoir and the power plant will be provided by means of a 8.5-m diameter concrete-lined power tunnel and two 5.3-m diameter vertical concrete-lined pressure shafts.

11. Current plans call for the power generated to be transformed underground from generator voltage (16 kV) to transmission line voltage (345 kV) and transmitted to the surface in high-voltage cable or bus through one or two vertical concrete-lined shafts. 345-kV breakers will be located in a small switchyard on the surface and the power transmitted by 345-kV overhead line to Ohio Edison's 'Star' substation, a distance of about 5 km.

GEOLOGY

12. The geology of the area in which the plant will be located is characterized by a flat-lying sequence of sedimentary strata of the Appalachian Basin. A generalized stratigraphic column is shown in Figure 3. The mine is located in the upper levels of the Columbus limestone at a depth of approximately 670 m below ground level. The limestone, which has a total thickness of about 76 m overlies strata of the Salina series. Overlying the limestone is a roughly 550 m thickness of shales thinly interbedded with some sandstones and limestones. Over most of the area of the site the shales are in turn overlaid by a stratum of the Sharon Conglomerate which provides a source of drinking water for the many household wells which ring the site.

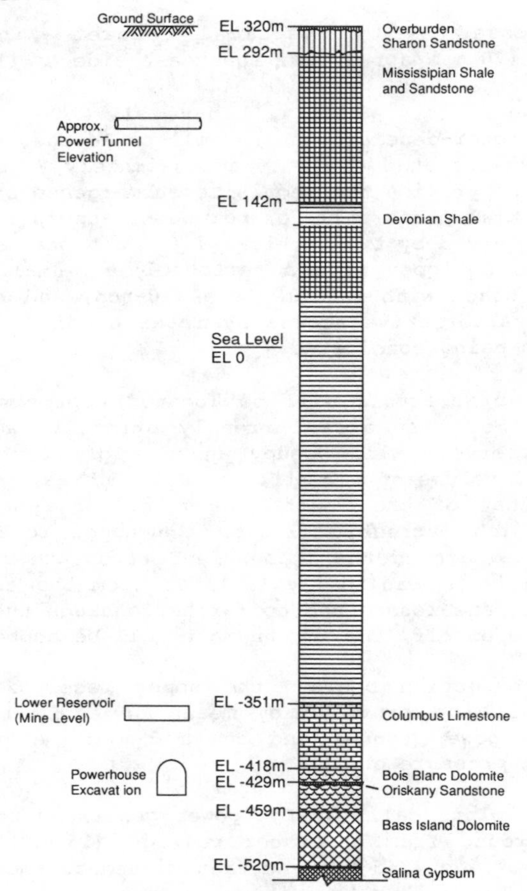

Fig. 3. Generalized stratigraphic section

13. Overburden generally ranges from 2- to 7-m in thickness, and is comprised of a sandy-silty till with occasional gravel lenses. The area in which the upper reservoir is to be located is crossed by an infilled bedrock valley to a depth of more than 50 m below the current ground level.

THE MINE

14. The Norton (originally Barberton) mine was developed by PPG Industries, Inc. for the production of high quality limestone for use in glass making, cement and other chemical processes. Development of the mine commenced in 1940 with the sinking of two 2.5-m by 5-m concrete-lined shafts to a depth of approximately 670 m below ground level in a stratum of the "Big Lime" series of the Columbus Limestone.

15. Excavation of the limestone continued until 1976, at which time, production was halted and the mine placed on "active standby" status. Approximately 9 million cubic metres of limestone was removed from the mine over this period using a room and pillar technique as shown in Figure 4. After some

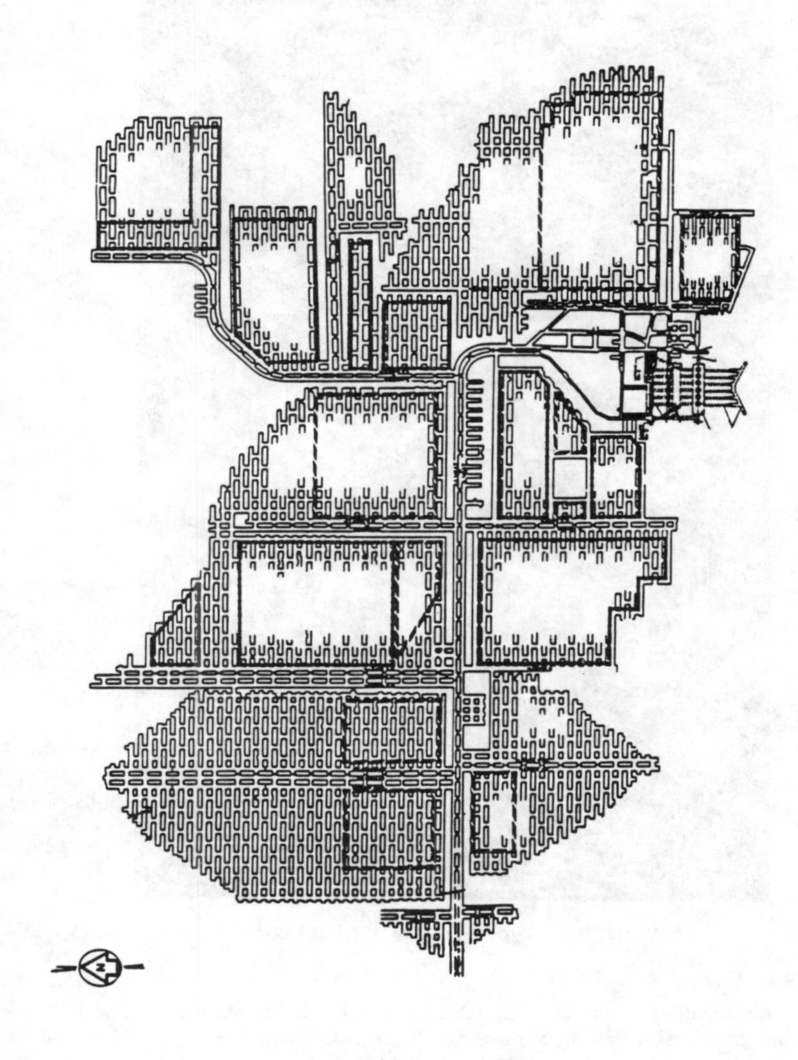

Fig. 4. Mine plan

initial experimentation, the room and pillar configuration was established on 23 m centers, with both room and entry widths set at 9.7 m, and an initial room height of 5.2 m. After full room depth had been established, the roofs were raised by stoping to partings at 8.5 m and 14 m, as shown in Figure 5.

Fig. 5. Typical 14-m high room

16. Although mining operations were discontinued in 1976, the mine has been maintained in "active standby" so that full access, and at least partial ventilation is still available, although most of the in-mine production equipment (crushers, conveyors, etc.) has been removed. A most remarkable aspect of the mine is that it is, and always has been, completely dry,

despite its size and depth. Although the mine has been essentially unattended since 1976, there is no evidence of any instability having developed over this period. The mine has required virtually no support measures throughout its life; the only areas in which support generally in the form of some pattern rock bolting has been applied are where roof spans have been increased to provide space for workshops and storage areas.

PROJECT PLANNING

17. In considering the use of the Norton Mine as the lower reservoir for a hydroelectric pumped storage facility, a number of unique opportunities and potential concerns have had to be addressed. These can be categorized generally in the technical, environmental and cost/financial areas.

Technical

18. From the technical aspects, dealing with the essential practical state-of-the-art feasibility of the project, there are, with perhaps the exception of the lower reservoir and the access to the power plant, no components of the project which are not well provided with adequate precedent. Although at the top end of current experience in reversible pump-turbine units, the proposed 250-MW reversible single stage units will be essentially conventional, having a rotational speed of 600 rpm.

19. For the lower reservoir, the key concern is, of course, its long-term stability under the repeated filling and emptying associated with the pumped storage cycle. Initial testing has shown the limestone to be only very slightly soluble in the water which will be used to fill the reservoir system. Bearing in mind that the project will operate in a closed cycle mode (no discharge to the external environment), it is anticipated that the water in the system will very rapidly become saturated with carbonates so that even minor solutioning will stop. Of more concern, perhaps, are the thin (up to 5-cm thick) shale partings which extend horizontally at between 0.7 m and 2 m vertical intervals throughout the mine. However, preliminary stability analyses based on Bieniawski's criteria (Ref. 5) indicate that even under conservative assumptions the long-term stability would appear to be assured. The considerable lateral extent (in excess of 100 km) of the limestone stratum and the thick overlying strata of shales will essentially guarantee the water-tightness of the mine. A detailed program of geotechnical investigation and stability analysis is currently underway to support the initial conclusions.

20. Initial studies have recommended that access to the power plant, which is located some 746 m below ground level will be provided through two vertical shafts, each approximately 7.6 m in finished diameter. One of these shafts will be equipped with a heavy hoist capable of handling loads of up to

150 tons, the other will contain a 120-man double-deck personnel elevator and various services including water, ventilation and electrical power. The heavy hoist will be equipped to operate at a higher speed for loads up to 20 tons.

21. The possibility of providing access to the power plant by means of an inclined spiral road tunnel rather than by the shafts has also been examined. It was found that although the unit excavation costs for the tunnel were lower than for the shafts, the significantly increased volume of material to be removed and the potentially longer construction schedule, inclined the selection toward the shaft alternative.

Environmental

22. As discussed earlier in this paper, one of the advantages that the early proponents of the underground pumped storage concept had postulated was its reduced environmental impact when compared to conventional surface-located projects, and development of this project has certainly borne this out. The only major structure located at ground level will be the upper reservoir, and because of the high head at which the plant will operate, it has been possible to keep the overall surface areas of this at full pool to no more than 80 ha whilst providing a full 10 hours of operation at 1500 MW of operation.

23. In selecting the location of the upper reservoir relative to the existing mine, a total of six alternative possible sites was evaluated. From the purely economic standpoint, the most attractive option appeared to be to use Lake Dorothy, the existing man-made lake located immediately above the mine. However, more detailed examination revealed a number of significant drawbacks to this selection; principal among those were the fact that the lake currently constitutes a significant recreational resource within the community and that use for pumped storage would completely destroy this. Added to this, the area and depth of the lake would have to be considerably increased to provide the required operating volume, with resulting destruction of surrounding areas of valuable wetland. When the cost of the necessary mitigation measures was factored into the evaluation process, it was concluded that a site located approximately 2 km to the west of the mine would provide a more satisfactory answer, particularly when the smaller catchment area of the more western site was taken into account, as discussed below.

24. The site selected for the upper reservoir is located in a mixed woodland/farmland environment, and is ringed on three of the four sides by single-family residential developments. The primary source of drinking water for the residences is wells drilled into the Sharon conglomerate stratum (Paragraph 12 above). The reservoir area is currently crossed by a small stream (South Run) with a drainage area of about 15 km^2. In

order to isolate the reservoir system from the external environment to the maximum extent possible, this stream will be diverted around the outside of the reservoir embankment structure so that there will be no natural streamflow discharge either into or out of the project reservoir system. In this way, potential environmental problems associated with point discharges from the project are avoided. At the same time, the need to provide a spillway structure is also avoided.

25. The stream diversion channel will be configured as much as possible to replicate the natural channel, with meanders, pool and riffle lengths, and shade trees to prevent heat build-up. A typical channel cross-section is shown in Figure 6. As construction of the upper reservoir will destroy approximately 4.5 ha of wetlands, provision has been made in the layout of the area around the reservoir to replace the same area by the provision of a number of weirs to form small impoundments. The balance of the area will be developed for recreational purposes, including recreational trails and picnic areas.

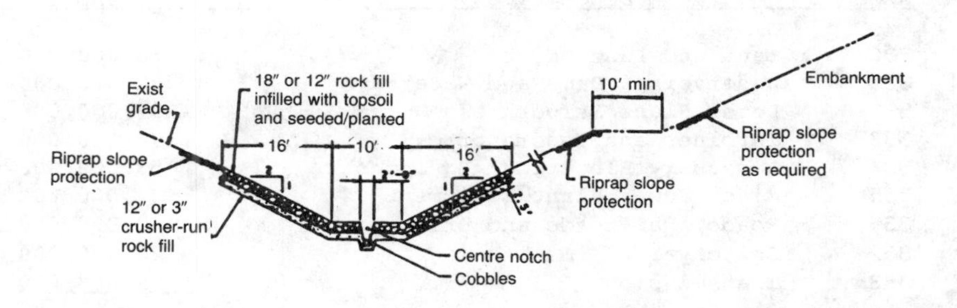

Fig. 6. Typical channel cross-section

26. A number of alternative possible sources of water to fill the reservoir system were considered; the principal source selected is from the public water supply system of the nearby city of Barberton. The total volume required is approximately 10 million cubic metres, with an annual make-up estimated amount to approximately 1.4 million cubic metres.

27. The relatively small volume of water contained in the reservoir system, combined with the proposed 1500 megawatts of installed capacity, and the high (approximately 30°C) rock temperatures of the lower reservoir, give rise to some concern as to possible fogging occurring at the upper reservoir under adverse meteorological conditions. Studies were made using a computer simulation of the temperature regime in the complete reservoir system to determine potential effects under a range of operating and meteorological regimes. These studies showed that despite concerns to the contrary, the principal parameter

affecting the upper reservoir temperature was the ambient meteorological conditions, and that although the water temperature would average perhaps two to three degrees above that anticipated under ambient conditions, the extent of additional fogging would be no more than two or three percent under the most conservative assumptions.

28. Because of the relatively short length of tunnels and shafts to be excavated, the quantity of spoil to be disposed of is small, with the majority being incorporated into the upper reservoir and roadway structures, leaving small quantities for on-site disposal.

ESTIMATED CONSTRUCTION COST AND SCHEDULE

29. The estimated construction cost for the project in mid-1988 dollars is summarized in Table 1, indicating an anticipated cost per kW installed of $662 US.

Table 1. Estimated construction costs: Summary by accounts

FERC ACCOUNT	DESCRIPTION	COST IN 1988 DOLLARS TOTAL
330	Land and Land Rights	$ 50,000,000
332	Reservoirs, Dams and Waterways	282,000,000
331	Power Plant Structures	134,000,000
333	Turbines and Generators	190,000,000
334	Accessory Electrical Equipment	53,000,000
335	Misc. Power Plant Equipment	30,000,000
336	Roads, Railroads and Bridges	7,000,000
352	Switchyard	22,000,000
353	Transmission	9,000,000
	SUBTOTAL	$ 777,000,000
	Contingency Allowance	111,000,000
	SUBTOTAL DIRECT CONSTRUCTION COST	$ 888,000,000
	Project Management, Engineering and Owner's Costs	105,000,000
	TOTAL PROJECT COST (MID-1988 $$)	$ 993,000,000 ($662/kW)

30. The currently planned construction schedule, based on the assumption that the FERC license will be granted in early-1991 and construction can commence in 1992 is shown in Figure 7.

PROJECT ECONOMY

31. Current projections by electric utility companies in Ohio indicate a potential need for peaking energy and capacity beginning in the mid- to late-1990's. Preliminary evaluations undertaken by Ohio Edison Company of the economy of the Summit Project relative to alternative sources of power (principally combustion turbine) within this time frame indicate that the

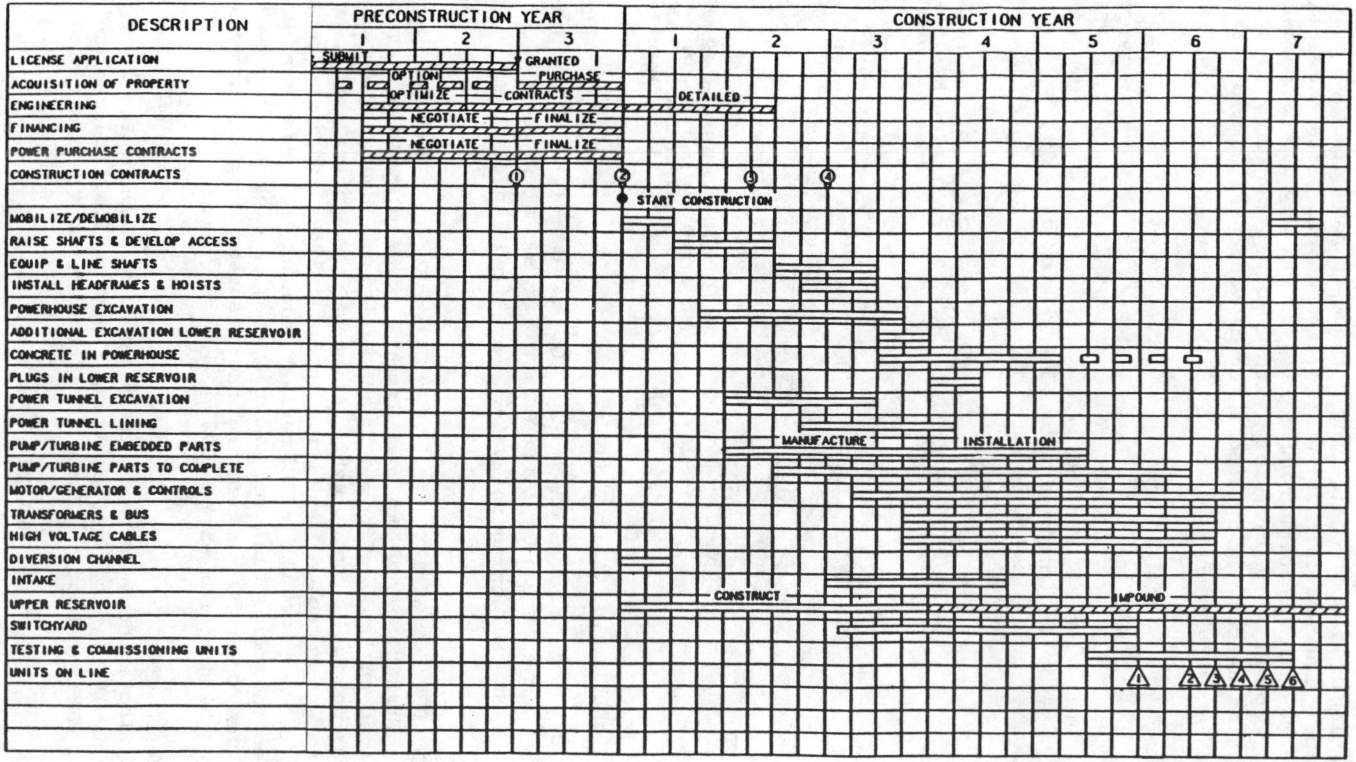

DESCRIPTION

PRECONSTRUCTION YEAR 1 2 3

CONSTRUCTION YEAR 1 2 3 4 5 6 7

LICENSE APPLICATION — SUBMIT — GRANTED
ACQUISITION OF PROPERTY — OPTION — PURCHASE
ENGINEERING — OPTIMIZE — CONTRACTS — DETAILED
FINANCING — NEGOTIATE — FINALIZE
POWER PURCHASE CONTRACTS — NEGOTIATE — FINALIZE
CONSTRUCTION CONTRACTS — ① ② ③ ④
START CONSTRUCTION
MOBILIZE/DEMOBILIZE
RAISE SHAFTS & DEVELOP ACCESS
EQUIP & LINE SHAFTS
INSTALL HEADFRAMES & HOISTS
POWERHOUSE EXCAVATION
ADDITIONAL EXCAVATION LOWER RESERVOIR
CONCRETE IN POWERHOUSE
PLUGS IN LOWER RESERVOIR
POWER TUNNEL EXCAVATION
POWER TUNNEL LINING
PUMP/TURBINE EMBEDDED PARTS — MANUFACTURE — INSTALLATION
PUMP/TURBINE PARTS TO COMPLETE
MOTOR/GENERATOR & CONTROLS
TRANSFORMERS & BUS
HIGH VOLTAGE CABLES
DIVERSION CHANNEL
INTAKE
UPPER RESERVOIR — CONSTRUCT — IMPOUND
SWITCHYARD
TESTING & COMMISSIONING UNITS
UNITS ON LINE — 1 2 3 4 5 6

CONTRACTS: ① PUMP-TURBINE/MOTOR GENERATOR
② CIVIL WORKS
③ AUXILARY MECHANICAL/ELECTRICAL SUPPLY
④ MECHANICAL/ELECTRICAL INSTALLATION

Fig. 7. Construction schedule

project is feasible. Based on the estimated project costs set out in Table 1, and considering a range of possible fuel costs, the studies showed that using levelized annual costs the break-even long-term capacity factor is a relatively low 8 percent, which compares very favorably with typical operating capacity factors of averaging 15 percent for conventional pumped storage facilities (Figure 8).

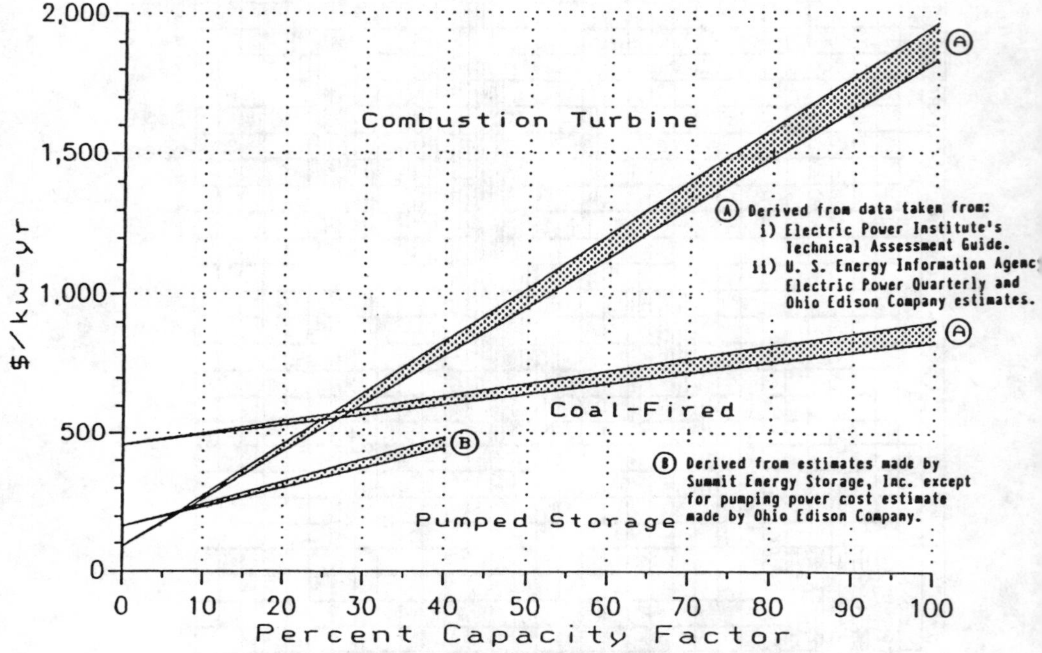

Fig. 8. Screening analysis: Levelized annual costs

CURRENT PROJECT STATUS

32. Summit Energy Storage, Inc., a privately owned company, submitted an application to the U.S. Federal Energy Regulatory Commission for a construction license in January 1989; the license was accepted for filing in April 1989, and is currently being subject to the FERC review process. In the meantime, engineering, financial and power system studies are proceeding with the objective of meeting the planned start of construction, scheduled for early 1992.

ACKNOWLEDGMENTS

33. The author wishes to thank Mr. Olof Nelson, President of Consolidated Hydro, Incorporated, the principal shareholder of Summit Energy Storage Inc., for permission to publish this paper.

REFERENCES

1. ISAAKSON, G., NILSSON, T. and SJORSTRAND, B. Pumped storage power plants with underground lower reservoir. VII World Power Conference, section 2, paper 160, Moscow, 1968.

2. RODGERS, F. C. Underground pumped storage developments. Joint Power Conference, September 1972.

3. WARNOCK, J. G. and WILLETT, D. C. Pumped storage underground. Symposium on Hydroelectric Pumped Storage Schemes. United Nations Economic Commission for Europe, Athens, Greece, November 1972.

4. ACRES AMERICAN INCORPORATED. Preliminary design study of underground pumped hydro and compressed air energy storage in hard rock. U.S. Department of Energy and Electric Power Research Institute, volumes 1-13, EPRI EM-1589, April 1981.

5. BIENIAWSKI, Z. T. Rock Mechanics in Mining and Tunneling. A.A. Balkema Publishers, 1984.

Panjiakon combined hydroelectric storage plant

C.S. CAO, Tianjin Prospecting and Design Institute

Synopsis
The Panjiakou Project consists of a large dam with 107.5 m height, an upper reservoir, a combined hydroelectric storage plant and a lower pond. The plant, consisting of one conventional unit and 3 pumped storage units with total capacity 420 MW, is the biggest combined pumped storage plant in China. The characteristics and unusual aspects are described in this paper.

General description of the scheme
The Panjiakou water control project is situated in the upper valley of the Luan River about two hundred kilometres north-east of Tianjin. As shown in Fig. 1, the project consists of two parts. The first part comprises the upper main dam (Fig. 2), the combined hydroelectric plant (Fig. 3), spillway and outlet works. The second part comprises he lower small dam, scouring sluices and a small power plant. The main dam creates a big reservoir which is the

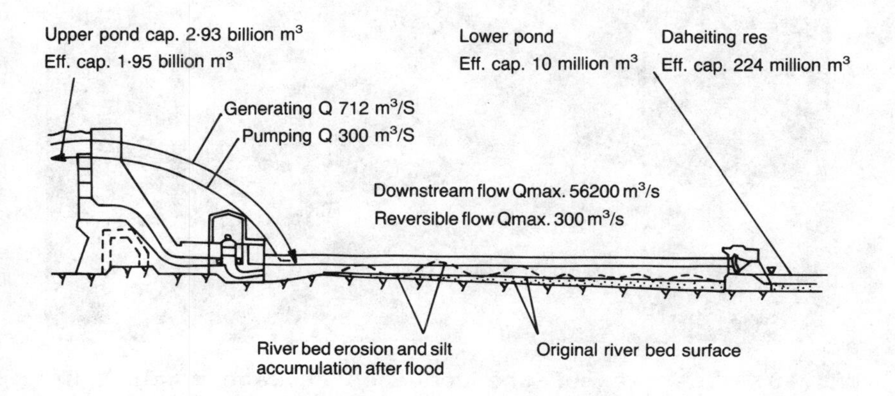

Fig. 1. The sketch of Panjiakou Storage Plant

Pumped storage. Thomas Telford, London, 1990.

Fig. 2. The down stream view of the upper main dam

Fig. 3. The view of the combined storage power house

upper pond of the combined hydroelectric storage plant, while the lower small dam creates the lower pond with a smaller capacity near 10 mm^3. The drainage area above the dam is 33 799 km^2. The average annual runoff is 2450 mm^3. The reservoir, created by this dam, gets a maximum capacity of 2930 mm^3. A capacity of 1950 mm^3 will be used for regulation of water for water supply.

The rocks forming the foundation and abutments for the dam are gneiss of excellent character, hard and strong.

The main dam is a concrete hollow gravity dam. The maximum height is 107.5 m, and the crest length is 1039 m. The central portion, consisting of 18 panels of spillway (15 m x 15 m) is designed to pass a flow of 54 000 m^3/s. There are four outlet works provided in the central portion of the dam at a lower elevation with a maximum discharge flow of 3000 m^3/s. The combined hydroelectric pumped storage plant is on the right side of the spillway adjacent to the downstream face of the dam. The plant consists of a conventional generating unit and three pumped storage units. The upper pond of the project has been designed as multi-year regulating reservoir, which is scheduled mainly for water supply and simultaneously also for the generation of electricity. Therefore the range of the reservoir's stage varies annually. In wet years the maximum reservoir level may reach 225.7 m. In dry years the minimum reservoir level may decline to 180.0 m. The lower pond is a reversible regulating reservoir with a relatively small volume. Its daily variations of water levels range from 139.4 m to 144.0 m. Because of the demand of the water supply of Tianjin and Tangshan, the discharge flow available for the power plant is small in January to February. The pumped storage units of the plant operate in pumped-storage manner from July to the next March, while from the April to June, they generally operate in the generating manner only.

The whole project will require more than 3 mm^3 of concrete and installing one generating unit and three pumped storage units. The construction work commenced in 1975 and is divided into two stages. Now the first stage construction work and the power house of the combined hydroelectric plant has been completed. The conventional unit started generating electricity in April 1981. The second stage construction including the lower small dam is under construction. The total volume of the concrete used in the second stage is about 300 000 m^3. The design of this project (1st stage) was awarded the golden prize by our state.

When the whole project is completed, it will be provided 1950 mm^3 (P = 75%) for water supply and will generate 580 GWh electricity per year. In the power system, it mainly works at the peak load and operates as the synchronous condenser.

The whole project consists of the upper dam, the combined hydroelectric storage plant, and lower dam including the lock and a small power plant. Due to the complexity of the water control project with multi-purpose requirements and multi-year regulating reservoir, the characteristics and unusual aspects of the project are stated as follows.

High discharge, high head, high velocity, reversible flow and the overflow dam with flaring piers

As shown in Fig. 1, down-stream of the dam is the lower pond created by the river channel 6 km long. The flow in the pond is reversible. In the flood season, the maximum downstream discharge from the reservoir may reach a very high velocity,

56 200 m^3/s. This is one of the highest discharges in the world. Due to its high discharge, high head and high velocity, the severe erosion and silting will elevate the downstream water level of the power plant and will diminish the effective volume of lower pond. While in pumping state, the water in the lower pond flowing upstream will probably carry the sand deposited in the tail-water channel of the plant into the draft tubes to cause some trouble in the ordinary operation of the plant. In order to solve the above problems, a large amount of model tests and research works have been fulfilled. A series of effective measures have been taken, such as optimizing the shape of the energy dissipater, improving the layout of the project, lengthening and changing the direction of the guide wall between the plant and overflow dam, and providing three wide flaring piers on the left end of the overflow dam. The overflow dam with the flaring piers is quite a new type of hydraulic structure (see Fig. 4). It plays a predominate function in the dissipation of energy. As shown in the upper right part of Fig. 4, the flow passing through the piers convergence from 15 m width to 10 m width, and the flows of neighbouring panels with high velocities impact each other. Creating very high water crowns behind the tails of the piers, it shows an effective energy dissipater. Its principal advantages include increasing the pressure on the overflowing surface, diminishing the backward flow velocity near the tow of the dam, improving the downstream flow condition and preventing river bed erosion and silt accumulation. Since the silt accumulation in the

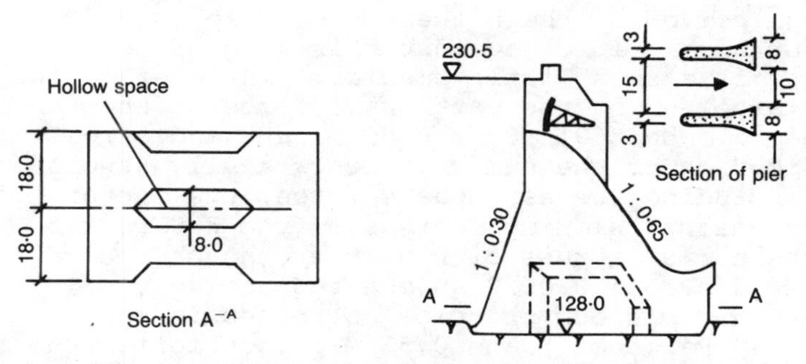

Fig. 4. The overflow dam with flaring piers

tail water channel has been alleviated to a great extent, the problem of sand carrying into the draft tube no longer exists. As the concentration of the water flow due to the flaring piers will induce some vibration of the overflow dam itself, only three flaring piers are provided in 18 panels of spillway.

Hollow gravity dam with wide and low cavities

As the earthquake intensity of the dam site is 7 degrees on the Mercalli scale, considering the height of the dam, the large capacity of the reservoir and the complex geologic structure of the reservoir area, a reservoir-induced earthquake may possibly occur in future. For the above reasons, the design earthquake intensity of the dam is 8 degrees. As it is more favourable for the earthquake resistance, the hollow gravity dam with rather greater cavity is adopted. As shown in Fig. 4 the hollow gravity dam is provided with wide and low cavities. The width of each block is 18 m, and the cavity is relatively wide at 8 m and relatively low, only one-third of the height of the dam. The principal advantages are diminishing the uplift pressure under the dam foundation and simplifying the structure of the upper two-thirds of the dam. For the convenience of inspection and maintenance in the cavities pumping-drainage installations have been provided. Under special circumstances in the future, these pumping drainage installations may also be operated permanently. Under this condition it may further reduce the uplift pressure under the foundation to raise the factor of safety against sliding of the dam if necessary.

After careful static and dynamic studies including the dynamic induction-spectrum method and dynamic structural model test, their results are very close. The frequency of the hollow gravity dam is little greater than the corresponding gravity dam, and the maximum period of the highest dam bloke is 0.28 sec. The earthquake force against this kind of hollow gravity is also a little smaller. The coefficient of enlargement of the earthquake force on the top of the dam is about 9. So the stability and stress distribution of the dam must be carefully studied. When including the earthquake force, the factor safety against sliding decreases from 3.0 to 2.5, and the normal stress of heel/toe changes from 0.42/2.10 mPa to 0.06/2.40 mPa respectively. The maximum and minimum principal stresses in the dam are 3.09 mPa (comp.) and 0.51 mPa (tension). Due to the high intensity of earthquake and large amount of discharge, the hollow gravity dam is preferable.

Power house installed with two kinds of units

The power house is installed with one conventional unit and three pumped storage units, these two kinds of units have quite different dimensions both in plan and in elevation. As shown in Fig. 5, the power house is at the downstream toe of the dam, and in the longitudinal section. These two kinds of units have different heights and widths (23 m/21 m), i.e. the power house section of the pumped storage units is thinner and higher than that of the conventional one. As the suction heads of these two kinds are -4.5 m/-9.4 m, the established elevations of these two turbines are 134.5 m/130.0 m. In Fig. 5, you may see clearly the upper and lower generator floors of these two units are at the same elevation 145.5 m and 141.0 m respectively. Besides this, the complexities of the plant also show in the complicated auxiliary machineries such as the converter, DC choke, and the controlling and ventilating systems. We see the complexity of the power house installed with two different kinds of units, but after careful consideration, the power house seems quite neat and coherent.

The wide range of water head and P/T with variable speeds

As stated before, the upper reservoir is designed for multi-year regulation. Generally, according to the hydrology of this region, there may be one wet year in every four or five years. Typically the first and fifth years could be wet years, and the others all dry or mean years. In the first year the water level rises to maximum water level, then in the following years, the water level may drop down step by step. Almost in the whole year of the first, and also in the second and fifth years the water level is high, and in the other years the water level is low. Therefore the conventional

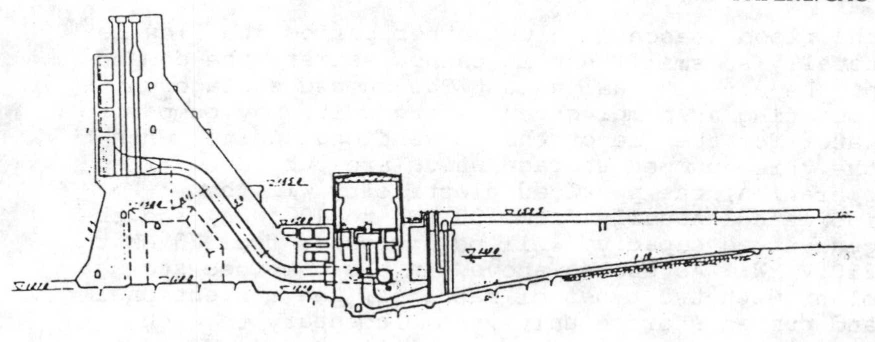

(a) Transverse section

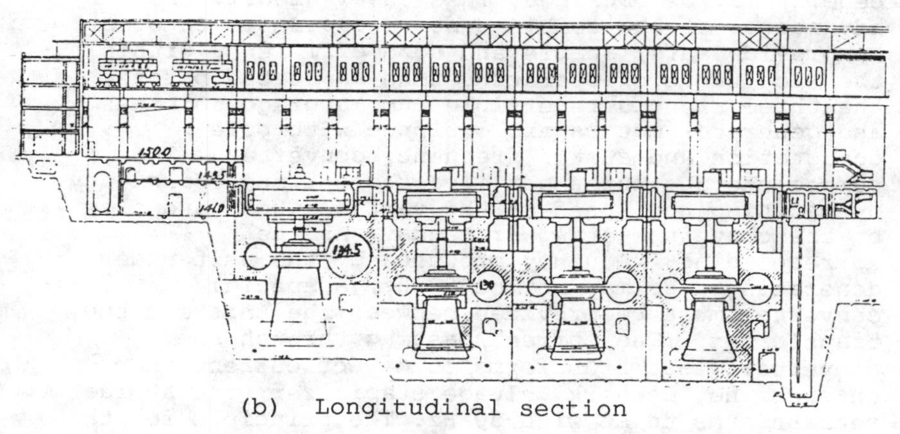

(b) Longitudinal section

Fig. 5. Section of the combined pumped storage plant

generating and pumped storage units may operate under high or low water head all the year round. The ratio of the minimum head (35.5 m) to maximum head (85.5 m) is very low, only 41%. For the above reasons, the conventional generating and pump-storage machines have to be safe and stable in operation with high efficiency at any head within the wide range of variation. As this multi-year regulating reservoir is mainly for inter-basin water supply, the generating of electricity must be subordinate to this goal, and at the same time the water supply shall also take care of the generation of electricity. Under the optimum reservoir operation, we should keep the water level as high as possible in order to store more water and to generate more electricity

During the high-flow period the pump-storage units operating only in generating state to complement the conventional unit, may operate both for producing peakload and for decreasing the waste of water in the flood season. In the other period the flow is usually so small that it cannot satisfy the demand of the conventional unit. The pumped storage units, operating as pumping-generating unit, may pump more water for the use of the conventional unit. When the three pumped storage units are put into operation, the peakload electricity will be increased annually from 130 MWh to 492 MWh, and the guaranteed capacity will be increased from 73 MW to 210.7 MW. As stated above, in the combined storage plant with two types of machines, the conventional and pumped storage units, complementary to each other, more electricity especially peakload electricity may be generated, and greater economic benefit may be obtained. This shows clearly the advantages of the combined storage plant. In order to provide safe, stable and good efficiency within the wide range of operating head as stated before, the three Francis reversible pump-storage turbines and generator motors are equipped with pole commutation and static frequency converter. The pumped storage units can give the optimum efficiency by matching the speed to the variable head and flow by the converter at lower heads. As the corresponding frequency produced by the unit under generating mode may not be 50 Hz, a special converter must be provided between the unit and the transformer. The current passing through the converter first transforms to direct current then through the DC chork voltage stabilizer, and at last retransforms to AC with 50 Hz. The principle for the pumping mode is the same. This is a quite effective and novel technique in hydroelectric plant. As the converter is put into service, it may increase the gross efficiency of the unit from 60% to 80% approximately under lower heads, and the guaranteed capacity of the plant may be increased by 15%. This kind of unit may be operated in optimum manner, either with pole changing at two speeds of runner or with variable frequency at any variable speed. It may also improve the cavitation-resistant condition and lead to a stable and safe operation.

The above three pump-storage units are supplied by TIBB and
DPEW of Italy and the guaranteed characteristics of the unit are shown in the following table.

Since the variable speed operation of pump-storage unit is the first application in the world, its feasibility has to be verified. In co-operation with the Scientific Research Institute of Electricity (Beijing), we have built up a digital-physical model of the G/M and a physical model of the SPC system. On the basis of which the

Table. Chart of speed power, flow and efficiency

Mode	Converter	Net head (m)	Speed r/min	Flow m\/min	Power MW	Efficiency %
	No	85.0	125.0	116.0	90.0	93.2
	No	70.0	125.0	114.9	86.8	83.7
	No	60.0	125.0	135.1	70.6	88.9
	No	45.0	125.0	118.5	46.6	89.1
	No	35.4	125.0	105.5	31.1	85.0
	With	45.0	115.9	117.9	46.6	89.7
	With	35.4	102.8	104.6	32.6	89.6
	No	86.2	142.8	81.7	76.4	90.4
	No	70.8	142.8	115.8	85.4	93.1
	No	65.4	125.0	72.6	51.5	90.3
	No	46.1	125.0	116.5	57.4	91.6
	No	37.1	125.0	120.0	57.2	76.3
	With	60.0	127.7	95.0	60.0	93.1
	With	45.0	115.4	95.0	45.1	93.0
	With	36.7	107.7	95.0	37.0	92.4

simulation test of the variable speed operation has been carried out. In this test, the working principle of the variable speed operation and the variable range of the parameters submitted by the seller for variable speed operation of every mode have been verified.

Summary

The civil works of this plant have been completed, the conventional unit of 150 MW started generating electricity in 1981 and 3 pumped storage units are just under installation, and the first unit will be commissioned soon. With a multi-year regulating reservoir, the pumped storage plant provides some special characteristics such as the wide range of water head of the units, the complexities of these two kinds of units existing in the same power house and the reversible flow with high velocity in the lower pond. After careful studies and liaisons with the relevant hydroelectric-machinery enterprises, the unusual aspects of this plant include some new and special structures of the control project and the plant, hollow gravity dam with wide and low cavities, overflow dam with flaring piers, and the reversible pumped storage units with the pole changing and a static frequency converter. Matching the wide range of water head (35.5 m to 86.5 m), these units with pole commutation and static frequency converter can give the optimum efficiency by matching the speed to the variable head and flow and will lead to a safe, stable and economical operation. This kind of hydroelectric machinery with variable speed is quite new and the application in the hydroelectric plant is the first time in the world. The Panjiakou Plant is the largest combined storage plant in China. With these two kinds of units, complementary to each other, more electricity may be generated and greater economic benefit may be obtained.

Construction of the new Koepchenwerk pump turbine plant at Herdecke, West Germany

Dr Ing J. KOHLI, Lahmeyer International Consulting Engineers, Frankfurt am Main

SYNOPSIS. The new Koepchenwerk pump turbine plant at Herdecke, near the city of Dortmund, was constructed between 1985-1989 to replace the old plant, which has been in operation for nearly 60 years. The new plant consits of an intake tower, a pressure tunnel and a shaft powerhouse. The upper reservoir, with a net storage volume of 1.53 mio m^3, and the Hengstey Lake, which is the lower reservoir, are incorporated into the new scheme in virtually unchanged condition.

INTRODUCTION

Construction of the new Koepchenwerk plant was considered and initiated when, in 1980, the spiral casing of one of the four storage pumps of the old plant was torn during pump-start. Investigation of the cause showed that the flange on the stayring was torn off from the inside due to material fatigue accentuated by notch effects. Subsequent inspection of the other three storage pumps revealed incipient surface cracks in the same places on the stayrings. On the basis of detailed cost/ benefit analyses for the construction of a modern pump turbine versus rehabilitation and overhauling the almost 50-year old equipment and superseded, the Rheinisch-Westfälisches Elektrizitätswerk AG (RWE), the owner, decided, in 1981, in favor of the installation of a modern reversible pump turbine plant.

The essential newly constructed components of the pump turbine plant are (Fig. 1):

- a 39 m high polygonal intake tower equipped with a cylinderical gate as well as appertaining inflow funnel with 64 m outer diameter, constructed in the southern part of the upper reservoir;
- a steel-lined pressure tunnel connected with the bottom of the intake structure and ending in the powerhouse shaft. The pressure tunnel has a net diameter of 4.75 m and a total length of 396 m;
- an underground powerhouse built in a 50 m long, 19 m wide and 47 m deep rock shaft in which the 150 MW

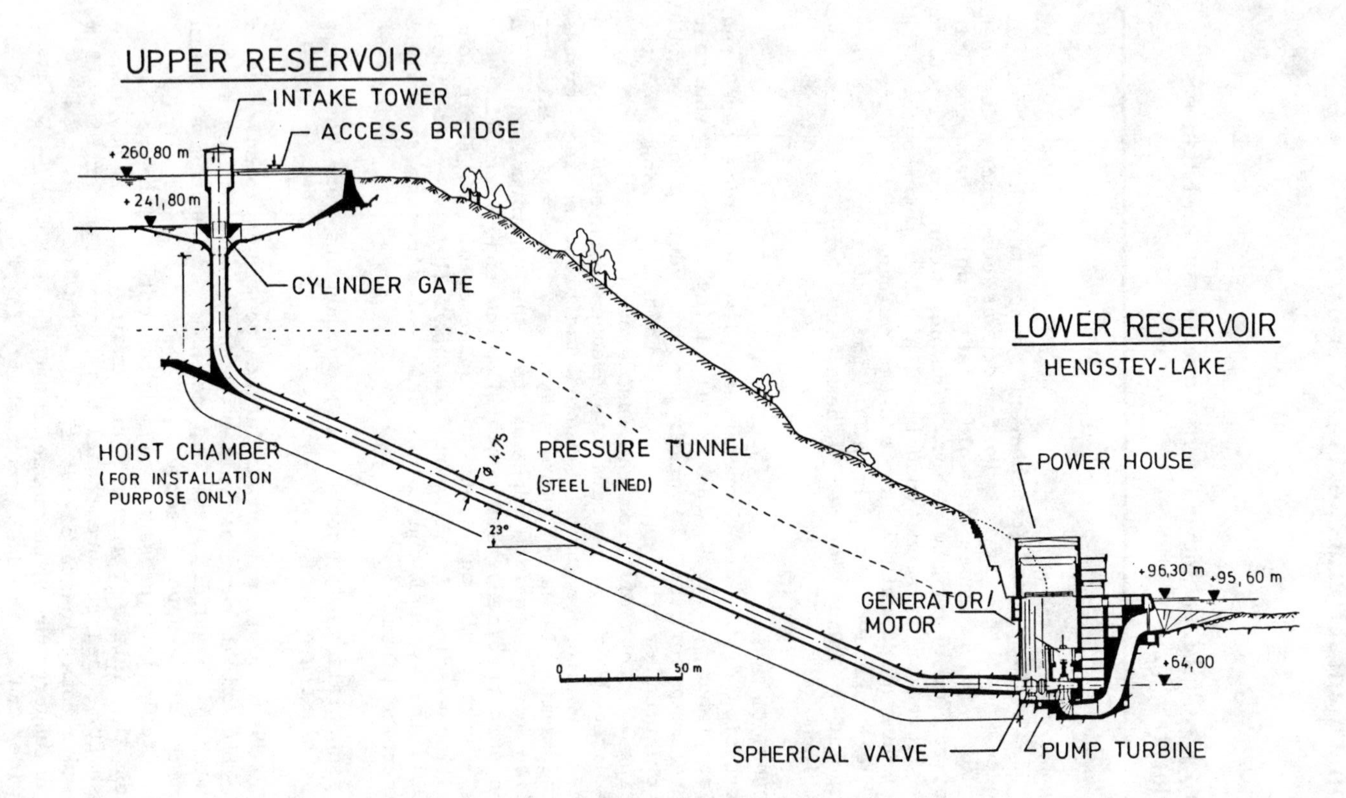

Fig. 1. Cross-section of new Koepchenwerk pump turbine plant

reversible pump turbine unit, 190 MVA synchronous generator-motor and auxiliary equipment are installed. The adjoining outlet structure is equipped with two roller gates and trash rack with rake machine;
- a superstructure consisting of a machine hall, office building, as well as operations and control building;
- transformer and air cooler foundations, railroad and auxiliary civil works.

2. GEOLOGY

The construction site of the new plant is situated south of the existing powerhouse, in the narrow strip on the lakefront between the Ardey Hills and Hengstey Lake.

Strata of the Upper Carboniferous Age, known as Vorhaller (Namur B) and Sprockhöveler (Namur C) strata, outcrop in the project area. The Vorhaller strata essentially consist of massive, moderately to fairly jointed stratified siltstone series containing isolated fine-grained sandstone intercalations. The Vorhaller beds on the other hand are composed of moderately jointed medium grained sandstones and siltstones.

3. GEOLOGICAL AND ROCK MECHANICS FIELD INVESTIGATIONS

The following program was carried out to investigate the rock mass conditions and to determine the rock mechanics parameters for the design of the pressure tunnel, power house and cut slope [1]:
- 16 core drilings of a total length of 720 m, Lugeon tests in selected areas of the civil structures;
- 20 dilatometer tests with the LNEC probe;
- 10 measurements of in-situ stresses by overcoring;
- excavation of approximately 30 m deep exploratory shaft of 3 m diameter in the powerhouse area to investigate the rock mass conditions, the joint systems, as well as to perform rock mechanics field tests and a test grout curtain;
- special geological mapping of access road rock slopes in the vicinity of the plant;
- rock mechanics laboratory tests on large-scale samples, 60 cm in diameter, and drill core material.

The existing rock slopes, up to 30 m high, enabled a detailed judgement to be made on the geometry of joints and the stability of the powerhouse pit and of the cut slope.

SYSTEMS OF PLANES OF SEPARATION

The joint directions as well as their spacing and extent were determined from the geological mapping of valley slopes along the access road and in the trial shaft. The statistical evaluation of data revealed three main systems of separation planes which are of importance to the stability of the cut back rock slopes, the vertical powerhouse walls and the pressure tunnel:

J1:N 12°E/75°E main (parallel) system of joints
J2:N 110°E/75°S secondary (diagonal) system of joints
J3:N 68°E/ 8°SE bedding (of minor importance)

A scattering of ±10° and ± 20°, respectively, has to be considered for the dip and strike of the joints.

Joint system J1 is the most prominent feature for the structural stability and strikes nearly parallel to the valley slope. System J2 strikes approximately perpendicular to the slope and thus nearly parallel to the powerhouse axis. It is much less marked and is only relevant for the wedge-type analyses of the parallel powerhouse walls. The bedding planes J3 are recognizable as latent and rarely developed joints. Their effect on the slopes and walls is small, but greater on the pressure tunnel.

From the results of the investigations cited above, the rock mechanics parameters [1] were selected for the stability analyses by the finite element method.

4. ALTERNATIVE ARRANGEMENTS

Taking into consideration standards of modern technology, cost and benefit analysis and experience made at similar plants, the owner opted, in the course of a prestudy, for a single pump turbine unit with vertical axis and capacity of 150 MW. The adjacent area south of the aforementioned was selected as construction site.

The following alternatives for the project components were considered to develop different competitive project layouts. As inlet structure, the intake tower with a cylindrical gate and the lateral inlet with roller gates were included in the study. The underground steel-lined pressure tunnel and the penstock along the hill surface represented the alternatives for the waterways. As alternatives for powerhouses, the shaft and cavern-type were to be compared. Based on these considerations the technical and economical aspects of the following four alternative project layouts were compared:

1. Intake tower with pressure tunnel and shaft powerhouse.
2. Intake tower with pressure tunnel, cavern powerhouse and tailrace tunnel.
3. Lateral inlet with pressure tunnel and shaft powerhouse.
4. Lateral inlet with slope penstock and shaft powerhouse.

Alternative 1. (Fig. 1), in comparison with all the others, proved to be the best as far as technology and costs are concerned.

Cavern powerhouse, alternative 2, proved to be approx. 6% more expensive compared with alternative 1. Moreover, it showed potential insecurities concerning the geological conditions due to the direction of the J3 bedding plane.

Alternative 4 with slope penstock would have been more expensive by approx. 10% and would have led to further impediments regarding protection of the environment, maintenance and security.

Finally, also, the lateral inlet of alternatives 3 and 4 was determined to be risky for the existing surrounding wall of the reservoir.

The selection of alternative 1 was therefore evident. Moreover, a skillful arrangement of the intake tower in the upper reservoir allowed the construction of a cofferdam between the intake of the old Koepchenwerk and the site of the new intake structure, so that the old power plant could continue operation with the bigger part of the upper reservoir during construction.

5. HYDRAULIC MODEL STUDY

Reasonable cavitation security and the technically safe practicable machine dimensions resulted in the choice of the rated speed for pump and turbine operation of 250 r.p.m. and setting height of 64.00 m a.s.l.

The model tests were performed on the basis of the selected shaft powerhouse, setting height of the pump turbine and the existing location of the downstream outlet to optimize the downstream outlet shaft of the pump turbine in connection with the draft tube [2]. Model draft tubes from various turbine manufacturers built on a scale of 1:20 were used. By means of a laser-doppler-anemometer the flow distribution and the twist in the cross-section of the pump turbine runner were investigated. Further, the downflow during turbine operation was tested. In this test a twist apparatus simulated different downflow twist angles on the runner level. By means of miniature measuring meters the flow speeds on the different levels in the outlet area were determined. Also, the turbulences in front of and behind the outlet trash rack and the flow in the outlet shaft extending to the pump turbine draft tube were observed by means of a paint sounding device. During testing different types of outlet shafts and especially different inclinations of the outlet shaft were examined, and the balance between good inflow conditions on the runner level during pump operation and sufficiently good outflow conditions on the trash rack section during turbine operation had to be compared. The result of this optimization showed a 73° inclination of the outlet shaft. The model tests also led to a constructive determination of the shaping of the embankment slopes and the foundation in the outlet structure.

6. CONSTRUCTION WORK

6.1 Upper reservoir and related structures. The upper reservoir, built in the late Twenties, is located on a 160 m high knoll in the Ardey Hills. Concrete gravity retaining walls, up to 21 m in height, enclose the basin. The proximity of the reservoir to the right bank of Hengstey Lake results in a very favorable ratio of about 1:2 between the gross head of 165 m and the horizontal projection between upper intake and lower outlet.

Because operation of the upper reservoir was required during the construction period, a cofferdam with double sheet pile walls, measuring 8 m in height and 150 m in length, was erected on the existing reservoir bed. The main components of the cofferdam are the 8.50 m wide concrete foundation slab, the lateral sheet piling and upper and lower anchor bars. The filling material's dead weight provided the stability of the cofferdam. In order to prevent uplift pressure, a three-fold protection was provided for between the reservoir bed and the concrete slab. Asphalt sealing at the joint of the reservoir bed/foundation slab forms the outside protection and, just behind this joint, on the upstream face a betonite border that would swell at the insurge of water was incorporated in the foundation slab. In addition, numerous relief apertures were provided in the slab.

The newly built intake tower is situated about 40 m from the outer wall of the reservoir and can be approached via a connecting steel bridge (Fig. 2). The total height of the concrete tower is approx. 39 m. The cross-section is dodecagonal on the level of the inlet openings. Above this level, the outside cross-section is octogonal. The inside cross-section is circular. The machine house that accommodates the driving equipment for the cylinderical gate is a steel construction.

The twelve inlet openings of the intake tower are equipped with trash rack. The cylindrical gate, the closing device, has a diameter of 5.5 m and a height of 4.75 m. The closing time in an emergency is 50 s.

In addition, a concrete ramp of approx. 170 m in length is installed as a bridge in the upper reservoir. The bridge functioned as a service ramp during construction. Now it will serve as an access road for inspection and repair work.

Fig. 2: Intake Tower and Upper Reservoir

6.2 Pressure tunnel and intake funnel. The inside diameter of the pressure tunnel was optimized to 4.75 m (Fig. 1). Thereby the startup time for the hydraulic system will be approx. 1.6 seconds, and the velocity of the water at maximum turbine performance 6.2 m/s. These figures facilitated the achievement of a favorable governor design of the power unit.

For safety reasons it was decided to steel-line the entire length of the pressure tunnel. Fine-grain steel with a 15 mm plate thickness was used on the upper reservoir end and 40 mm on the powerhouse shaft end. Claws or steel rings for protection against buckling of the steel lining due to water pressure from the outside were not installed. The selected smooth pipe section can bear an outside water pressure of up to 5 bar. Pressure-release valves that open at a pressure of 3 bar were built in. Concrete of B25 quality, as stipulated by the German building code, was used to backfill the steel plate. The thickness of the backfill is 30 cm (inclusive of shotcrete protection).

For the excavation of the rock tunnel, 5.35 m in diameter, separate stability design concepts were developed for the vertical, horizontal and inclined stretches. The assumption of rock wedges to determine the rock stability of the vertical stretch was based on the experience made during the cut slope excavation. Theoretical wedges with heights of 5 to 16 m required a stabilization through rock bolts and shotcrete. SN type bolts every 5.2 m^2, in lengths of 2.9 m and 2.4 m, were set up in a regular cross-section. Stabilization of the vertical stretch in this manner is sufficient even if the rock characteristics should be less favorable due to intercalations.

Because of the limited importance of wedge failure in the tunnel walls in comparison to possible failure of the roof after excavation, the stability design for the horizontal and diagonal stretches was developed based on finite element calculations. In addition to using rock bolts as dowelling for the joint faces, the zones experiencing overstress in the tunnel walls and the danger of roof caving were investigated.

In the design, three different rock classes were defined, which were stabilized with a shotcrete sealing incorporating steel mesh reinforcement mats of 1.88 cm^2/m steel sections and anchors varying between 4 and 8 anchors per 1.5 m tunnel length.

After initial progress in the excavation of the inclined stretch, it was found that a "head protection" consisting of mats of reinforced steel mesh and double-wedged anchors was sufficient. A shotcrete sealing was then used only where local conditions required it.

After the steel-lining of the pressure tunnel was completed, excavation on the area of the intake funnel could begin. The sound silt/sandstone rock was primarily excavated with chisel excavator. Blasting in the lower depths was necessary to loosen the rock mass; measures to secure the slope in the funnel area were not necessary.

6.3 Cut slope. The confined area between the shoreline of Hengstey Lake and the mountain slope necessitated a back-cutting of the natural slope, in order to provide the necessary space for the power plant shaft. The back cut is approx. 18 m in depth, having a new slope inclination of 3:1 and an intermediate berm.

The cut slope stretches over a length of 170 m and has a max. exavation height of 30 m. The principal joint system J1 runs virtually parallel to the slope, and the secondary joint system J2 vertical to that. The stability of the slope was therefore determined by rock wedges formed by these joint systems. This required a systematic stabilization of the northern flank with shotcrete d = 10 cm, and SN type bolts (l = 6 m and 8 m) with a bolt for every 3.3 m^2 below and 4.4 m^2 above the berm. Due to the more favorable orientation of the separation planes with regard to the southern flank of the cut slope, the density and length of the bolts could be reduced and a shotcrete sealing waived.

Fig. 3: Open Joints in the Cut Back Slope Area

The northern flank of the slope evidenced up to 8 m of overburden, composed of weathered rock and loose soil. The rock underneath showed exposed joints - approx. parallel to the slope - with openings of up to 18 cm in width (Fig. 3).
Here, the cavities were filled through 15 m subhorizontal bore holes in the first series and 30 m in the second. The record of injection takes indicated that the open joints were 10-15 m deep in the slope. In addition, 18 prestressed anchors, P = 600 kN and l = 30 m, were set up. Because of possible further slope deformation through stress relief after excavation, these prestressed anchors were prestressed with 450 kN only. To date, however, no recognisable movement of the slope has been observed.

6.4 Shaft powerhouse and related structures. The powerhouse shaft is 50 m long, 19 m wide and 47 m deep (Fig. 4). The cross-section on the mountain side is a semi-circle and a rectangle on the lake side. The lakefront portion downstream of the turbine/generator unit is divided into ten floors. The electrical and mechanical auxiliary equipment is installed here. The outlet shaft, downstream of the draft tube, is inclined at an angle of 73° and ends at the outlet structure in Hengstey Lake.

To protect the shaft excavation against insurges from Hengstey Lake, an almost semi-circular 160 m long cofferdam

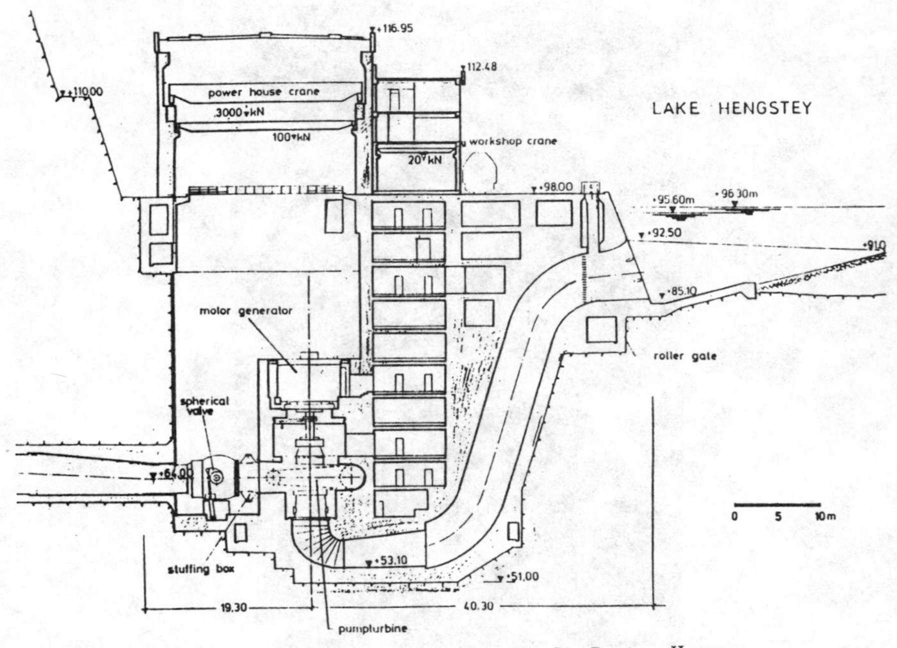

Fig. 4: Section Through Shaft Power House

with double sheet pile walls was built. At the same time, a grout curtain of 25 m depth was injected around the construction area between the flanks of the cofferdam and the cut slope. Since the topmost 10 m of overburden involve, to a large extent, a landfill originating from the construction of the old power plant, a large amount of cement was needed - about 1,000 t of solid material for 8,500 bore meters. Due to the positive results of the grouting the seepage water in the powerhouse shaft has declined to approx. 3.2 l/s.

When excavating a deep rock shaft, it is common practice to start with the construction of a concrete rim. The height and width of the rim depend on the thickness of the loose soil overburden and the dimensions required when used as a control gallery. In this case, a two-story construction with a width of 4 m and a height of 9 m was decided upon. The concrete rim was erected on sound rock and anchored back in the surrounding rock mass by 74 pieces of 750 kN prestressed anchors 15 m long.

Because the old plant was to remain in operation, as well as the transformer installation behind it, the construction of a bridge near the lake side was necessary prior to the excavation of the shaft. Stabilization of the two rock abutments was carried out through 30 m long, horizontal prestressed anchors of 1000 kN in a 2 m x 2 m grid. Each abutment was fortified with 16 anchors.

To prevent a wedge failure in the rock embankment, the vertical shaft walls naturally needed stronger stabilization than the slope embankment. In addition to 15 cm of shotcrete and wire mesh reinforcement of 2.57 cm^2/m, 6 and 8 m long SN type bolts 28 mm in diameter were arranged in a 1.3 m x 1.3 m grid on the northern and cut slope flanks as well as in a 1.8 m x 1.8 m grid on the southern and lake side flanks.

When the shaft was sunk, however, two weak zones were encountered that stretched over the entire 19 m width of the shaft and, in the excavation zone, a combined thickness of 1 m was evidenced. As in the area of the curved section of the shaft 28 horizontal prestressed anchors, P = 750 kN and l = 15 m, in a 2 m x 2 m grid were provided in the design, these weak zones did not pose any problems for stability of the rock walls.

After the shaft excavation was completed, the erection of the draft tube and embedding in concrete could begin. The concrete works for the outlet shaft and lakefront powerhouse structure with ten floors followed immediately. On the other hand, the concrete foundation for the pump turbine unit (Fig. 5) could begin only after the erection of the 3000 kN powerhouse crane in the machine hall and concreting of the two columns of the hall, founded on the ground floor of the powerhouse.

DRAINAGE SYSTEM

To effectively reduce the uplift in the surrounding underwater area of the shaft site, an extensive drainage system was built [4]. It consisted of 80 mm drainage bore holes of up to 10 m in length that were drilled into the rock at an upward inclination of 10°. The boring grid measures 2.90 m x 4.50 m. The drainage borings discharge into vertical slotted collecting pipes 100 mm in diameter. The distance between the neighboring pipes is 2.90 m. These pipes stretch from the gallery at elevation 89.84 to the drainage gallery at elevation 56.60. A filter screen was installed between the collecting pipes and the adjoining rock surface.

A water quality analysis shows that a decrease in the cross-sectional areas of the borings through iron staining can, to the greatest extent, be ruled out. The collecting pipes can be cleaned from elevation 89.84.

SUPERSTRUCTURE

The machine hall covering the powerhouse shaft is 44.4 m in length, 24 m in width and 18.9 m in height. Equipped with a 3000 kN and a 100 kN crane, it served as erection bay during the construction period and will now be used as mechanical workshop. The office building, 51.6 m long and 10.2 m wide, has workshops on the ground floor and offices on the two

Fig. 5: Pump Turbine Foundation

upper floors. The neighboring operation and control building, 20.3 m in length and 15.4 m in width, consists of a basement and three further floors. It accommodates the static frequency converter, passage for bus bars, control room and other technical auxiliary equipment.

7. MECHANICAL AND ELECTRICAL EQUIPMENT

The main components in the powerhouse are the spherical valve, the pressure equilizing sleeve, the pump turbine and the generator/motor [3].
The spherical valve is rigidly flanged at the steel lining of the pressure tunnel. It is designed for a pressure of 28 bars. The spherical valve has an inside diameter of 3.3 m, a total height of 5.1 m and a total weight of 167t. The force-balancing sleeve is situated between the spherical valve and the spiral casing of the pump turbine. It allows a longitudinal shifting of the valve in the order of ± 20 mm. This reduces the hydraulic thrust on the concrete machine block.

The rated capacity of the pump turbine is 153 MW, the rated speed 250 r.p.m., the rated turbine flow 105.3 m^3/s, and the rated pump flow 101.7 m^3/s. The runner of the pump turbine is made of chrome-nickle cast steel. It has an external diameter of 4.6 m and seven fixed blades. The setting height of the pump turbine was fixed at 64.00 m above sea level for a cavitation-free operation, i.e., 31.60 m under the lowest tailwater level.

The motor-generator has a rated output of 190 MVA as generator, a rated power factor of 0.85 and a rotor inertia moment of 1530 tm^2. The diameter of the rotor is 6 m and it weighs 300 t. A static frequency converter, which is installed in the control building, allows startup of the pump operation in 80 s.

8. MONITORING SYSTEM

A comprehensive monitoring system was installed as early as possible to pursue a continuous control of the stability of the underground works during the whole construction period. The installed instrumentation consists of [4]:

- 14 horizontal multiple extensometers for the powerhouse shaft and 5 in the cut-back slope;
- 5 vertical multiple extensometers for the shaft;
- 4 inclinometer boreholes for the shaft and 2 in the cut slope;
- 28 geodetic measurement points for the powerhouse shaft and on additional 8 points in the cut slope;
- 23 tension measurement devices for the total of 151 prestressed anchors;
- 5 convergence measuring cross-sections in the pressure tunnel.

The redundancy of monitoring of the rock deformations was ensured by the geodetic measurements and by the extensometer records. Monitoring records collected in short regular intervals were compatible with the deformations predicted on the basis of the prior field investigations and design calculations [5]. The effects of the weak zones encountered during the shaft excavation were evidenced by the differences in the magnitude of the deformation of the north and south wall of the rock shaft. However, the maximal perpendicular displacement of the rock was 12 mm and safe excavation conditions were monitored during the execution period.

The monitoring system will remain further in function, although the records will be carried out in larger intervals, and will provide early indications whether any changes do occur in the underground conditions.

9. CONCLUSION

The new Koepchenwerk Pump Turbine Plant at Herdecke represents the tremendous progress made during the last five decades in the field of hydropower technology in general and in underground construction in particular. Four units of the earlier plant have been replaced by a single unit, augmenting the generating capacity by 18 MW and the overall efficiency by more than 10 percent. The modern methods of geotechnical investigations, design and construction enabled a safe and economical execution of underground and other civil works. The built-up area of the new plant is less than 60 percent of the earlier installation. During the construction period the old plant remained in operation, although with limited capacity.

REFERENCES

1. SCHENK, V.; HÖNISCH, K.; SCHEIBE, H.J. (1986). Geotechnical Investigations for Koepchenwerk. Water Power & Dam Construction, July 1986.
2. ELS, H. (1987). The Hydraulic Design of the Outlet Works for Pump Turbines in a Shaft. Water Power & Dam Construction, October 1987.
3. HARTMANN, E.; HELMER, F.; LATRILLE, W. (1987). Replacement of the Existing Pumped Storage Plant Koepchenwerk, Herdecke/Germany, by a New 150 MW Pump Turbine Plant. Water Power & Dam Construction, Conference on Uprating and Refurbishing Hydropower Plants, Strassbourg, France.
4. BOGENRIEDER, W.; KOHLI, J. (1988). Entwurf und Ausführung der Tiefbauarbeiten für die 150 MW Pump turbine Koepchenwerk. 8. Nationales Felsmechanik-Symposium in Aachen, April 1988.
5. HÖNISCH, K.; KERN, D.L. (1989). 3 D Back Analysis for an Unusually Shaped Powerhouse Shaft in Jointed Rock. Third International Symposium on Numerical Models in Geomechanics, Niagara Falls, Canada, May 1989

ACKOWLEDGEMENT

The kind permission of Hydropower Department, RWE, Essen, to publish this paper is gratefully aknowledged.

Discussion

J.G. WARNOCK, ***LE Energy, Edinburgh, UK***

The Author refers to early 1970s examples of pumped storage projects based on underground reservoirs. The original ideas were presented at the World Power Conference in 1968 by Swedish engineers. Mr Willett and his colleagues (of whom I was one) saw a

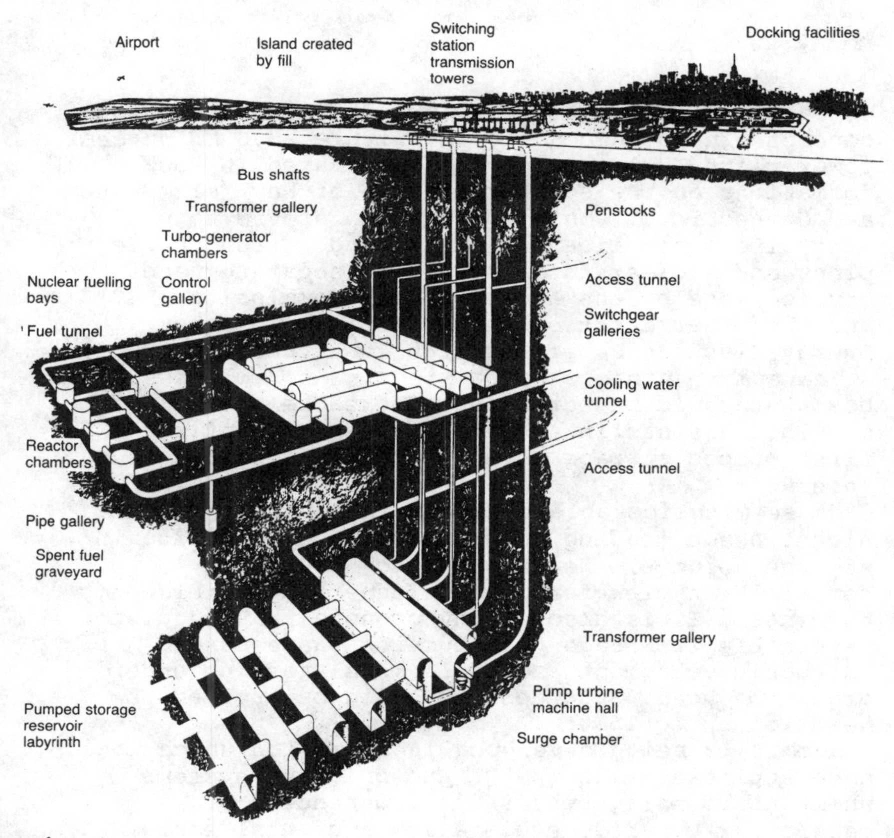

Fig. 1

Pumped storage. Thomas Telford, London, 1990

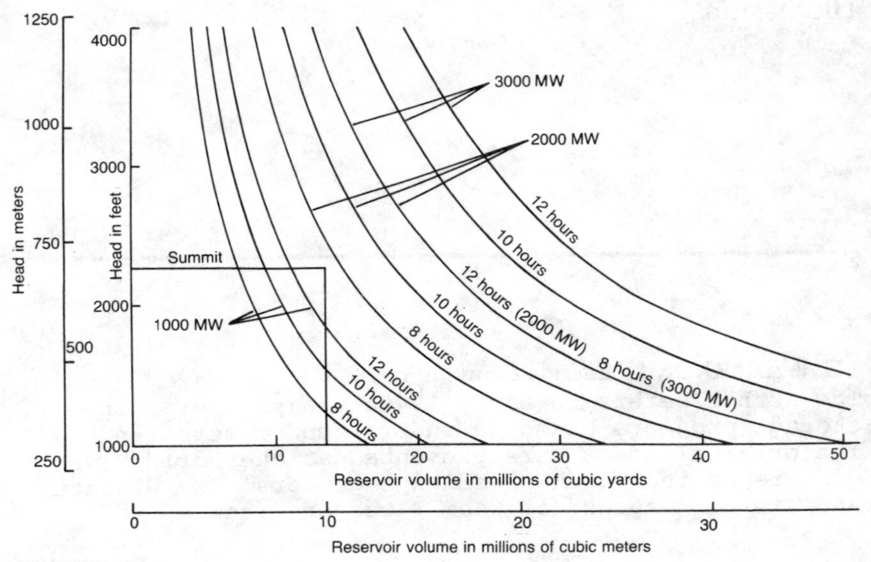

Fig. 2

potential solution for New York State. This first concept linked pumped storage with a 2000 MW nuclear power plant, all underground, excavated in rock formations on the southern coasts of New York State and Connecticut. This combination proved to be excessively complicated and the underground nuclear plant and nuclear facilities were separated and studied, one by the Atomic Energy Commission (AEC) and the other later by the US Department of Energy/Electric Power Research Institute (DoE/EPRI).

The summit project dealt with in Mr Willett's paper has what could be termed a comfortable head of 670 m. This assists in achieving acceptance of the first pumped storage plant with an underground reservoir.

We see considerable benefits in adopting very much higher heads leading, of course, to a reduction in storage volumes. Heads of up to 1000 m are certainly in view for single runner reversible pump turbines. It is also worthy of note that regulated reversible two-stage pump turbines have taken through development and scale model testing under prototype head. Such designs could be applied to heads of up to 1500 m.

Summit is being developed in a fashion which could have application in the UK and on other systems where third party ownership (independent of a particular utility) could allow a general energy storage service to be provided.

T.H. DOUGLAS, ***James Williamson & Partners, Glasgow, UK***

Could Professor Cao describe the flow conditions at the downstream side of the flared piers and, in particular, any problems with the turbulence at the ends with the constrained flows.

Dr Kohli described the control of the hydraulic system being effected by one cylinder gate with a closing time of 50 s. What further emergency protection was provided either for operation or maintenance of the below-ground power station from the upper reservoir?

E.T. HAWS, ***Rendel Parkman, London, UK***

With respect to Professor Cao's reference to the flared spillway piers, I assume the effect is major air entrainment in the downstream napple. As air entrainment is extremely difficult to simulate in the hydraulic laboratory, is the surface of the downstream napple predicted safely? I assume the training walls are very high.

With a dam more than 100 m high, storing 10^9 m^3 of water, there is a reasonable chance of reservoir induced earthquake. Has this phenomenon been experienced?

D.C. WILLETT, ***Author***

The principal ingredient which makes the Summit site so attractive is, of course, the presence of the existing limestone mine at a depth and with a configuration appropriate to pumped storage development using a relatively small extrapolation of current machine designs. As Mr Warnock notes, there may be merit in considering higher heads and multistage machines to reduce the necessary storage volume. I suspect, however, there will be a notable reluctance on the part of privately funded developers to accept both the technical risks involved in the use of unproven equipment and the cost risks associated with new major excavations at the depths Mr Warnock is proposing. In this era of privatization of the electric utility industry, the disappearance of the publicly funded utility willing and able to fund the up-front costs involved in proving up this advanced technology regretably makes it seem unlikely that advances will be made in this direction.

C. S. Cao, ***Author***

As Mr Douglas and Mr Haws note, a large amount of air entrainment and turbulence may occur. When the water passes through the flared pipes, the width of the flow is diminished and its depth is increased.

When the flows of the neighbouring panels impact at high velocity large water crowns are created behind the flared piers. The very high water crowns with a large amount of air entrainment are very favourable to energy dissipation. The principal advantages include increasing the pressure on the overflow surface, diminishing the backward flow velocity, improving the flow condition downstream and improving river bed erosion to a great extent. The above phenomena are observed in different model tests we have made over a period of more than ten years.

Although air entrainment can not be fully simulated by models it is understood that the energy dissipation will be developed more thoroughly than that shown in the models. As the concentration of the water flow due to the flared piers will induce some vibration of the dam itself, only three piers are provided in 18 panels of spillways.

J. KOHLI, ***Author***

A spherical valve is additionally provided for operation and emergency protection of the below-ground power station from the upper reservoir (please refer to Fig. 1).

Experience and design features of motor-generators in pumped storage plants

J.J. SIMOND and H. VÖGELE, Asea Brown Boveri Ltd, Switzerland

Synopsis

Pumped-storage plants use electrical machines which can work not only as motors, but also as generators. A synchronous machine cannot start up by itself, therefore to run it up, special starting assistance is required.

In this paper, we shall report on experience with these aids: pony motors, asynchronous starting, and low-frequency starting.

With extreme variations in the turbine or pumping heads the turbine cannot run at constant speed. A suitable solution is offered here by the pole-changing machine; the two speeds permit adaptation of the operation to the storage level so that the plant can be run at the best possible conditions.

By using static converters to adapt the machine output to the network frequency, the turbine can operate under optimum conditions while running the set at variable speed.

Experience and design features will be described on the basis of machines which are now in service.

1. General Design Criteria

Modern pumped-storage power stations have to meet strict requirements in regard to operation and reliability. The specification for a motor-generator reflects the hydraulic situation and the network relationships, together with a number of other important factors, as follows [1]:

Hydraulic Situation:

Under this, we consider the effective power, the rated speed, tripping speed, the inertia constant, the type of construction, the rotation (in one or both directions) and the possibility of converter operation with variable speed.

Network Relationships:

This has to do with the apparent power, reactive power, the stability and control conditions (X_d, X_d'), the voltage range, the operating modes, switchover times, the frequency of starting, and starting methods in relation to allowable voltage drops, and other network data.

Ancillary Specifications:

Included here are such factors as space requirements, loss index, limits on temperature rise, and standards to be used.

For the planning of a pumped-storage power plant, a guideline describing the technical feasibility of 3-phase salient-pole machines is of great help. A family of curves illustrating the maximum economical ratings for various speeds and types of cooling is given in [1]. The lower limit shown for water-cooled machines, however, also depends on other factors such as the inertia constant and the loss index.

The hydraulic conditions at the site often do not provide a constant head, but one which varies over a wide range. In a case like this, two operating speeds must be available so that the pump-turbine can be operated with the best possible efficiency. A pole-changing salient-pole machine will have, depending on the combination of speeds, either two interwound windings [2] or a stator winding which can be switched over from one configuration to another [3]. The rotor poles will have at least two different pole configurations [2, 4]. The extra price for a pole-changing synchronous machine cannot be stated in general terms, because it depends largely on the speed combination used. Under favorable conditions, it will amount to an extra cost of 15 to 20 %. In each individual case, an economic analysis which takes into account all of the system components and main influence factors must be the determining factor.

The aim of having an optimum, continuously variable operating speed for the pump-turbine, so as to adapt to large changes in head, can now be reached even more effectively by using a frequency converter between the motor-generator and the network. The rapid, even spectacular, progress in power electronics in recent years makes it possible to use frequency converters for ever increasing power ratings [5]. The unit consisting of the frequency converter and synchronous machine must, of course, be investigated, designed, and optimized as a complete system [6]. Mutual interconnections and interreactions of an electrical and mechanical nature between the network, the frequency converter, and the motor-generator must be considered most carefully.

2. Starting Methods

For starting a reversible motor-generator in the pumping mode, any one of a number of methods can be used [7], as follows:

- starting motor(pony motor),
- asynchronous starting, direct on-line or with line-side control equipment such as reactors or starting transformers,
- back-to-back starting (or full-frequency starting),
- partial-frequency starting,
- synchronous starting with static converter.

In any given case, the optimum selection of the starting method will depend on:
- the network conditions (maximum allowable starting current),
- the total inertia constant,
- the countertorque (starting with the turbine full of water or dewatered),
- the desired runup time, and the frequency of starting,
- the space requirements for equipment, including cabinets.

In the following pages, we shall briefly describe and comment on the more important starting methods.

2.1 Starting Motor

This method is now seldom used, being increasingly replaced by static starting equipment. The starting motor is an asynchronous motor of the wound-rotor type coupled to the main machine; the rotor circuit is connected to a liquid resitor [8]. This method of starting is suitable only when the countertorque is relatively low, otherwise the costs and the space requirements cannot be justified.

2.2 Asynchronous Starting

In comparison with the other starting methods, asynchronous starting of a synchronous machine is faster, simpler, and cheaper, but on the other hand it leads to higher stresses in the machine. The approximate maximum unit ratings - disregarding the network conditions - for which asynchronous starting is possible are given in [1] in relation to the rated speeds and inertia constants. When necessary, the load on the network can be reduced by using certain line-side control equipment, but this will also affect the starting characteristics.

Torque Conditions in Asynchronous Starting

As is known, a synchronous machine when in the asynchronous state produces two torque components, namely a mean component m_a and a torque m_p oscillating at double slip frequency. Whose value and variation with speed depend largely on the rotor design. Fig. 1 shows typical curves for machines having laminated poles and a damper winding, and for those with solid poles.

A machine with solid poles or with a damper winding of poorly conducting material (e.g. brass or bronze) often has a saddle in the torque curve (see Fig. 1) for low slip values. A common

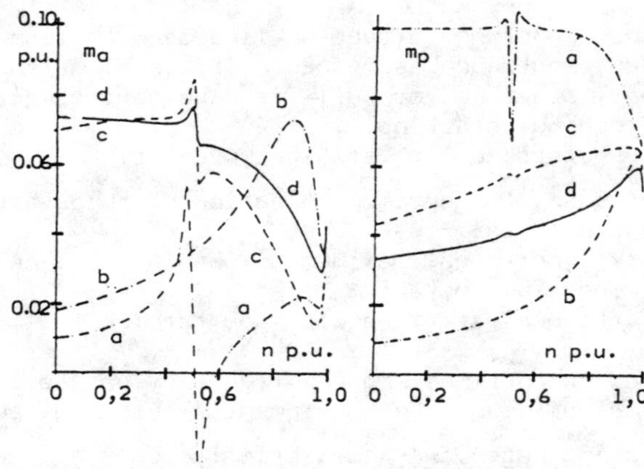

Fig. 1 - Influence of the rotor construction on the mean torque m_a and the oscillating torque m_p during the asynchronous starting
Synchronous machine 6020 kVA, 6.6 kV, 1500 rpm
Starting voltage 1620 V
a = laminated poles with grid damper winding
b = laminated poles with complete damper winding (Cu)
c = laminated poles with complete damper winding (Bz)
d = solid poles with pole connexions

way of improving this is to connect an additional resistance in the field circuit so as to increase the asynchronous torque at low slip. This resistance also reduces the so-called Görges' phenomenon at half speed, and reduces the oscillating torque, but only at low slip.

Various Possible Circuits for Asynchronous Starting

The simplest method is direct starting, in which the machine to be started is connected directly to the network. In this case, however, the starting current is very high, and causes a voltage drop in the network which many exceed the allowable limits.

If a reactor is inserted in the circuit then the starting current is reduced.The amount of the asynchronous torque, however, is also reduced. This means that a compromise must be found so that the starting torque remains high enough, while the starting current remains within the allowable range for the machine and for the network.

It is also possible to use a starting transformer with a ratio $k < 1$. The main advantage of this method over reactor starting is that for the same network current the machine current is greater, by ratio $1/k$, and the starting torque is higher, by ratio $(1/k)^2$. In cases where the starting current is very strictly limited, it is also possible to work with a tapped starting transformer.

Heat Energy in the Rotor

During an asynchronous start, an amount of heat energy E appears in the rotor circuits of the synchronous machine; the amount is determined as follows:

$$\frac{E}{S_n} = H + \int_0^{t_a} m_g \, s \, dt \qquad (1)$$

Where S_n = rated apparent power [kVA]
H = inertia constant [kWs/kVA]
m_g = countertorque [p.u.]
s = slip [-]
t_a = runup time [s]
E_{kin} = HS_n = kinetic energy [kVAs]

From equation (1) it can be seen that at very low levels of countertorque m_g, the heat energy in the rotor is practically equal to the kinetic energy E at synchronous speed, and hence is independent of the starting voltage, runup time, and rotor design. If, however, m_g is not negligible (e.g. when starting with the pump full of water), then as shown by equation 1 the relative starting heat E/S_N can be considerably higher than H. In other words, the amount of heat in this case depends on the runup time and the torque curve, i.e. on the rotor design.

The heat capacity of a rotor with solid poles is much greater than one with a cage-type damper winding. Consequently the solid-pole machines not only have the advantage of higher starting torques, but also, for thermal reasons, are very suitable for difficult cases of asynchronous starting.

It is not so much the total amount of starting heat in the rotor which is responsible for the stresses in the machine; the stresses are more a result of the heat distribution in the rotor body. Thus the rotor construction must be selected not only to give sufficient heat capacity, but also to avoid as far as possible any larger concentrations of heat.

The design phase for a machine to be used with asynchronous starting requires very precise modeling of the machine. The active parts of the rotor, and in particular the solid iron parts, must be very carefully simulated [9]. It is essential to check the rotor temperature rise, specially in regard to any local hot spots [10], which could be largely eliminated by suitable design measures (grooved pole-shoe surface, transverse expansion slots in the pole, suitable pole connections). It is only under these conditions, and with the backup of long, substantial experience, that asynchronous starting can be successfully applied in units of over 200 MVA [11].

2.3 Synchronous Starting

Machines which have laminated poles and damper windings are known for the fact that they are not well suited to absorbing the heat energy released during a heavy asynchronous runup. In addition, there are plants for which the power network conditions and the specifications covering voltage dips make it very difficult or even impossible to use asynchronous starting. In such cases, synchronous runup, with its various possibilities, becomes more attractive. This method is particularly suitable for pumped-storage power stations where there are a number of units. The modern techniques of static excitation or frequency converters permit full automation, and at the same time offer great flexibility and safety in operation [12].

Fig. 2 shows the main circuit used in a synchronous runup system. The driven, excited generator G produces a voltage and a current in the stator circuit which in turn produce a torque in motor M, so that the motor begins to rotate.

With this method, two basic cases can be differentiated [13]:

- Full-frequency starting: the motor is excited while still at standstill, and runs up in synchronism practically from the start.
- Partial-frequency starting: at first, the motor starts asynchronously, then after reaching the same speed as the generator it is excited and continues the runup in synchronism.

Neither of the two basic cases cause any significant stress in either the network or the machines themselves.

The success of a synchronous runup not only depends on the machine data (reactances, inertia GD^2, breakaway torque or countertorque) and the connections (conductors, transformers),

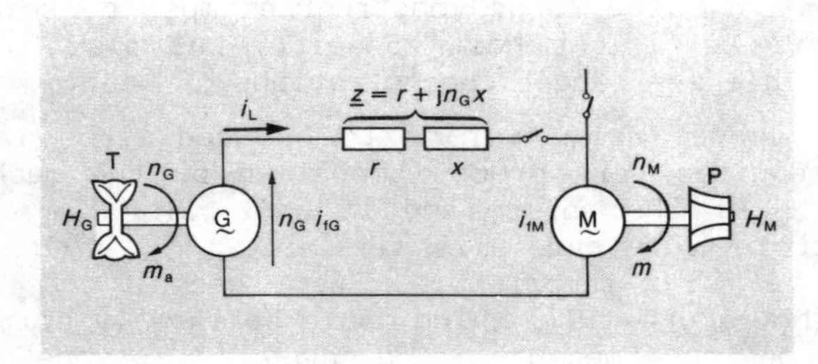

Fig. 2 - Basic circuit configuration for frequency starting
M = Motor, G = Generator, P = Pump, T = Turbine
H_M, H_G = Total inertia constants
n_M, n_G = Speeds
i_L = Line current
i_{fM}, i_{fG} = Excitation currents
m_a = Generator prime mover torque
m = Electromagnetic torque of the motor
$\underline{z} = r+jn_G X$: Line impedance

but also depends significantly on a number of parameters which can be selected as desired, i.e. on:

- the settings of the excitation systems,
- the drive torque on the generator,
- the initial speed of the generator (in partial-frequency starting).

These parameters must be optimized in each individual case, thus a comprehensive knowledge of the physical relationships involved is essential [13]. These starting procedures can be simulated, so that even in the preliminary design phase reliable predictions can be made and suitable optimizing steps can be taken. Synchronous runup in two existing plants are illustrated in Fig. 3 and 4.

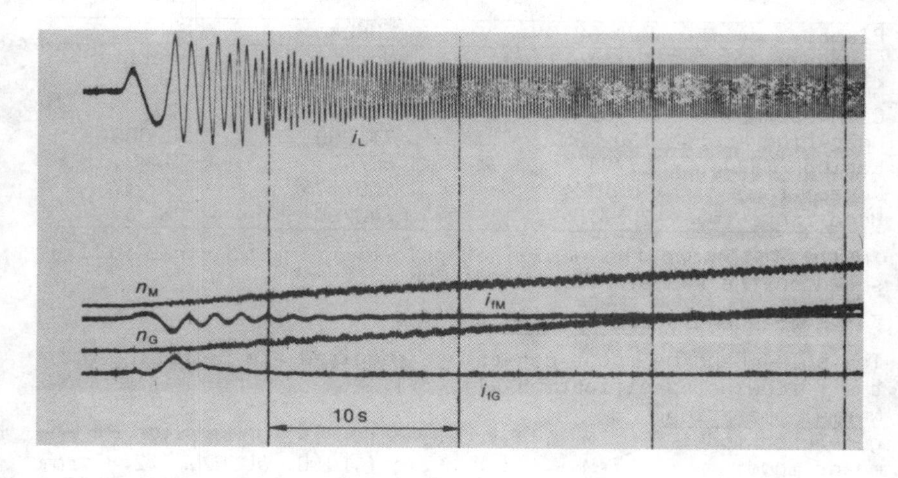

Fig. 3 - Full-frequency starting
Motor/generator ratings 64 MVA/32 MVA
Connection via unit transformers. For notations see Fig. 2.

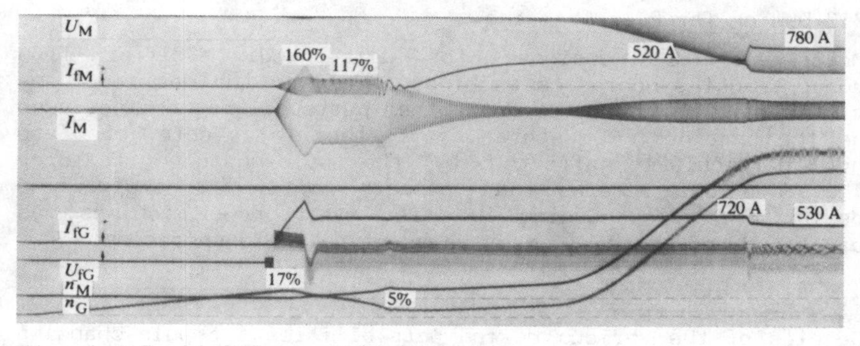

Fig. 4 - Partial-frequency starting
Machines: each 169 MVA, 13.8 kV, 600 rev/min
Connection via unit transformers
U_M, I_M = Terminal voltage and stator current of motor, for remaining notations see Fig. 2

3. Pole-Changing Synchronous Machine

Pole-changing synchronous machines have particular application when, for hydraulic reasons (e.g. cavitation, efficiency), large variations in the turbine or pumping heads cannot be handled by a single-speed machine.

The reference list (Table 1) shows several pole-changing machines designed and manufactured by ABB.

All of these pole-changing machines have given completely satisfactory service from the time they were installed.

Table 1 Pole-changing synchronous machines

Plant	OVA SPIN	JUKLA	MALTA
Country	Switzerland	Norway	Austria
No. of units	2	1	2
Commissioned	1970	1974	1977
Operating hours	130 000	73 000	38 000
Rating (MVA)	27/21	48/40	70/42.5
Speed (rpm)	500/375	500/375	500/375
Head range (m)	205/70	240/60	198/63

In the following the design of pole-changing machines will be shown on the example of PAN JIA KOU.

3.1 Motor-Generators for PAN JIA KOU

The pole-changing motor-generators required are be designed to the following specification [14, 15]:

Normal operation

generator mode: 98 MVA, 13.8 kV, p.f. 0.9, 50 Hz, 125 rpm
motor mode: 97.7 MVA, 13.8 kV, p.f. 1.0, 50 Hz, 142.86 rpm

Converter operation

generator mode: 66.48 MVA, 12.8 kV, p.f. 0.88, 48.8 Hz, 122 rpm
motor mode: 70.91 MVA, 12.61 kV, p.f. 0.88, 45.5 Hz, 130 rpm

3.2 Design for Pole-Changing

Motor-generators designed for pole-changing differ from conventional synchronous machines mainly in the design of the stator winding and rotor. All other parts of the machine, such as the bearings, the stator frame, the stator core, etc. are not particularly affected by the pole-changing feature. Therefore our remarks will concern mostly the engineering design for pole-changing in large synchronous machines, as practised and proven in many years of reliable operation.

3.3 Rotor Pole Design

Details of the various design possibilities for pole-changing on the rotor side are given in reference [2]. For the PAN JIA KOU machines, the following has shown itself to be the optimum arrangement:

- 6 pole groups, each consisting of 8 poles. Each pole group itself is symmetrical, and contains 3 narrow poles and 5

wide poles; the pole pitch (center to center distance of pole shoes) is non-uni-form (Fig. 5a).
- The 48-pole arrangement comprises all of the poles; the polarity changes from each pole to the nexxt, i.e. the poles have alternate polarities (Fig. 5b).
- For changeover to the 42-pole arrangement, every second pole group changes its polarities, and in each series of 3 narrow poles the middle one is disconnected and remains inactive (Fig. 5c).

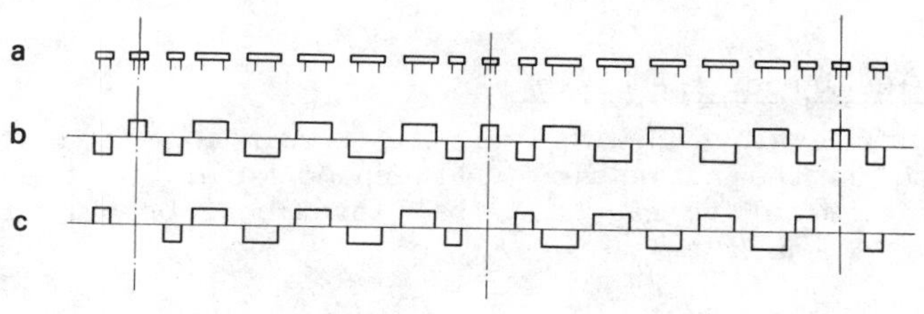

Fig. 5 - Schematic arrangements
a: Pole arrangement
b: Pole field corresponding to 48-pole
c: Pole field corresponding to 42-pole

The final pole arrangement and relevant field diagram is shown in Fig. 6.

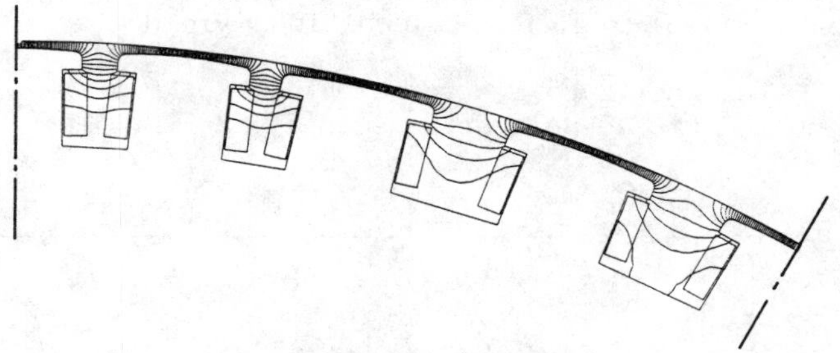

Fig. 6 - Pole arrangement and pole field for half of a pole groupe (48-pole connection)

3.4 Stator Winding

For the reason explained in [2], the machines will be provided with two separate windings for the two numbers of poles. In order to obtain the most favorable conditions in regard to the number of stator slots and the winding factors, both windings are designed as fractional-slot windings, accepting the presence of subharmonics in the armature reaction (parasitic forces).

Both windings are designed electrically as single-layer windings. The two bars in each slot belong to different windings. Adjoining bars of a winding are positioned alternately in the bottom and the top of the slots. Both windings together have the appearance of a normal two-layer winding, though mechanically there is an essential difference in that one bar in each slot carries no current, which increases the risk of thermally induced relative movement of the two coils.

Calculation and experiments have both shown that firm fixing of the windings to each other can prevent this relative movement. The resultant tensile and compressive stresses in the individual bars are within acceptable limits. At their ends, the windings are fixed together by an impregnated bandage.

3.5 Evaluation of the Design

The final evaluation of the stator windings and the pole changing on the rotor side will depend mainly on the quality of the terminal voltage and the vibrational behaviour of the stator.

3.5.1 Terminal Voltage. The international electrical standards require that the telephone harmonics (THF) for the machine not exceed 1.5 %. This is checked realitively easily, using a very flexible computer program for exact calculation of the air-gap fields for any given arrangement of poles. The variation of terminal voltage with time, and hence the THF factor, can be determined by Fourier analysis of the relevant pole-field curves (Fig. 7) and the corresponding winding factors for the relevant stator winding. Our calcualtions yield the following results:

for the 42-pole arrangement: THF = 0.52 %
for the 48-pole arrangement: THF = 0.45 %

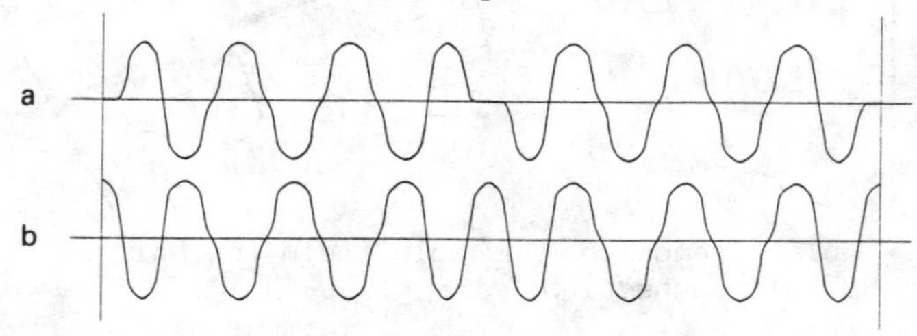

Fig. 7 - Calculated pole field curves
a: 42-pole connection
b: 48-pole connection

3.5.2 Vibrational Behaviour of Stator. The harmonics spectrum of the pole-field curve is generally much wider in pole-changing synchronous machines than in conventional machines. There are also characteristic subharmonics. In addition, certain compromises have to be made during the design of the two stator windings, in regard to the number of slots per pole and phase and the relevant winding pitch. Hence the spectrum of the harmonics of the armature reaction field is also much more apparent here than in the conventional case.

The values of the exciting parasitic forces are obtained from the Maxwell stress $B^2/2\mu_0$; these forces act on the stator core and on the poles, in both the radial and the tangential directions. The term B represents the resulting air-gap flux density (sum of pole field plus armature-reaction field). It is clear that in pole-changing synchronous machines with their distinct harmonics in the pole field and in the armature reaction field, the spectrum of parasitic forces will also be greater. By using a computer program, the amplitudes, frequencies, and number of nodes of these forces can be calculated.

The vibrational behaviour of the stator has been checked in regard to these exciting forces, using a finely detailed mechanical computer model for the stator. The calculations show, that the vibration amplitudes lies below 15 μm zero-to-peak.

4. **Static Frequency Converter for Adjustable Speed Control**

By using a static frequency converter in the system, it is possible not only to prevent the machine from falling out of step, but also to drive it at any desired speed up to twice its synchronous speed [5, 6, 15]. The control range can be from 10 % to 200 %, which can be applied in either the pumping mode or the generating mode of hydro-electric sets, the only limitation being the mechanical design of the set.

4.1 Design and operation of a Static Frequency Converter

A static frequency converter consists mainly of:
- a line-commutated converter, which rectifies the line current to dc current and controls its value,
- a machine-commutated converter, which changes the dc into an ac current having a variable frequency corresponding to the required machine speed,
- a dc intermediate circuit containing a reactor for smoothing the dc and for decoupling the two systems,
- a control circuit for monitoring and controlling the converter,
- a static excitation circuit for controlling the voltage and flux in relation to the speed.

The purpose of the machine-commutated converter is to feed this dc voltage to the synchronous machine so as to produce a rotating field in the stator.

The semiconductors in the machine-commutated converter are connected in pairs so that at any given time, two phases in the stator are energized in order to produce the field. With suitable commutation on the semi conductors, the field can be made to rotate and the rotor will follow.

The significant point here is that the instant of commutation is given by the position of the rotor itself, which ensures that the rotor neither hunts nor slips.

4.2 Operating Modes of the Static Frequency Converter

For the pump-turbine set, the following operating modes of the static frequency converter are provided:

a. Pumping mode, with variable speed. Here the line-commutate converter is controlled so that the pump is driven at the desired speed. The line-commutated converter then operates as a rectifier, the machine-commutated converter as an inverter. The voltage and reactive power required for commutation are delivered by the network or by the over-excited machine.
The set is run-up to speed with the watered turbine and against closed guide vanes. When the required speed is reached, the vanes are opened and the turbine speed is controlled according to the pumping head and power drawn.

b. Turbine mode, with variable speed. For this mode, the function of the two converters is reversed: the line-commutated converter acts as an inverter, the machine-commutated converter as a rectifier. The electrical energy is produced at a variable frequency corresponding to the set speed, and is changed into electrical energy at the line frequency and is fed into the network.

c. Run-up mode. Here there are two kinds of operation possible:
I. Synchronous condenser operation. With the turbine dewatered, the set is run up to synchronous speed by the converter and is synchronized with the network. Following this, the converter is disconnected and the generator can deliver re-active power to the network.
II. Pumping operation at fixed speed. This is similar to the previously described condenser mode. The motor-generator is run up to synchronous speed and is synchronized. After disconnection of the static frequency converter, the guide vanes are opened and pumping operation at synchronous speed is started. Previous dewatering of the turbine is not necessary.

4.3 The Design of the Pan Jia Kou Static Frequency Converter

The frequency converter control at the Pan Jia Kou station [15] will initially be applied only for one pump-turbine set at a time, at a reduced rating. The main data for the converter are:

Rated load	60 MW
Rated voltage	13.8 kV
Rated current	2800 A
Rated frequency	0-50 Hz
Control range	5-50 Hz

Voltage matching on the machine side is provided by series connection of thyristors. Each leg of the ac bridge contains 14 thyristors in series, of which one thyristor is redundant, i.e. if one thyristor fails, the converter can continue to operate at rated load.

Thyristor firing and monitoring is performed by light emitting diodes and pulse transmission by fiber optics.

Bibliography

[1] K. BALTISBERGER, M. CANAY, H. VOEGELE and M. WIMMER. Motorgenerators for pumped storage schemes. Brown Boveri Rev. 65 1978 (5) 280 - 291

[2] H. VOEGELE. Pole-changing techniques in high-rating pumped storage machines. Brown Boveri Rev. 61 1974 (7) 327 -331

[3] J. CHATELAIN and M. TUX XUAN. Pole-changing synchronous machines. Cigré, Symposium Rio de Janeiro 1983 contribution 220 - 07

[4] K. BALTISBERGER. Motorgenerator sets for the upper stage of the "Malta" (Austria) pumped storage scheme. Brown Boveri Rev. 62 1975 (7/8) 348 - 355

[5] F. PENEDER, R. LUBASCH and A. VOUMARD. Static equipment for starting pumped storage plants, synchronous condensers and gas turrbine sets. Brown Boveri Rev. 61 1974 (9/10) 440-447

[6] J.-J. SIMOND. Modelling of the system synchronous machine, frequency converter and three-phase network with simulation of its dynamic behaviour. etz Archiv 10 1988 (12) 375 - 379

[7] M. CANAY. Methods of starting synchronous machines Brown Boveri Rev. 54 1967 (9) 618 - 629

[8] K. HIRT and M. WIMMER. The Drakensberg pumped storage plant Brown Boveri Rev. 70 1983 (7/8) 268 - 279

[9] M. CANAY. Extended synchronous machine model for the calculation of transient processes and stability. Electric. Mach. a. Electromech. 1 1977 (2) 137 - 150

[10] J.-J. SIMOND. Transient temperature rise in the solid-iron poles of salient-pole machines. ICEM, Budapest 1980

[11] M. CANAY. Asynchronous starting of a 230 MVA synchronous machine in Vianden 10 pumped storage station. Brown Boveri Rev. 61 1974 (7) 313 - 318

[12] F. PENEDER and V. SUCHANEK. Static frequency changers for driving and running up high-power synchronous machines. Brown Boveri Rev. 67 1980 (9) 524 - 529

[13] B. BOSE, M. CANAY and J.-J. SIMOND. Frequency starting in pumped storage plants, possibilities and operation. Brown Boveri Rev. 70 1983 (7/8) 295 - 302

[14] M. JAQUET. "The Pan Jia Kou Pumped Storage Station, Part I: Hydraulic Equipment" Fourth ASME International Hydro Power Fluid Machinery. Symposium, December 1986

[15] H. VOEGELE and K. WEBER. "The Pan Jia Kou Pumped Storage Station. (Pump-turbine operated with two speeds and with variable speed). Part II : Electrical Equipment" Fourth ASME International Hydro Power Fluid Machinery. Symposium, Anaheim California, Dec.1986.

[16] A. MEYER and R. LUBASCH. "Digital Simulation of Converter-Fed Synchronous Motors" CEA Spring Meeting, March 20/24, 1983, Vancouver, Canada

Operation of pumped storage power stations

P. GUICHON, Electricite de France

ABSTRACT :

Faced with an economic situation that has evolved little since 1984, Electricité de France (EDF) has enhanced the operation optimization of its generation-transmission system. The result is increased use of hydroelectric power plants and, in particular, pumped storage stations (STEP).

A statistical review between 1984 and 1986 shows a substantial increase in the use of STEP's, together with a growth in French network output due to pumped storage and improved kinetic qualities in turbining and in pumping. The appreciable drop in the cost of fossil fuels in 1987 tends to decrease and even to stop the progression of these growth indexes.

The Monitoring and Planned Maintenance Method (MECEP), developed and applied for many years by EDF's Hydroelectric Generation Department, has made it possible to constantly lower the number of incidents during the startup of pump turbines (rate of successful startups in the main STEP's in 1987: 98 %).

In addition, improved automatic control systems have resulted in fewer incidents and in reduced connection time. The required maintenance time has also been reduced owing to the use of specialized tools and robotized repair techniques.

INTRODUCTION

The power demand of the French network varies at a 1:2 ratio on a yearly basis, with daily fluctuation ranging from 30 to 50 %.

So as to cope with these load variations at all times, one must be able to make use of generating facilities with rapid startup capabilities and high flexibility.

In addition to this technical aspect, there is the economic aspect where the periodic, but especially random, character of marginal production costs makes operation optimization an extremely complex matter.

Pumped storage stations (STEP) meet these two types of concerns.

The installed capacity today is 5000 MVA.

DESCRIPTION OF FRENCH PUMPED STORAGE STATIONS

In 1973, the overall capacity of pumped storage stations in FRANCE was 172 MVA. Fifteen years later, this capacity has been multiplied 30 times.

This strong growth is characterized by three periods:

- The first period is between 1973 and 1981: this is the period of the construction of pumped storage stations with daily energy transfer. This concerns all power facilities with a storage or pumping capacity allowing only pumping in off-peak hours or at certain times of the day.

Two variants were built:

. STEP's without gravity inflow or hydraulic links with other stations
. STEP's connected hydraulically to a series of power facilities conditioning their operation to a great extent.

- The second period between 1982 and 1984 is more favourable to uses of STEP's with a reservoir capacity making filling possible during the weekend and turbining during the peak periods of the week. These are known as power stations with weekly transfer.

- The last construction phase from 1985 to the present day provides the possibility of using STEP's with seasonal reservoirs. It should be noted that these power stations also perform weekly and daily pumping operations.

The table below presents the characteristics of pumped storage stations in France today :

	UNITS	LAC NOIR	VOUGLANS G4	REVIN	STE CROIX G2	LA COCHE
Kind of pumping cycle	-	D	D	D	W	C/D
Commissionning year	-	1938	1973	1974	1976	1977
Upper reservoir capacity	hm^3	3.9	419.5	6.9	301	2
Lower reservoir capacity	hm^3	2.3	1.7	6.9	8.1	0.4
Maximum gross head	14 m	125	97	247	83	927
Total capacity	MVA	100	72	800	64	340
Maximum turbining flow	m^3/s	100	75	411	70.6	40.5
Maximum pumping flow	m^3/s	56.3	72.5	284.4	70.6	32.2
Types and number of reversible units (3)	- -	T 4	R 1	R 4	R 1	R 4
types of startups (4)		T	AS Watered Runner	Pony Motor	AS Watered Runner	DA

(1) D = Daily transfer, W = Weekly, S = Seasonal, M = Combined gravity-STEP

(2) For combined gravity-STEP schemes, the capacities and flows concern reversible equipment only

(3) Type of generating unit : T = Units with 3 turbines, R = Reversible units

(4) Type of startup: T = Turbine (ternary unit), DA = Direct asynchronous startup, AS = Asynchronous - Synchronous, FTC = Frequency thyristor converter

The table below presents the characteristics of pumped storage stations in France today :

	UNITS	LE CHEYLAS	MONTEZIC	POUGET G5	GRAND MAISON 2	VIEUX PRE	SUPER BISSORTE 2
Kind of pumping cycle	-	C/D	W	S	C/S	S	C/S
Commissionning year	-	1981	1982	1982	1987	1988	1988
Upper reservoir capacity	hm^3	4.9	30	9.3	134	50	37.5
Lower reservoir capacity	hm^3	4	20	2.8	19	0.5	1.2
Maximum gross head	14 m	261	423	445	926	77	1159
Total capacity	MVA	540	1000	41.5	1360	12	680
Maximum turbining flow	m^3/s	234	252	10.2	140.8	19	56
Maximum pumping flow	m^3/s	170	206.4	6.6	135	17.1	50.8
Types and number of reversible units (3)	- -	R 2	R 4	R 1	R 2	R 8	R 4
types of startups (4)		FTC	FTC	AS	DA	Back to back	Back to back Watered Runner

(1) D = Daily transfer, W = Weekly, S = Seasonal, M = Combined gravity-STEP

(2) For combined gravity-STEP schemes, the capacities and flows concern reversible equipment only

(3) Type of generating unit : T = Units with 3 turbines, R = Reversible units

(4) Type of startup: T = Turbine (ternary unit), DA = Direct asynchronous startup, AS = Asynchronous - Synchronous, FTC = Frequency thyristor converter

USE OF PUMPED STORAGE STATIONS

The ability of STEP's to take constant part in matching generation and consumption, or to provide the network with immediate backup under incident conditions, requires the utilization of facilities in the 5 following areas:

1 - Energy Transfer

This was the basic function at the outset. Since then, however, the difference in cost between oil and coal fuels has continued to drop. This downturn phenomenon makes its effects felt particularly at the marginal production costs level.

The result is a reduction of the economic valuation of the energy transfer function and, in particular, for daily cycle STEP's.

This can be illustrated by the development of the number of hours of pump operation for a daily cycle power station (REVIN: 800 MVA) and a weekly cycle power station (MONTEZIC: 1000 MVA).

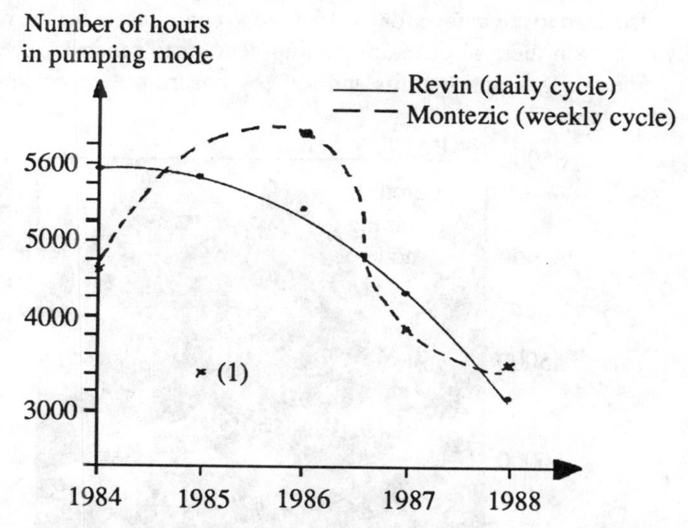

(1) Reinforcement work on the Montezic headrace works lasted 5 months and greatly modified the operation results so that it is not necessary to take this point into account.

Despite this, fuel saving in 1985 amounted to some 50 million French francs for each power station.

2 - Power Reserve

The STEP's play an active part in spinning reserve, the purpose of which is to cope with unforeseen generation or consumption situations.

This results in savings on the startup cost and on the cost of operating at reduced power the additional thermal units required to create this reserve.

The fast startup capabilities of STEP's, thanks to centralized control facilities (Hydroelectric Control Centres) makes them particularly well adapted to this type of use.

The saving obtained, compared with what it would have cost to use thermal facilities providing the same service, came to 40 million French francs in 1985 for pumped storage stations such as MONTEZIC or REVIN.

3 - Load Follow

The implementation of these power stations with high load variability is also used to advantage for load follow to compensate for the relative rigidity of certain thermal generation equipment, or to minimize the constraints of certain generating units and thereby ensure the generation-consumption balance.

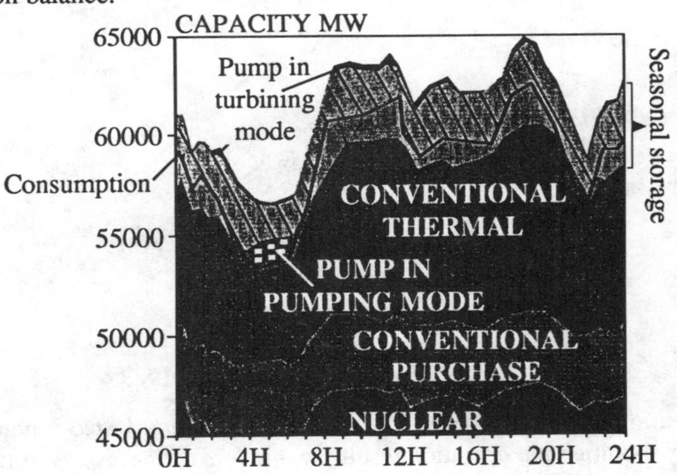

4 - Frequency - Power Remote Control

About 10 % of the total STEP capacity contributes to controlling the frequency and exchanges with foreign countries.

5 - Synchronous Condenser

When no active power is consumed or produced, these facilities can also operate in synchronous condenser mode and gradually provide or absorb the reactive power for voltage map maintenance.

The synchronous condenser mode operation time of STEP's generally varies between 12 % and 20 % of the overall operation time.

STEP's era less used in pumping mode, but are used for the same number of hours per year in synchronous condenser mode.

This function is, in fact, experiencing some growth.

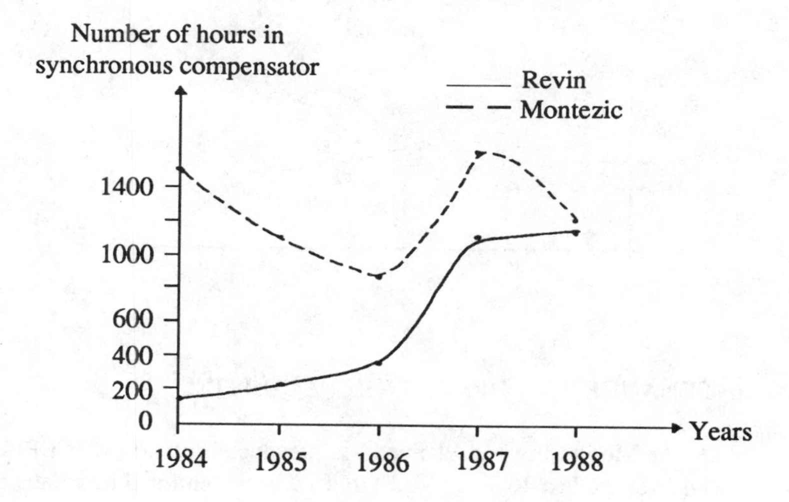

THE RELIABILITY OF PUMPED STORAGE STATIONS

The reliability of STEP's is an essential criterion of the performances to be met so that they fulfil their mission.

Reliability is characterized by the annual power availability rate and the rate of incidents during transitions (change of operating states, including startups).

The criterion selected to measure the availability rate of hydroelectric facilities is the ratio of the sum of the rated capacities of unavailable equipment to the total rated capacity. This calculation is performed every Wednesday at 8 a.m.

The rate of incidents during transitions is the proportion of the number of faults during transitions to the total number of transitions.

For all French STEP's, equipment availability over the past five years, apart from maintenance, is97 % +/- 1 %.

The rate of incidents during transitions criterion shows a steady downward trend.

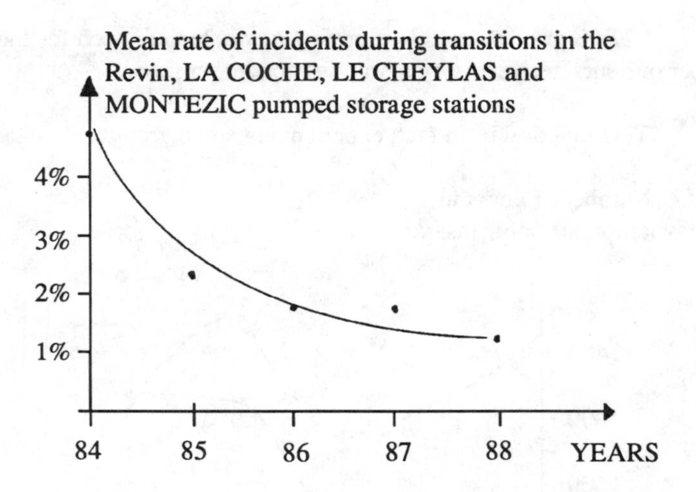

MAINTENANCE AND OPERATION FACILITIES

The Monitoring and Planned Maintenance Method (MECEP) has been used over the last 20 years for EDF's 1,21 conventional hydroelectric units and related equipment.

The evolution of the mean incident rate is proof of the pertinence of this method. In addition, maintenance costs for all hydroelectric power stations have been practically constant over the past eight years despite an increase of 25 % in installed capacity.

At the same time, expenses incurred for fault occurrence have witnessed a downward trend.

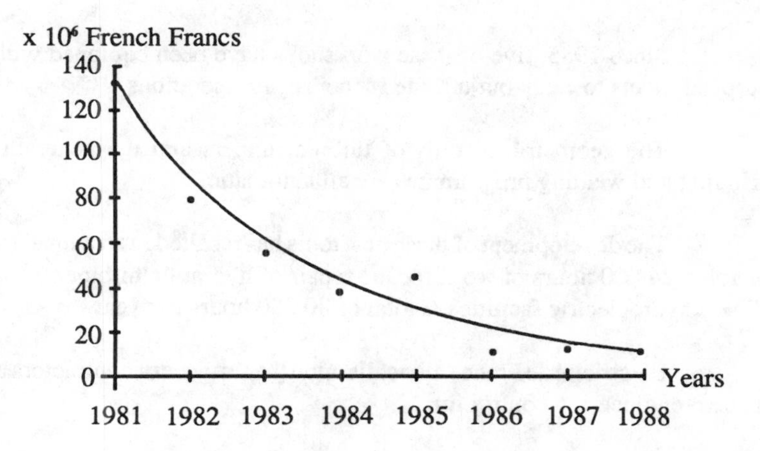

These results have been obtained owing to operational structures which are able at all times to receive the technical backup of specialists.

To make enhanced results possible and use to account the experience acquired in the field of maintenance, the computerization of data collection and assessment by an expert system is in progress.

1 - Maintenance Structures

Maintenance is organized around hydroelectric power station groups: light, local maintenance structures grouped into hydroelectric generation subareas (136 groups - 39 subareas in FRANCE).

The staff of the power stations perform operation, troubleshooting and routine checks, as well as day-to-day maintenance of computer, electrical and mechanical equipment.

For example, GRAND'MAISON power station (1360 MVA in pumped storage plus 680 MVA for seasonal gravity inflow) is operated by only 38 persons, and MONTEZIC power station (1,000 MVA) by 23 persons.

The subarea possesses the complementary human and technical skills to optimize the structure, skills that can be called upon at short notice for major maintenance operations.

In the field of hydromechanical equipment, both staff and equipment (high-capacity machine tools, high-performance tools) have been grouped in workshops.

Since 1985, five of these workshops have been equipped with all-purpose robots to carry out turbine runner repair operations.

The geometric records of turbine runners and the generation of grinding and welding programmes are all automatic.

The development of these new tools has resulted in an annual saving of some 24,000 hours of work for the repair of hydraulic turbine runners in EDF's hydroelectric facilities, a total of 80,000 hours per year.

Electricité de France also calls upon the services of contractors under its own engineering control for:

- major modernization and rehabilitation programmes;

- overhaul of generators or transformers by specialized national repair companies;

- civil engineering work entrusted to regional firms or to national public works companies.

The subareas are grouped into 9 hydroelectric generation regions called Hydrogeneration Areas.

2 - Technical Structures

The Hydrogeneration Area has specialized engineers who provide the operational structures with their technical skills in civil, mechanical and electrical engineering, conventional or computerized control, and in hydraulics, hydrology and environmental sciences, etc.

Measuring and monitoring teams also take part in electrical tests of control and protection equipment.

The Hydroelectric Generation Department coordinates the services of these specialists and liaises with the other departments of Electricité de France, in particular with the Research and Development Division. Whenever required, it can also mobilize the appropriate means of the Company and form the liaison with contractors and manufacturers.

The Hydroelectric Generation Department's General Engineering Branch at Grenoble has been instrumental in enhancing this technical assistance organization. It has the necessary personnel and equipment to implement specific techniques for the control of hydroelectric schemes which require adapted equipment and tried and tested methods. Highly experienced teams perform measurements and analyse subsequent results by means of powerful digital computer facilities. The General Engineering Branch enjoys a reputation for scientific competence which allows it to serve as a virtually independent expert in many fields within the Company, and also with the administration and with manufacturers, and abroad.

CONCLUSION

The drop in the use of STEP's observed over the past few years should gradually slow down and show a reverse trend as soon as the optimum level of all French generating facilities is reached.

This relative decline in the use of STEP's in no way detracts from the great advantage offered by these installations in the way of immediate and firm power availability.

This is why EDF is continuing its efforts to improve the performances of pumped storage stations by making use of operation feedback and state-of-the-art techniques.

Pumping in a tidal power plant: experience at La Rance and main aspects of the turbine design

J.P. FRAU, Electicité de France, and P.Y. LARROZE, Neyrpic, France

SYNOPSIS. Like conventional hydraulic power plants, pumping can be used in tidal plants. Since 1966, LA RANCE was operated according to a double effect scheme, with pumping. 20 years of operation show that pumping increases energy generation quite significantly. Contrary to conventional pump storage schemes where the efficiency is about 0.7, LA RANCE power plant shows an efficiency of more than 1. This is because the basin is ower filled at low heads, thus offering an extra head available at much higher heads during the generating cycle. ELECTRICITE DE FRANCE developed a dynamic programming method to optimize the income and the various operating modes of the plant. To suit the various cycles at best, up to date hydraulic design will feature :

- adjustable runner blades designed for high specific discharges and allowing good pumping efficiencies while maintaining top level efficiencies in the turbine mode,
- adjustable of fixed guide vanes, designed for high discharge,
- a unit layout featuring a direct driven bulb unit (in open forebay or in closed conduct) or a pit turbine with step-up gear for lower outputs,
- an automatic draft tube gate for safety and reliability during starting, stopping and all other numerous transients involved in a power plant.

A TWENTY FIVE YEARS EXPERIENCE

Characteristics of structures

Although they have been often mentioned, we must rapidly recall the basic data of this plant.

1. Civil structure. From the left to the right bank :
   - a 13 m wide lock
   - the power station : 390 m long, 53 m wide, 33 m high
   - the 160 m long rock-fill dyke, or inactive dyke
   - the 115 m long dam equiped with six 15 X 10 m sluice gates

The active parts of the structure are therefore the power

station and the dam where the 6,000 cu.m/s of the estuary go through.

2. Electromechanical equipment
   (a) - 24 horizontal bulb units
   - rated output : 10 MW
   - turbine runner diameter : 5.35 m
   - number of mobile blades : 4
   - mobile distributor : 24 blades
   - rated speed : 93.75 rpm
   - discharge : 275 cu.m/s
   - generator : 10 MVA - 3.5 KV

   (b) 3 transformers, 80 MVa - 3.5/220 kV
   Each group of 4 units discharges on the tertiary of a "block" transformer directly connected to the 220 kV network through an oil cable under 3.5 bar

3. Dam gates. Six (15 x 10 m) sluice gates operated by oil servomotors are installed on the right bank of the dam for emptying and filling of the basin.

Operation of LA RANCE Power Plant

Many engineers have been interested for long in tidal energy and, more particularly, in the operating cycles of a tidal power plant. Without knowing the results and requirements, the most complete cycles were selected. The choices for LA RANCE Power Station were based on R. GIBRAT's studies and resulted from experiments and well-determined economic and technical conditions.

They have led today to a simple and cost-efficient operation. This optimum use of tides involves two cycles.

(a) Single-effect cycle (with or without pumping) (Fig.1)
One-way use of the tide : from the basin to the sea
- filling-in of basin
  Reverse orifice operation of the units (machines not connected to network), dam gates in open position
  Direct pumping operation can be used depending on power demand and energy costs
- stopping
- emptying of basin
  Direct flow through turbine - Stopping of machines for H = 1.20 m.

(b) Double-effect cycle (with or without pumping) (Fig.2)
Optimum use of the tide.
- filling-in of basin
  Reverse flow through turbine with machines stopped when water level difference reaches 1.70 m
  Completion of basin filling-in in reverse orifice operation, the dam gates being open

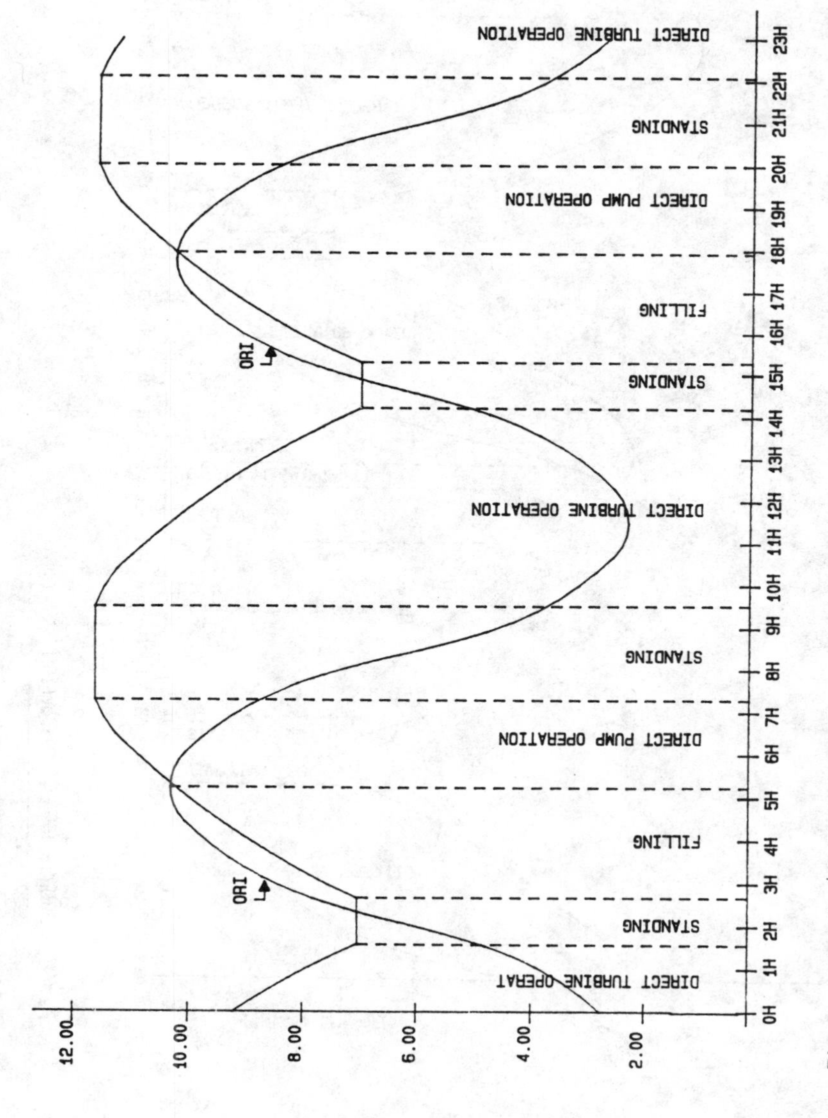

Fig 1 : SINGLE EFFECT OPERATION

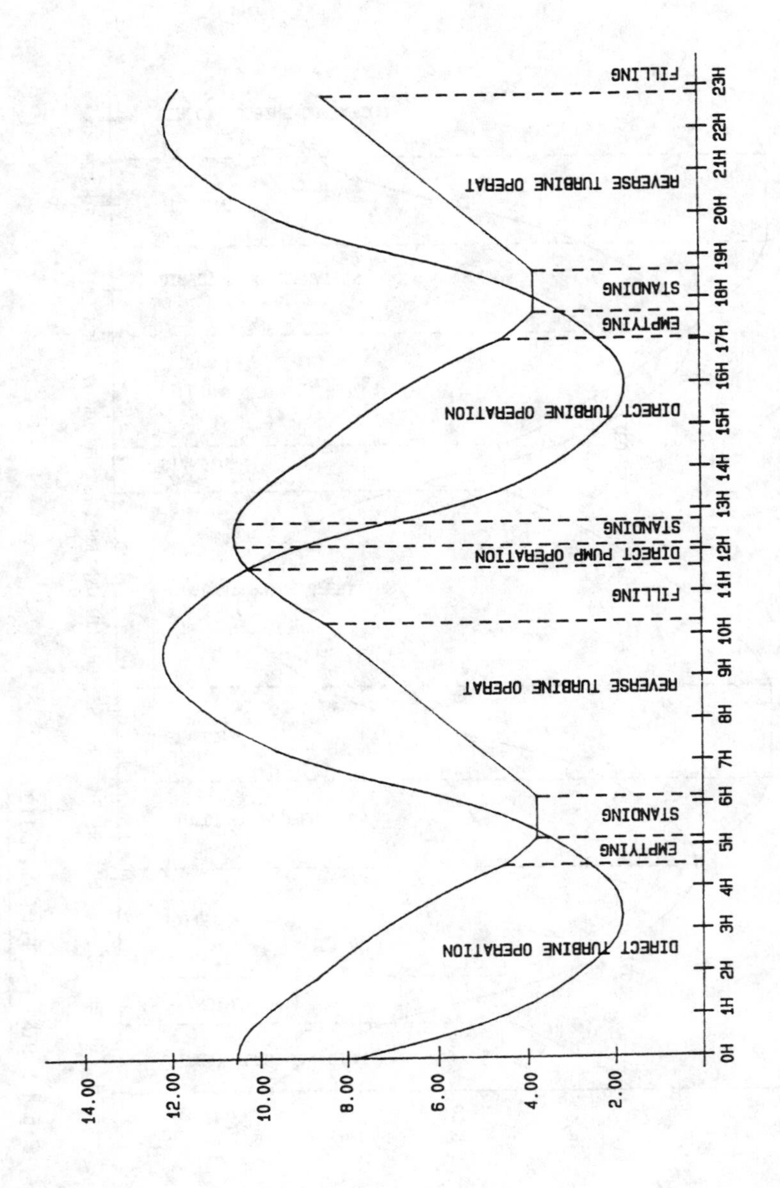

Fig 2 : DOUBLE EFFECT OPERATION

Pumping operation can be used depending on the power demands or energy costs.
- stopping
- emptying of basin
  Direct-flow through turbine and, on completion of the cycle, direct orifice operation and dam gate opening to speed-up the basin emptying.

4. Evaluation - Results. (Fig.3) A brief analysis of the life of this unique tidal power plant in industrial operation permits to differentiate several periods.

(a) tests and starts-up : until the end of 1968
(b) research of energetic optimum : 1969 to 1974
(c) a necessary repair : 1975 to 1982
(d) incident-free operation since 1983

A FINER ANALYSIS OF THE OPERATION RESULTS OBTAINED SINCE 1980 IS GIVEN BELOW

YEAR	GROSS PRODUCTION GWH	PUMPING CONSUMPTION GWH	DURATION (TOTAL OF 24 UNITS) IN %					
			TD	TI	PD	P1	O	TOTAL HOURS
1980	503	10	64	5	3	0,1	28	116 800
1981	570	62	61	2	16	0	21	134 700
1982	607	88	59	2	18	0	21	151 000
1983	610	98	57	6	17	0,1	20	155 900
1984	609	107	58	3	18	0	21	157 500
1985	612	118	58	5	19	0	18	158 110
1986	595	99	58	3	20	0	19	156 391
1987	578	91	57	5	19	0	19	156 077
1988	575	80	57	5	20	0	18	160 360

It can readily be noted that the design data (efficiency) favour direct operations : direct turbine and direct pumping. Direct pumping accounts for about 60 % of the total operating time, but it is of course the best normal way of using the machine !
After 1982, direct pumping corresponds to about 20 % of the operating time ; this shows the importance and economic advantage of this function.
Orifice operation is a natural operation and is only complementary to dam gate opening to pass the maximum amount of water into the estuary.
Runner distributor and blades are locked in fully-open position for a maximum discharge with unit speed lower than the rated speed. Advantage is taken from speed operation to start the machines in direct pumping mode. This method allow to avoid stresses induced by asynchronous starts-up.

Reverse operations are very seldom used. Reverse pumping has been entirely discontinued and reverses the operating time. The owner has carefully chosen between the loadings detrimental to the equipement and the expected gain. However,

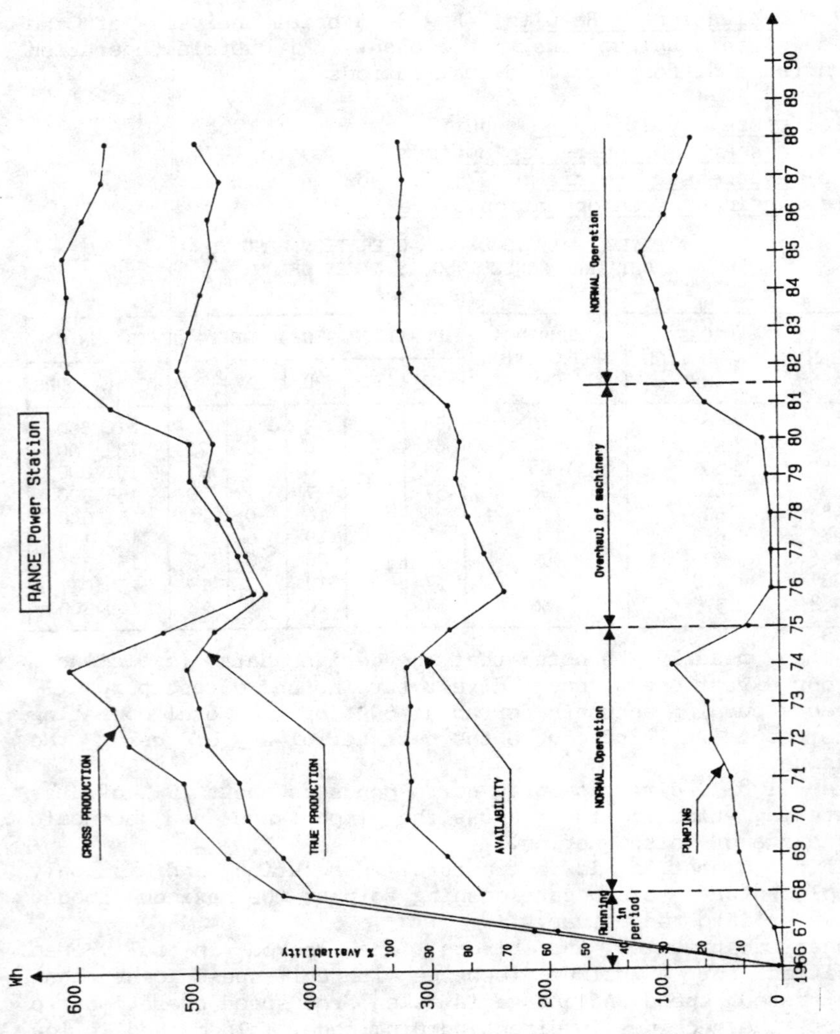

Fig 3 : Production, Consumption and Availability of Generating Sets

it must be stressed that operation in reversed-flow through the turbine may give a significant contribution to power production when carried-out at a well-chosen hour. This explains the 8,000 hours of reversed-flow operation.

5. Modernization of control system. In view of the interesting operation results obtained and due to the programmable controller becoming antiquated, E.D.F. decided to modernize the control system with the following complementary targets :

- ensuring fully-automatic and unattended operation,
- improving control performance.

The modernization work was carried out in 1987 and 1988.

Pumping mode - Advantage and specific features

The managers of the Production/Transport/Consumption system may tell you how interesting LA RANCE power station is. The power at the station terminals is today very small compared to the installed power of the French network, and it is therefore used for district production. However, it must be kept in mind that the operation of the power station is based on the research of maximum return.

It is of interest for both the designers of new tidal power stations and the owners to know :

- the choice of the best type of operation regarding power production (single or double effect with or without pumping) ;
- the modulation of marginal costs which may be the ratio of energy cost between on-peak and off-peak hours.

The analysis of the operation results has shown the interest of pumping mode. Today, it accounts for nearly 20 % of the operating time of the machines.
Contrary to conventional plants, the analysis of the operation results of a tidal power station does not allow to demonstrate the advantages of pumping. However, the use of the optimization program permits simulations which can help the Engineer in its task. In May 1986, during the third International Symposium on Wave, Tidal, Otec a small scale hydro energy (Brighton-England), Messrs Hillairet and Weisrock (E.D.F.) presented these simulations and their results.

Based on data issued from simulations made at that time, we can speak to you of :

- the interest of pumping regarding energy efficiency in a tidal power plant ;
- the valorization of production.

6. Pumping : energy efficiency (Fig.4). Contrarily to conventional pumping power stations, the downstream level of a tidal power plant varies in time. During direct pumping operation (sea to basin) the downstream level is maximum while the low-tide period is used for turbine operation.It readily appears that such special feature leads to an original situation where energy efficiency is much higher than in conventional situations (for which efficiencies remain smaller than 0.8).

Let us illustrate this with the results of LA RANCE power plant, considering a tidal amplitude from 4 to 12 m. The simulation studies have given for each tidal amplitude and on a constant-cost basis :

ET1 = net energy of single-effect turbine operation, without pumping ;
ET2 = net energy of single-effect turbine operation, with pumping ;
EP2 = energy of single-effect pumping.

The study of these types of operation allows to calculate the pumping energy efficiency which is defined as the ratio between power generation difference with and without pumping and single-effect pumping.

$$\frac{ET2 - ET1}{EP2}$$

Table 1 and curves 4 show that this efficiency varies from 1.5 to 2 depending on tide amplitude. This result may appear somewhat surprising but the effects of tide on downstream level variation - and therefore on head variation - must be considered. However, it should be noted that maximum is reached for amplitude of 8-11 m.

7. Valorization of the production. A series of simulations conducted over two periods (spring tide : coefficient 76.4 and neap tide : coefficient 69.03) and based on modulation ratio valorization from 1 to 9.81, have led to the following conclusions :

- the difference between single-effect with pumping and without pumping varies from 10 % to 20 % depending on the valorizations used ; the double effect with pumping results in a gain of 2-10 % against single-effect with pumping ;
- a highly contrasted valorization (9.81) does not affect too much the quantity of produced energy ;
- energy pumping efficiency being very high, it may lead depending on the modulation ratios to a very high efficiency in value, when pumping is made at given hours.

Valuation	Springs fortnight			Neaps fortnight	
	const. costs	Val. 3	GRETA	const. costs	Val. 3
Modulation ratio	1	5.70	8.81	1	5.70
Single operation without pumping, T_1 (GWh)	22.43	21.61	21.35	19.80	18.99
Single operation with pumping actual (GWh)	24.60	22.84	21.68	22.31	
Single operation with pumping, T_2: produc. GWh	28.40	28.76	28.31	26.36	27.52
P_2: consump. GWh	3.80	5.93	6.63	4.05	8.06
Energy yield of pumping %	157	121	105	162	106
Value yield of pumping %	157	176	192	162	170

Tabl 1 : Energy yield of pumping

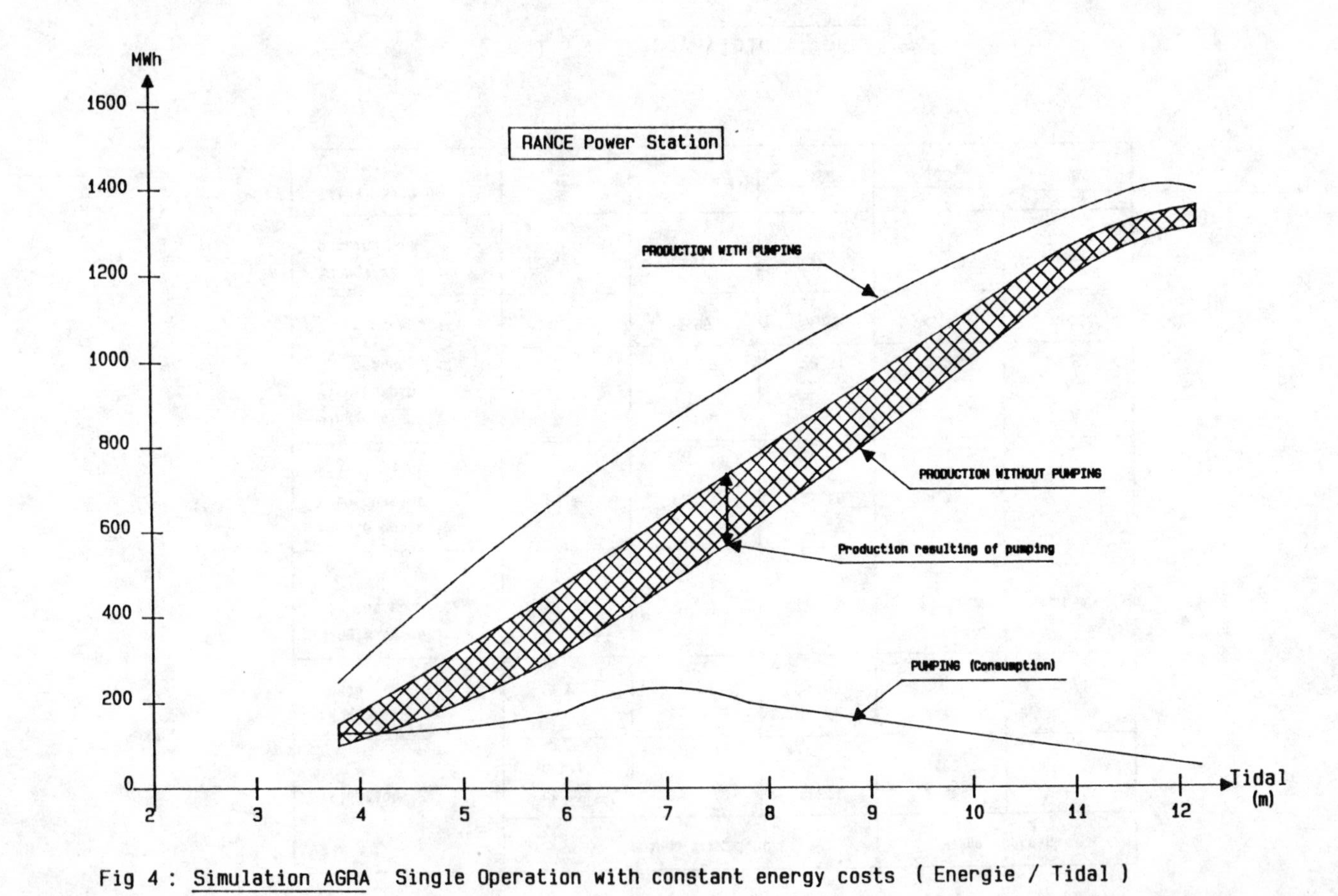

Fig 4 : Simulation AGRA Single Operation with constant energy costs (Energie / Tidal)

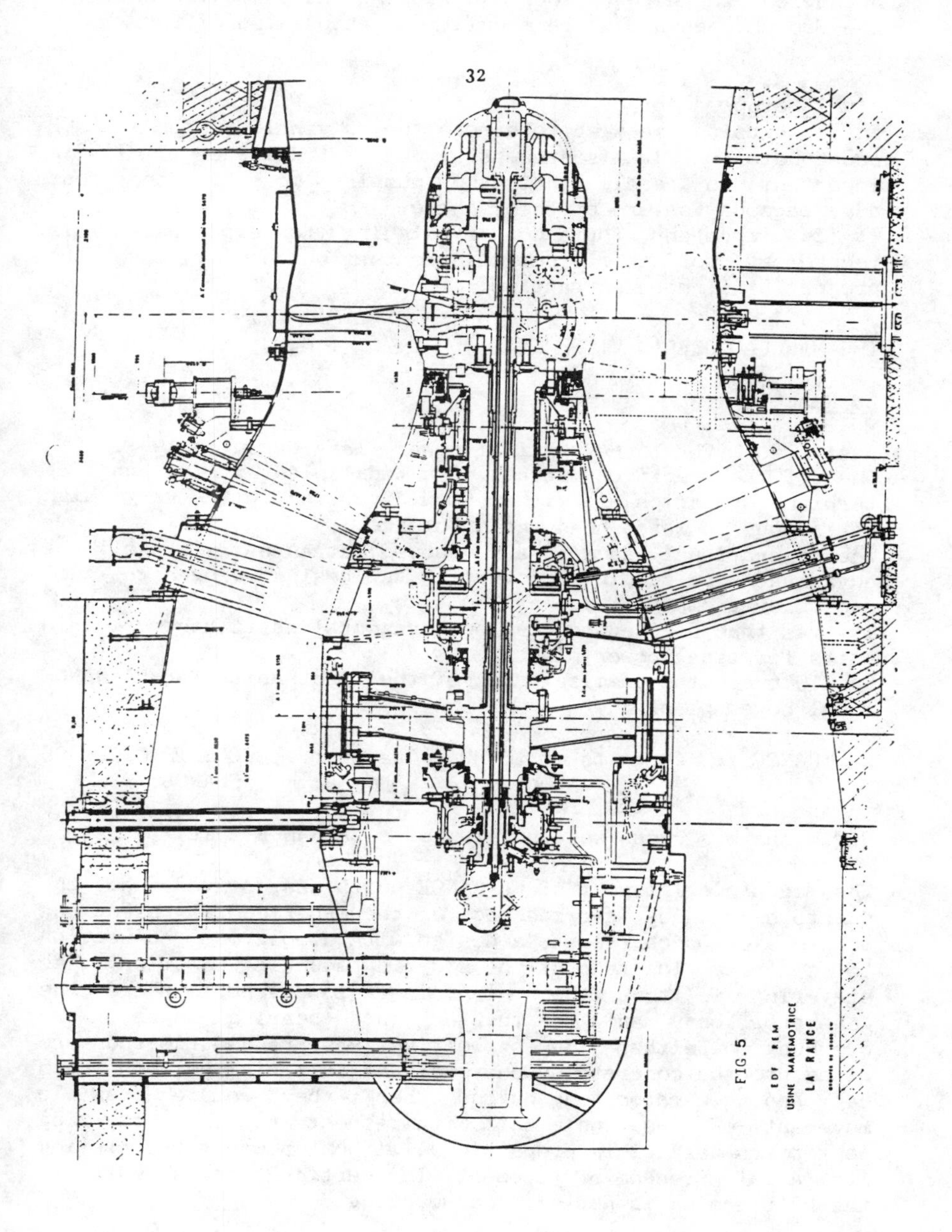

FIG.5

These results permit to choose the best suitable solution in countries where power source distributions are different from the French ones and where modulation ratios also differ.

A unique operation

In conclusion, we wish to stress the originality of LA RANCE power station in its successful use of tidal energy. It is important to recall that the pumping mode is the most advantageous feature of this scheme.
It is important for a country wishing to experiment this technology not to forget this feature which will make the project still more attractive.

THE MANUFACTURER'S VIEWPOINT

La Rance

Design work for LA RANCE tidal power plant began in 1943.
One of the specific features of a tidal plant is its range of turbine operation in either flow direction, with very high variations in water head and discharge.
The operational studies also showed, at an early stage, the advantages of pumping to overfill the basin during high tide and to over-empty it during low tide.
It was thus decided to develop horizontal-shaft upstream bulb units for tidal power plants.
So, 1966 saw the commissioning of the first units for LA RANCE tidal power plant :

LA RANCE : 24 units of 10 MW - Customer : ELECTRICITE DE FRANCE.

Fig.5 shows a cross-section view of a LA RANCE unit.

The mechanical layout of LA RANCE units was derived from the numerous studies carried-out for the experimental plant at Saint-Malo which led to a design with almost one connection to concrete located at the stayring mounting section. The stay-ring is thus the only main rigid support where the generator body and the turbine guide-bearing support are overhung on either side, by bolting. Prestressed tie-rods are fixed to the concrete by one end, and to a ring between nose cap and generator body by the other ; they reduce possible movement of the whole unit with respect to the stay-ring center to negligible proportions, without however restricting its axial freedom of movement. The vertical access shaft to the bulb casing is not a rigid mounting.

The main advantage of this arrangement are as follows :

- the connections of supports to concrete have little effect on the strain in main bulb unit components : body and supporting cone ;

NEYRPIC BULB UNITS WITH RUNNER DIA ABOVE 3 M

NAME OF THE POWER STATION	DATE OF COMMISSIONING	NAME OF THE RIVER	RATED HEAD (m)	UNIT		RATED FLOW (m3/s)	RATED CAPACITY PER UNIT (MW)	TOTAL INSTALLED CAPACITY (MW)	RUNNER DIAMETER (mm)	RUNNING SPEED (Rev / min)	
				NUMBER	TYPE					TURBINE	GENERATOR
FRANCE											
CAMBEYRAC	1957	TRUYERE	10,75	1	VRKI	55	5,5	5,5	3100	150	150
WADRINAU	1957	MOSELLE	4,5	4	PFKM	36,4	1,5	6	3050	107	750
ARGENTAT	1958	DORDOGNE	16,5	1	VRKI	100	14,4	14,4	3800	150	150
BEAUMONT MONTEUX	1959	ISERE	12,5	1	BRK	79,7	9,1	9,1	3800	150	150
SAINT MALO	1959	TIDAL	4,8	1	BRKI	227	9	9	5800	88,3	88,3
LA RANCE	1966	TIDAL	5,75	24	BRKI	192	10	240	5350	93,8	93,8
PIERRE BENITE	1966	RHONE	7,95	4	BRK	284	20	80	6100	83,3	83,3
GERSTHEIM	1967	RHINE	9,8	4	BRK	244	23	92	5600	107	107
BEAUCAIRE	1970	RHONE	15,3	6	BRK	258	35	210	6250	93,8	93,8
STRASBOURG	1970	RHINE	14,5	6	BRK	252	29	174	5600	100	100
GERVANS	1971	RHONE	12	4	BRK	363	30	120	6250	93,8	93,8
GOLFECH	1973	GARONNE	15,5	3	BRK	164	23,2	69,6	5100	125	125
SAUVETERRE	1973	RHONE	9,3	2	BRK	400	33	66	6900	93,8	93,8
AVIGNON	1973	RHONE	10,5	4	BRK	350	30	120	6250	93,8	93,8
GAMBSHEIM	1974	RHINE	13,2	4	BRK	244	24,5	98	5600	100	100
CADEROUSSE	1975	RHONE	9,15	2	BRK	400	32,5	65	6250	93,8	93,8
CADEROUSSE	1975	RHONE	9,15	4	BFH	400	32,5	130	6900	93,8	93,8
PEAGE DE ROUSSILLON	1977	RHONE	11,5	4	BRK	400	40	160	6250	93,8	93,8
VAUGRIS	1980	RHONE	5,65	2	BRK	350	18	36	6250	75	75
VAUGRIS	1980	RHONE	5,65	2	BFH	350	18	36	6900	75	75
CHAUTAGNE	1980	RHONE	14,7	2	BRK	350	46,6	93,2	6400	107	107
BELLEY	1981	RHONE	14,7	2	BRK	350	46,6	93,2	6400	107	107
LA CROUX	1981	TARN	13,8	2	BRK	75	9,5	19	3250	200	200
BREGNIER CORDON	1983	RHONE	10,8	2	BRK	350	35	70	6250	93,8	93,8
SAULT-BRENAZ	1986	RHONE	8	2	BRK	350	23	46	6250	85,7	85,7
CAMBEYRAC	1987	TRUYERE	10	1	BFK	110	10	10	4250	107	107
ARGENTAT	1989	DORDOGNE	15,7	1	BRKM	124	17,2	17,2	4100	138	600
SAINT EGREVE	LC	ISERE	12	2	BRK	250	23,4	46,8	5600	100	100
HUNGARY											
TISZA 2	1973	TISZA	10,7	4	BRK	136	7,2	28,8	4300	107	107
INDIA											
SONE LINK CANAL	UC	SONE	3,7	6	BFK	61	2	12	3220	120	120
KOREA											
PALDANG	1972	HAN	11,8	4	BRK	200	21,2	84,8	5200	120	120
PORTUGAL											
CRESTUMA	1987	DOURO	11,3	3	BRK	375	40,2	120,6	6800	83,3	83,3
USA											
ROCK ISLAND	1978	COLUMBIA	12,1	8	BRK	481	53	424	7400	85,7	85,7
W T LOVE	1982	OHIO	9	3	SFK	330	25	75	6100	90	90
WORUMBO	1989	ANDROSCOGIN	8,2	2	BRK	128	9,5	19	4250	120	120
WHITE RIVER	UC	WHITE RIVER	3,4	3	PRKM	233	6,7	20,1	5500	67	720
ALLEGHENY 3	DS	ALLEGHENY	3,4	2	PRKM	202	6,1	12,2	5800	65	450
USSR											
KISLOGUBSKAYA	1967	TIDAL	1,3	1	BRKMI	32,9	0,4	0,4	3300	72	600
YUGOSLAVIA											
CAKOVEC	1979	DRAVA	18,5	2	BRK	250	42,2	84,4	5400	125	125

SYMBOLS UC UNDER CONSTRUCTION
DS DESIGN STAGE

TYPE
1ST LETTER V= DOWNSTREAM BULB
B= UPSTREAM BULB
P= PIT TURBINE
S= BARGE
2ND LETTER R= ADJUSTABLE DISTRIBUTOR
F= FIXED DISTRIBUTOR
3RD LETTER K= MOVABLE BLADES
H= FIXED BLADES
4TH& 5TH LETTERS M= STEP-UP GEAR
I= PUMPING MODE

TABLE 2

- body fixing flanges to stay-ring and cooling nose do not transmit high loads. This facilitates durable sealing of these flanges against sea water, which is an essential requirement for satisfactory operation.
- connections between supports offer a good resistance to sea-water corrosion ;
- the concrete intake duct has a smooth simple shape with only one opening of small dimensions at the vertical access shaft.

The runner, with the blade servomotor, and the generator rotor mounted on a shaft line rest on two bearings. Vertical loads are transmitted to concrete :

- by the stay-vanes connecting the bulb casing to the duct wall,
- by four prestressed radial tie-rods.

The hydraulic thrusts applied to the runner during the various operating conditions are taken by a double thrust bearing between turbine runner and generator and transmitted to concrete through the stay-ring.
Assembly and dismantling of unit components are carried out through the turbine pit.
This brief description illustrates the state of technology, twenty years ago, on which further development of the bulb units is based. Since then, NEYRPIC who had already manufactured 29 units with diameters of 5350 mm and over, either alone or with outputs ranging from 18 to 54 MW and runner diameters up to 7400 mm. See Table 2.
We intend now to review the progress made in bulb units up to now and to analyze the new possibilities of using such machines for water heads under 10 m.

Equipment under 4-10 m head with high turbine discharges
The profitability of hydro schemes under 4-10 m heads and with high turbine discharges requires to develop an economic overall design leading to cost-savings in the civil structures and electro-mechanical equipment.

8. Use of units with simplified regulating system. The electro-mechanical equipment for low-head schemes accounts for a large amount of the total investment ; a significant cost reduction may be obtained through the use of simplified machines with satisfactory performances.
On a fixed-blade runner, power can be controlled through the wicket gates, with a fixed-vane distributor, load variations are obtained by changing the runner blade pitch angle.
Power stations with almost constant heads and discharges can be equipped with machines designed with no regulating device; they have a constant output at synchronism speed, for a given head. Start-up and shutdown are ensured by means of a gate generally located downstream of the runner.

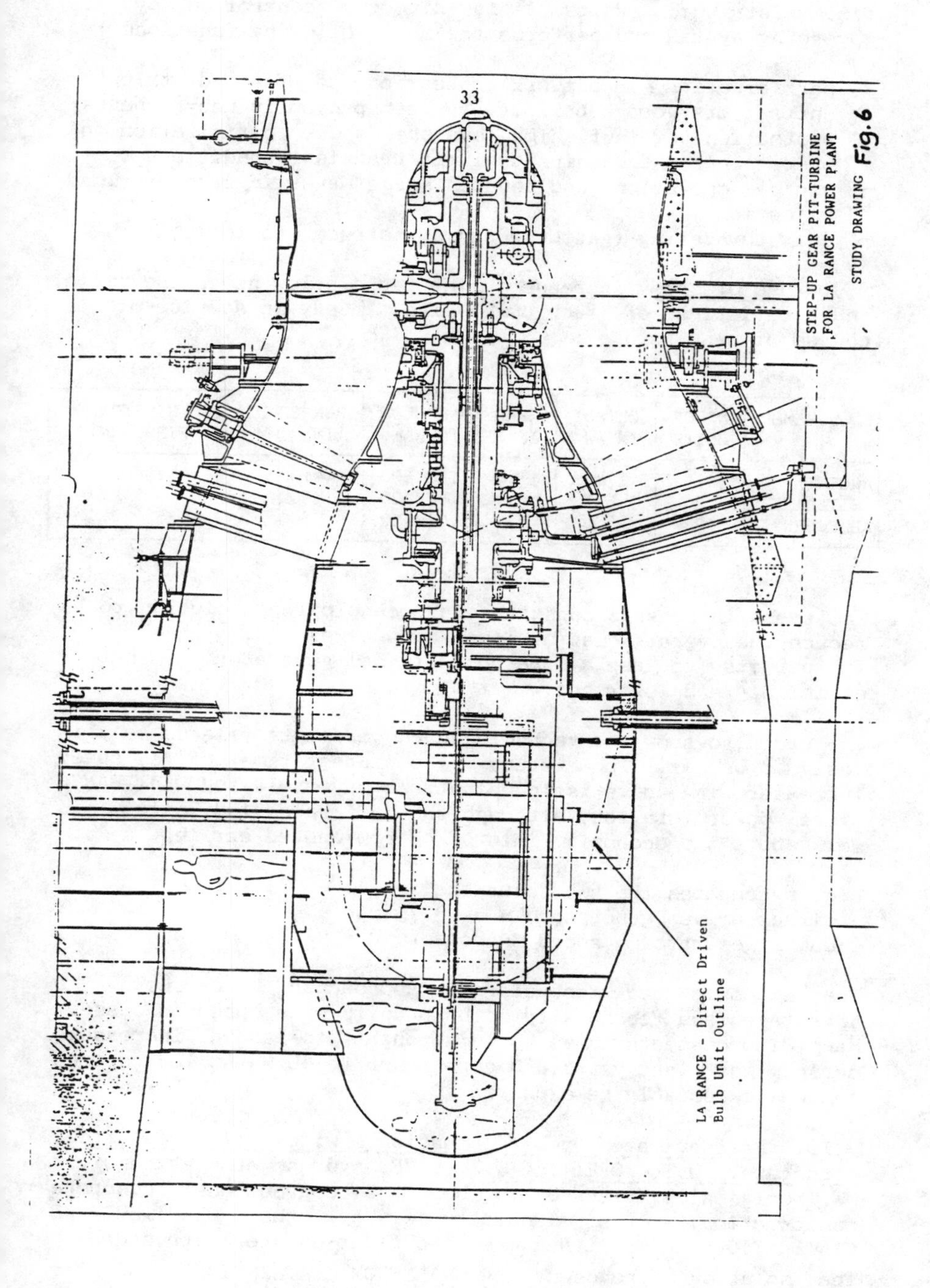
33
STEP-UP GEAR PIT-TURBINE
FOR LA RANCE POWER PLANT
STUDY DRAWING
Fig.6
LA RANCE - Direct Driven
Bulb Unit Outline

The downstream gate - main safety system for fixed-distributor units - is directly controlled by the governing system and performs the main following functions :

- participation in network connection of fixed-distributor units ; its good behaviour has been proved in many schemes;
- protection against high overspeeds ; optimization of operation relationships to limit transient conditions ;
- sluice operation and reduced surge waves in case of load rejection ;
- unit dewatering (gate used as downstream stoplog).

9. Use of speed increasers. The table below gives the main characteristics of such bulb units already in service or in course of manufacture :

SCHEME	N° OF UNITS	OUTPUT IN MW	HEAD IN M	SPEED IN RPM	RUNNER DIA IN MM	COMMISSIONING FIRST UNIT
ARGENTAT	1	17.2	15.7	138/600	4100	1989
WHITERIVER	3	6.7	3.4	67/720	5500	
ALLEGHENY 3	2	6.1	3.4	65/450	5800	

The speed increasers constitute a technological solution to reduce the overall cost of low-head schemes.
It permits to use a more conventional generator of better efficiency and lower cost.

In order to show the technologic requirements related to this design, we will now study the integration of a speed increaser in an existing plant. We have selected LA RANCE tidal power station, although the unit speed (93.75 rpm) is well above the economic limit of 80 rpm quoted earlier.

Fig. 6 compares the two solutions :
- direct-driven generator, n = 93.75 rpm
- quick generator + speed increaser, n = 92.25/500 rpm

The economic and technologic approaches have shown the advantages and feasibility of bulb units with speed increaser. Many improvements have been brought to speed increasers during the last years and outputs up to 40 MW at low speeds can now reasonably be expected.

10. Prefabricated power station with float-on steel structure: W.T. LOVE. The W.T. LOVE hydropower station of 72 MW, originally planned to be of a conventional concrete construction, was manufactured by the CHANTIERS DE L'ATLANTIQUE in Saint Nazaire in a float-on steel structure.
The solution proposed by ALSTHOM ATLANTIQUE was to manufacture in a building yeard a metallic floating structure

CENTRALE W.T. LOVE

W.T. LOVE GENERATING STATION

SITUATION DES ÉQUIPEMENTS PRINCIPAUX

MAIN EQUIPMENT LOCATION

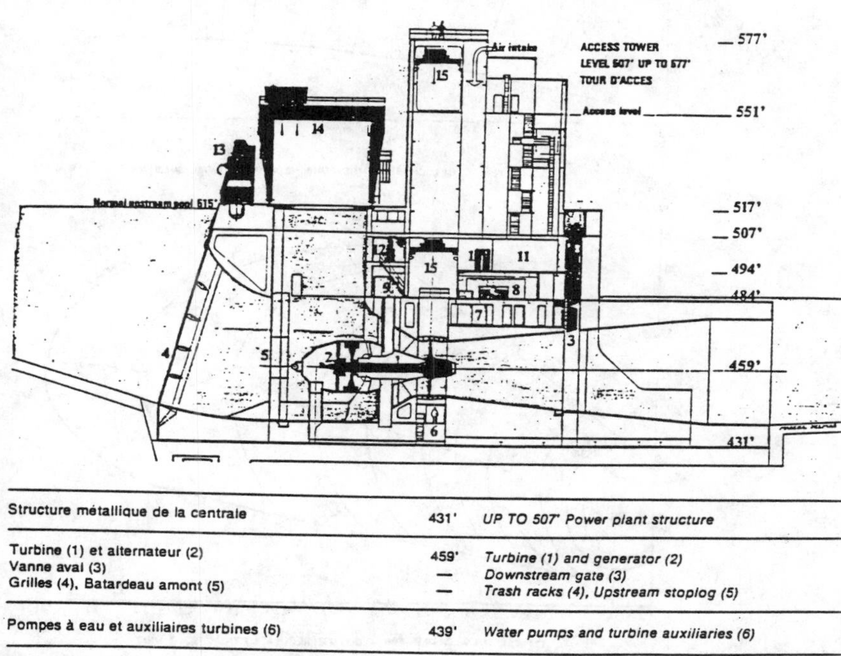

Structure métallique de la centrale	431'	*UP TO 507' Power plant structure*
Turbine (1) et alternateur (2) Vanne aval (3) Grilles (4), Batardeau amont (5)	459' — —	*Turbine (1) and generator (2)* *Downstream gate (3)* *Trash racks (4), Upstream stoplog (5)*
Pompes à eau et auxiliaires turbines (6)	439'	*Water pumps and turbine auxiliaries (6)*
Armoires relayage (7) et transformateurs auxiliaires	474'	*Relaying equipment (7) and auxiliaries transformers*
Salle de contrôle et de commande (8) Auxiliaires turbines et alternateurs (9) Transformateurs principaux et poste SF 6	484'	*Control room (8)* *Turbine and Generator auxiliaries (9)* *Main power transformers and SF 6*
Locaux protection incendie (10) et conditionnement air (11)	494'	*Fire (10) and air conditionning rooms (11)*
Diesel de secours (12)	497	*Standby diesel generator (12)*
Dégrilleur (13) et portique démontage (14) Ponts roulants (15)	517' —	*Trash rack rake (13) and gantry crane (14)* *Overhead travelling cranes (15)*

FIG.7

Fig. 8

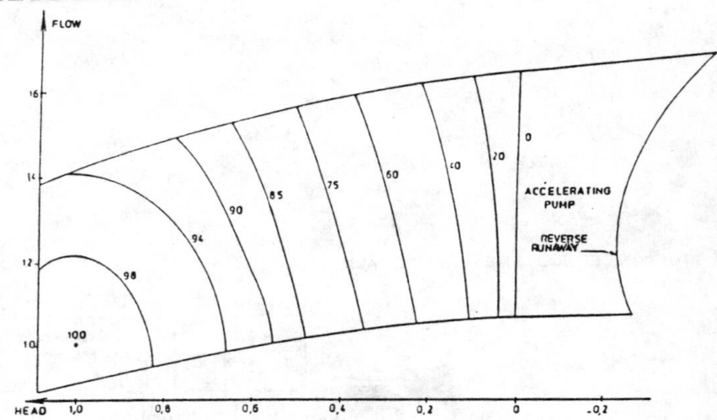

TYPICAL HILL CHART IN THE PUMP MODE FOR A TIDAL BULB UNIT

Fig. 9

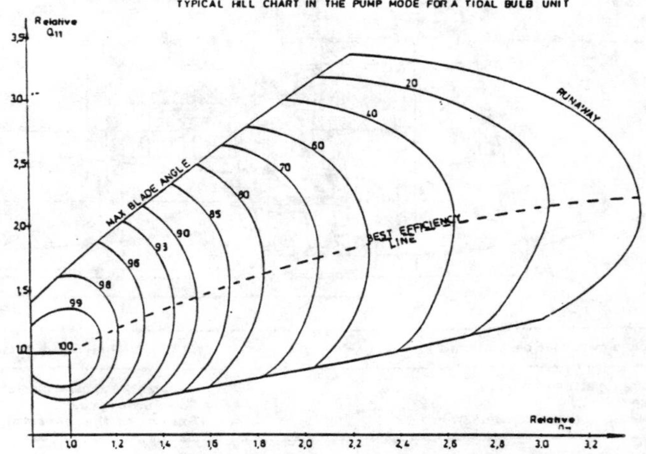

TYPICAL HILL CHART FOR A DOUBLE REGULATED TIDAL BULB UNIT

Fig. 10

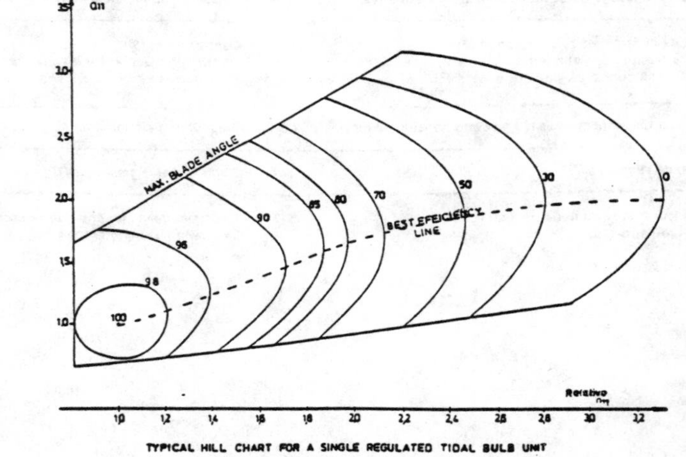

TYPICAL HILL CHART FOR A SINGLE REGULATED TIDAL BULB UNIT
(FIXED DISTRIBUTOR, ADJUSTABLE BLADES)

where the whole equipment will be mounted (see Fig. n°7). This "prefabricated" structure was then floated to site, landed and anchored in its final position by pouring of concrete.

A few specific features of tidal units

11. Pumping. During the lifetime of a power station, the economic operating conditions may differ from those prevailing at the design stage.
The flexibility of a double-effect power station with pumping, such as LA RANCE, allows easy adaptation to different economic data.
The optimization of a hydraulic design specific to tidal units is based on the improvement of pumping performances while maintaining turbine performances quite similar to river bulb units.
This optimization is integrated by NEYRPIC in its development program for new hydraulic designs.
Fig. 8 shows the expected characteristics in pump operation.
Pumping makes it necessary to use a mobile-blade unit as power consumption increases rapidly.
Moreover, since headlosses - other than those specific to the pump-turbine - increase with the amount of pumped flow, it appears interesting to operate the units at partial blade opening in pumping mode.
Pumping should also be considered in the optimization of the hydraulic design in order to define the shape of the water intake (basin side) which is used as draft tube in pumping mode.
Regarding direct pump operation, it must be noted that it represents nearly no additional cost in civil structure or electro-mechanical equipment for a mobile-blade unit at constant speed. It should be taken advantage of the flexibility and valorization offered now by such units and which may be still improved in the future.
This interest may vanish as concerns reverse pump operation since the reservoir volumes are smaller and the setting of units for this operating mode may require excessive submergence of the power station.
In a tidal power plant, turbine operation covers a wide range of heads. Through over-filling of the basin by pumping, the maximum head can nearly reach the maximum tide amplitude. On the contrary, it is desired to produce energy up to the lowest heads possible. The use of adjustable blades is quite appropriate for a very large head variation.
Another characteristic of tidal power plants is that during one cycle each head occurs twice with a different downstream level (viz. the sea level).
For the low level, it is of course the cavitation which limits the operation. Contrarily, for the high level, the possibility of blade over-opening is quite interesting.

Chosing adjustable or fixed wicket gates can only be decided after a detailed economic study.
Close to the maximum turbine output, similar performances are obtained with the two solutions. For smaller specific discharges (operation under high head), the adjustable wicket gates give, of course, a better efficiency. The comparative efficiencies of both solutions are illustrated on Fig. 9 to Fig.10. As an idea, the best efficiency of a fixed wicket gates, adjustable blades turbine is 95 % to 98 % of that of a double regulated turbine, depending on the hydraulic design and the size of the turbines.

12. Use of variable speed. Operation of tidal power units in generating mode is carried out within a very wide head range, from the maximum to the minimum head, that is the head for which the maximum runaway conditions correspond to the synchronization speed.
The operation under high heads is generally carried out within the good efficiency area while, for low heads, the efficiency decreases ith the head.
Unit speed deceleration as the head decreases permits to improve the turbine efficiency and power generation during the cycle.
The same speed reduction can be a gain not only in normal pumping operation but also when the suction head is greater than the delivery head.
For tidal power units where the powers to be transferred are in the region of some ten megawatts, the most interesting solution is that of a generator connected to a static frequency converter (SFC).
The CSF is a kind of rectifier transforming the alternating current produced by the generator into direct current. This direct current is then transformed into alternating current by an inverter at the network frequency.
As this system can be reversed, it can be used in pumping mode.
Static frequency converters have given satisfactory results at powers of some megawatts for the supply of rolling mill motors at variable speed, and as pump starters for pump-turbine units. In this case, the power attained is about 20 megawatts (MONTEZIC and CHEYLAS power plants).

13. Evolution of turbine and dam design. This evolution must allow for mass production to reduce costs and delivery times as well as interests on capital during the construction.

CONCLUSIONS

The design and construction of various prototype plants prior to LA RANCE scheme have permitted to develop large-size river bulb units representing a cost-effective equipment for low-slope and high-discharge rivers.

Further developments and improvements of river bulb units permit, in turn, both to reduce the cost and improve the design, operation and maintenance of future tidal power units. Owners and manufacturers always wish to capitalize on technologic innovations while maintaining hazards to a reasonable level ; the few evolutions shown in this study bring a contribution to these advancements in a technique aiming at the best service at an optimum cost.

The unit can be secured to concrete, like in a conventional power station, or inserted into a prefabricated metallic structure (W.T. LOVE) or into a concrete structure (cells of ARROMANCHES artificial harbour).
These structures equipped only with the built-in parts of the upstream and downstream isolating gears can be submerged. They offer a large flow section to the tide, thus reducing the very strong currents upon cut-off of the estuary. All these structures being installed, erection of downstream stoplogs or gates allows immediate operation of the units as and when they are commissioned. Under these conditions, further time saving can be obtained by reducing the erection time of the units.
For the SEVERN tidal power station, we have studied a solution enabling to position the fully-assembled units (about 1 500 tons) into the dam prefabricated structures. After lowering down, connection of the unit to the prefabricated structure is ensured by securing and concreting its built-in parts to the matching elements already fitted to the structure.

14. Provisions against corrosion. For LA RANCE power station, provisions have been made against the various types of chemical, electrochemical and biological corrosion, through the combination of the following methods:

- use of naturally corrosion-resisting materials,
- use of appropriate protective coatings with respect to the nature of substrate and corrosive environment,
- in addition, cathodic protection of submerged surfaces.

Aluminium bronze offering good mechanical properties - although lower than martensitic stainless steel - has mainly been used for the remaining half of runner blades.

We believe that for future tidal power units fitted with cathodic protection, the use of naturally corrosion-resisting materials may be limited to :

- 17-4 Mo martensitic steel for runner blades,
- 18-18-3 austenitic steel for throat ring hydraulic surfaces, shaft seal and sealing components.

In other words, and disregarding surface coatings, the cathodic protection enables to make almost the same provisions for tidal power units as for river power units.

B I B L I O G R A P H Y

G. GLASSER - F. AUROY. Les études des groupes de LA RANCE - Revue Française de l'Energie p 183 - Septembre/Octobre 1966 - special issue.

AGRA - Le nouveau modèle de gestion de l'usine marémotrice de LA RANCE

P. SANDRIN - E.D.F. Bulletin de la Direction des Etudes et Recherches Série B Réseaux électriques, Matériels électriques- n°3 - 1980

M. BANAL - A. BICHON - Tidal Energy in France : LA RANCE Tidal Power Station - Conference in Cambridge - September 1981

S. CASACCI - J. BOSC - Evolution des turbines KAPLAN et des groupes axiaux pour l'équipement des installations de forte puissance et basses chutes - La Houille Blanche n°5/6 - 1982

J. LE TUTOUR - J. BOSC - A hydropower station manufactured in a shipyard - NEYRPIC Technical Review n°2 - 1983

J. BOSC - L. MEGNINT - Evolution des groupes axiaux pour l'équipement des installations marémotrices - La Houille Blanche n°8 - 1984

L. MEGNINT - J. ALLEGRE - Present design of tidal bulb units based on the experience for LA RANCE tidal power plant and on-the-river bulb units - Conference in Brighton - May 14-16th 1986

P. HILLAIRET - G. WEISROCK - Optimizing production from LA RANCE tidal power station - Conference in Brighton - May 1986

J. BOSC - L. MEGNINT - Evolution des groupes bulbes à partir de l'expérience de LA RANCE et des installations de rivière - Conference in Saint Malo - November 6th 1986

Y. STROHL - Hydraulic and mechanical design of turbines of tidal power plants - Conference in New Dehli - 16-18th 1988

J.P. FRAU - LA RANCE tidal power station : 20 years of operation - Conference in New Dehli - February 1988.

Discussion

E. GOLDWAG, *GEC Alsthom Power Plants Limited*

A clarification of the point regarding energy efficiency in pumping is very important because the percentage improvement attributable to pumping at La Rance is seen by some as a target for the schemes currently being considered in the UK. Thus although the work undertaken on the Severn indicated that an improvement in output as a result of pumping could be as high as 9-11%, with comparable improvements on the Mersey scheme, there were those who claimed that twice this percentage was achieved at La Rance. This is, of course, in line with para 7, valorization of production. However, it is the aspect presented by the modulation ratio valorization that is causing some difficulties. Are we to understand that when this ratio is set to a value such as, 5.7 or 9.87, the costs of energy used for pumping and the value of generation per unit of output are in the ratios of 1/5.7 and 1/9.87? If this is the case then, of course, pumping is very cost-effective and the differences between La Rance and the proposed UK schemes could be satisfactorily explained.

Further, it is said that the 'pumping mode is the most advantageous feature' of the La Rance scheme. Is this view based on the very high values of 'the valorization' used? The appraisal of the UK schemes using a fairly flat tariff structure suggests that the value of pumping is roughly in keeping with additional output with some further small benefit arising out of re-timing. What are the Author's views on this matter? Also are the very advantageous rates a matter of internal accounting or would they be available to private generators?

E.T. HAWS, *Rendel Parkman, London, UK*

It is good to hear from M. Frau of a commissioned tidal power project when in the UK we are still studying the Severn and Mersey. We hope to learn from our French colleagues.

I was interested in Mr Goldwag's comments on the values assumed for imported and exported energy. For the Mersey we have had to do sensitivity checks on this, as imported energy will be at half hour competitive rates in the new competitive industry. The La Rance experience and economics appear to be based on cheap imported energy for pumping.

The problems for tidal power projects in the UK are high capital cost and economic assessment by discounted cash flow. Is this the reason for no second tidal power project in France? Could M. Frau give information on plant maintenance, particularly the major rebuild programme of 1976?

C.P. STRONGMAN, ***Merz and McLellan, UK***

At the design stage the performance of the surge chamber has considerable attention from the civil engineers, plant engineers and the client, the object being to achieve the necessary dynamic behaviour of the plant. Has the Author anything to report about the hydraulic performance of the surge chambers at Cruachan and Foyers in relation to the dynamic behaviour of the plant, particularly in view of the enhanced dynamic capability mentioned in Paper 10.

R.B. KYDD, ***City of Cape Town***

What has the experience of operating the pumped storage schemes indicated with regard to inspection and maintenance requirements?

Please comment particularly on frequency of activities and extent of work involved indicating if the tunnel system is drained and if so, why it was considered superior to the use of an ROV, as a much reduced outage time would result?

K. SØRLI, ***Kuaerner Hydro Power A/S, Oslo, Norway***

Referring to Dr Vögele's diagram showing MVA/rpm and the limits of the different generator cooling methods, it seems possible to select a combination of 340 MVA and 500 rpm motor/generator unit. Based on natural motor inertia, the questions are: Does Dr Vögele find these parameters comfortable and will the given data indicate a generator design well within proven generator experience?

ERLACHER, ***TIWAG, Innsbruck, Austria***

What is the influence of slip-ring pole forces on starting torque and starting circuit?

In the schedule of aspects for selection of starting

methods, I missed the number of units (four units together with only two static frequency converters would be possible).

The difference between air cooling and water cooling shown in the diagram is surely too low.

How many slip-rings are necessary for a pole changing machine and what is the Author's experience with slip-rings and brushes?

C.P. STRONGMAN, *Merz and McLellan, UK*

With regard to the pole-changing motor-generator, can a description be given of the method of attaching the different sizes of pole to the rim and how the rim design takes account of the lack of symmetry of the poles attached to it?

R.B. KYDD, *City of Cape Town*

Could Mr Barbet please explain how the cost of failures is composed and details of how such significant reductions as 130 million francs in seven years could be achieved?

J.L. BARBET (for P Guichon)

In reply to R.B. Kydd, the cost of failures is composed of

(a) expenses of replacements or repairs
(b) payment of external contracts
(c) internal costs, such as hours of work and workshop presentations.

(Costs of unavailability are not included.)

The spectacular downward trend of the cost of failures is due to many factors

(a) increased reliability of the most important storages recently installed
(b) use of the monitoring and planned maintenance method
(c) voluntary maintenance toward the reduction of the failures.

Meanwhile, maintenance costs for all hydroelectric power stations have been practically constant over the past eight years, in spite of an increase of 25% in installed capacity. A better understanding of the ageing units helps.

Manufacturing controls and experience have increased the quality of maintenance, which also helps to prevent failures.

The monitoring and planned maintenance method helps to increase knowledge.

H. VÖGELE, *Author*

In reply to K. Sørli and R. Erlacher on cooling systems, the diagram 'Maximum possible output as a function of speed' gives only a rough idea concerning main machine data and adequate cooling systems. In the case of a real project we have to investigate this problem in depth taking into account all relevant requirements such as inertia of the machine, runaway speed (maximum possible diameter of the rotor) power factor, voltage range, operational conditions (load variation, thermal elongations) and the very important costs of the machine.

In reply to R. Erlacher and C.P. Strongman on pole changing, in switching over pole groups and disconnecting poles we need five sliprings (OVA SPIN, JUCKLA, PANJIAKOU, table 1 of the report).

If pole groups are switched over and poles are switched in parallel we need four sliprings (MALTA, table 1 of the report).

In all these machines (after many years in successful operation) no problems concerning sliprings and brushes occurred.

Pole-changing machines will have at least two different poles and pole core pitches. In the case of a laminated rim this could lead to problems concerning the arrangements of the rim slots (hammer head or dovetail) to attach the poles. The final arrangement of the slots (equally spaced) will be a compromise between the magnetical behaviour of the flux density in the air gap and the requirements concerning ventilation and mechanical conditions and must be found individually for the individual project.

In reply to R. Erlacher on starting methods, I fully agree with Mr Erlacher to extend the schedule of aspects for selection of starting methods by the number of units.

Pumped storage in South Africa

J.H. HENDERSON, BSc, and B.W. GRABER, Dipl Eng, Eskom

SYNOPSIS. Eskom, the electricity supply utility of the Republic of South Africa has 44 550 MW in operation or under construction. Eskom formulated a Vision of an extended electrical grid system in conjunction with exploiting the large hydro potential in the North to foster the economic development of the Southern African Region. Eskom has two pumped storage schemes of 1400 MW in operation which in addition to performing pumped storage duties they also pump water for the Water Supply Authority. These schemes are important for the optimum operation of Eskom's system and are utilised in excess of 50% of the time and their performance compares favourably with other schemes in the world.

1. ESKOM, THE ELECTRICITY SUPPLY UTILITY OF SOUTH AFRICA

Southern Africa with its vast mineral and other resources is a region possessing enormous growth potential. The Republic of South Africa has a sophisticated first world economy, supported by an electricity utility which is the fifth largest in the Western world in terms of installed capacity. Eskom generates more than 97% of the electricity used in South Africa, representing 60% of the electricity used on the entire African continent.

Eskom's total installed capacity as at December 1988 was 33 176MW, subdivided as follows:

- 19 coal-fired power stations : 28 916 MW
- 1 nuclear power station : 1 930 MW
- 3 gas turbine installations : 390 MW
- 2 conventional hydro stations : 540 MW
- 2 pumped storage schemes : 1 400 MW

Power Stations for a further 11 374MW are under construction, raising the capacity to 44 550MW. This power is supplied to the Republic of South Africa via a distribution network 201 802 km in length.

Pumped storage. Thomas Telford, London, 1990.

2. ESKOM'S VISION

Eskom functions in an environment which is undergoing rapid social, political and economic change. We believe that our organisation has a role to play in the change process and we have accepted the challenge to be a positive pro-active force in helping to shape a future Southern Africa in which all its diverse peoples can enjoy prosperity and peace. How we see our role in this change process has been distilled into what we call our "vision", which is expressed in three affirmations,

* Eskom is recognised internationally as a top utility.

* Eskom makes electricity available to everybody.

* Eskom supports the development of neighbouring countries by facilitating the creation of a regional network and thus optimising the use of the energy resources of the subcontinent.

Firstly as a utility, Eskom can already claim to be in the top class of power utilities. Production costs are amongst the lowest in the world with an average price of 2,3 US cents/kWh sold in 1988, furthermore, the technology employed is abreast of world developments. Eskom is pioneering the development of technologies in dry cooling and in the use of low grade coals. Kendal and Matimba power stations, each generating about 4000MW from six generating units, have the largest dry cooling systems in the world, while the Lethabo and Matimba power stations represent major advances in the combustion of low grade coal. The performance of Eskom's plant is also comparable to that of the leading utilities, with an average availability of 82,7% for the large coal-fired stations in 1988.

While statistics indicate that Eskom is already a top utility, we still have a long way to go in the development of our people. A large part of our work force is drawn from a third world environment where educational standards are low and an industrial culture is underdeveloped. The major challenge facing Eskom is to uplift the competence of all its people. This is an enormous task, but essential if we are to succeed in the future. Many programs which address this challenge are currently underway which cover primary, secondary and tertiary education, as well as extensive in-house development and training.

The second part of Eskom's vision is to extend rural and urban electrification, thus making electricity available to the twenty million South Africans who currently do not have access to domestic electricity. Eskom is establishing partnerships with business and local communities to mobilise the necessary resources to implement these plans. Eskom is also developing appropriate technologies to provide affordable, but safe electricity, and considerable success can be reported such as the ready board with light socket, overhead bundle conductors and budget energy controllers by magnetic cards. In addition to increase electricity sales, the realisation of this objective will enhance the quality of life for all South Africans, and stimulate economic growth by creating demand for other products.

The third part of Eskom's vision, and perhaps the grandest, is to facilitate the establishment of a Southern African electrical grid. This plan enables the harnessing of the vast hydro potential to the north of South Africa, thus providing cheap electricity to the benefit of the entire subcontinent. Not only will this stimulate trade and generate wealth in some of the most impoverished countries in the world, but also scarce capital can be redirected from costly fossil and nuclear investments to other employment and wealth creating ventures. It is also hoped that this stimulus to trade will open doors to other forms of exchange and cooperation. Eskom is exploring all avenues in pursuit of this vision and already exports electricity to Botswana, Lesotho, Mozambique, South West Africa/Namibia, Swaziland and Zimbabwe. Figure 1 on the next page shows the existing and possible future electrical grid system in Southern Africa which shows the merit of building hydro stations in the northern countries to supply their own needs and export surplus power to South Africa, earning income for their development. This hydro potential is substantial, estimated at over 5000 MW on the Zambesi River and 50 000 to 100 000 MW on the Zaire River.

3. ESKOM'S PUMPED STORAGE SCHEMES

Eskom has two pumped storage schemes in operation, Drakensberg and Palmiet. The pump turbines are of the vertical shaft single stage reversible Francis type for both schemes:

	DRAKENSBERG	PALMIET
(a) Capacity, MW	4 x 250 = 1000MW	2x200 = 400MW
(b) Head range for generation, m	400,0 to 448,5	245,5 to 285,6
(c) Head range for pumping, m	436,4 to 476,0	264,7 to 306,2
(d) Rated speed, RPM	375	300
(e) Method of pump starting	pony motors	Static frequency converter
(f) Operating cycle	weekly	weekly

(g)	Time required to fill upper reservoir, hours	39	33
(h)	Cycle efficiency,%	72,7	77,9
(j)	Duration of construction until commissioning of first set; years	6,5	4,5
(k)	Location of power station:	underground cavern	individual shafts
(l)	Commissioning of first set	May 1981	May 1988

At the time of decision to build Drakensberg, there was only one scheme in operation with a higher pumping head, namely Numappara in Japan with a maximum pumping head of 528m, but its capacity of 230 MW was slightly less. Eskom embarked therefore right on the forefront of pump-turbine

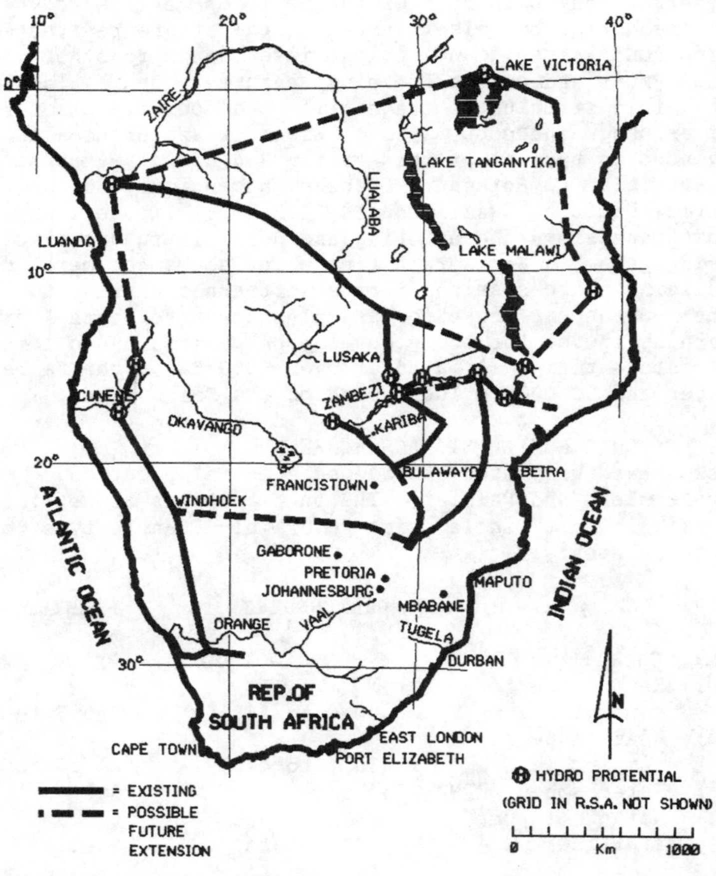

Fig. 1. Existing and possible future electrical grid system in southern Africa

technology. In order to obtain a high-quality engineering and performance of these pump turbines, we selected the two most promising manufacturers to build and test a homologous scale pump turbine model and to perform an advanced design, incorporating finite-element techniques. The successful tenderer was selected afterwards based on the witnessed model tests, the adequacy of the advanced design and the adjudicated tender price. Both manufacturers were paid for this initial work. At the time of commissioning, Drakensberg had the highest head in the Western world.

As Palmiet had many reference pump turbines in operation, the standard tendering procedure was adopted and model tests and advanced design work were performed after contract award.

4. MULTI-PURPOSE APPLICATION OF PUMPED STORAGE

Both the Drakensberg and Palmiet Pumped Storage Schemes are multi-purpose schemes and were built and financed jointly by Eskom and the Department of Water Affairs. In addition to performing pumped storage duties for Eskom, water is pumped on behalf of the Department of Water Affairs. This Department is a governmental organisation being responsible for all water matters in the Republic, and in particular the adequate water supply to consumers. The multi-purpose application of pumped storage between the electricity supply and water supply authorities is unique in the world.

At Drakensberg water is pumped from the Tugela River Basin into the Vaal River Basin to augment the water supply for Pretoria-Johannesburg-Vereeniging area. At Palmiet, water is pumped from the Palmiet River to augment the water supply of greater Cape Town.

Of particular interest is the multi-purpose application of Drakensberg. The industrial hub of South Africa is the Pretoria-Johannesburg-Vereeniging area producing 56% of the Gross Domestic Product and accommodates 29% of the population of South Africa. Consequently an assured water supply is of paramount importance. Originally, water was supplied from within the local Vaal River Basin. However, as a result of expansion, more water is required than is always available from the Vaal River and the shortfall has to be imported from elsewhere. Investigations showed that the most economic source for such additional water is from the Tugela River. Consequently a pumping scheme was commissioned in 1974 to transfer 330 000 cubic meters per day to the Vaal River Basin. However, to meet the growing demand, the supply from the Tugela had to be increased to 950 000 cubic meters per day in 1981.

A first solution appeared to be an extension of the existing pumping station from 30,8MW to 74,5MW. However, before this solution was adopted, investigations were carried out for a pumped storage scheme which would augment the water supply by allowing only a part of the pumped water to be returned for generation. The investigations showed that large financial benefits would accrue to both Eskom and the Department of Water Affairs as follows:

FIGURES IN %	COSTS TO DEPARTMENT OF WATER AFFAIRS	COSTS TO ESKOM	TOTAL COSTS TO COUNTRY
1. Each organisation builds own scheme	30	70	100
2. Joint scheme	23	56	79
3. Saving in capital	8	13	21

The 21% saving is about $105 million in 1989 money and the reduction in costs to Eskom is 18%. These costs are life-time costs and the cost saving by the Department of Water Affairs were apportioned equally between the two organisations. As a result of these benefits it was decided to build the Drakensberg scheme as a joint venture.

The water pumped by Drakensberg is pumped into a dam with a storage volume of 2 656 million cubic meters (Sterkfontein) inside the Vaal Basin near the watershed. This water is kept in reserve and is only released (at a high rate) under extreme conditions when the dams in the Vaal Basin are virtually empty. This allows the dams on the Vaal River to be depleted in excess of the previously dependable yield, thus providing more storage for absorbing periodic floods. This improves the water supply situation as follows:

	MILLION CUBIC METER PER ANNUM	%
1. Unaided Vaal River System	1 560	100
2. Import from Tugela	281	18
3. Additional yield from Vaal Basin	531	34
4. Total yield from extended system	2 372	152

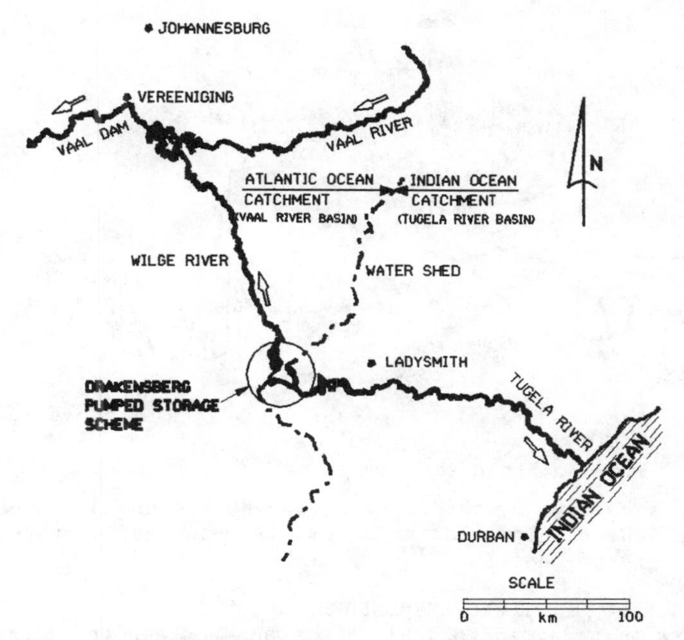

Fig. 2. Tugela-Vaal water transfer scheme

The yield from the local Vaal River System from natural inflow is increased by 34%, or each cubic meter of imported water increases the local yield by 1,9 cubic meter.

South Africa experienced a serious drought from 1979 to 1986 and the cumulative natural flow in the Vaal River during these 8 years was only about 30% of the long term average flow. By using Tugela water pumped by the Drakensberg Pumped Storage Scheme, serious water restriction could be avoided. From 1983 to 1987, a total of 1612 Mm^3 Tugela water was released which supplied about 43% of the demand during this period.

Figure 2 shows the location of the Drakensberg Pumped Storage Scheme in relation to the watershed between the Indian Ocean and Atlantic Ocean catchment areas.

As previously mentioned, the Palmiet Pumped Storage Scheme is also a multi-purpose scheme between Eskom and the Department of Water Affairs. As shown on Figure 3, the original proposal by this Department was to build a 16,5 km long tunnel from the Hangklip Dam to an underground pumping station to supply the waterworks. Investigations showed that it is more expedient to pump the water over the watershed by a pumped storage scheme. This was facilitated as such a scheme was required by Eskom in this part of the country. The portion of the water pumped for water supply will be used for hydro generation, firstly by a 5MW Pelton

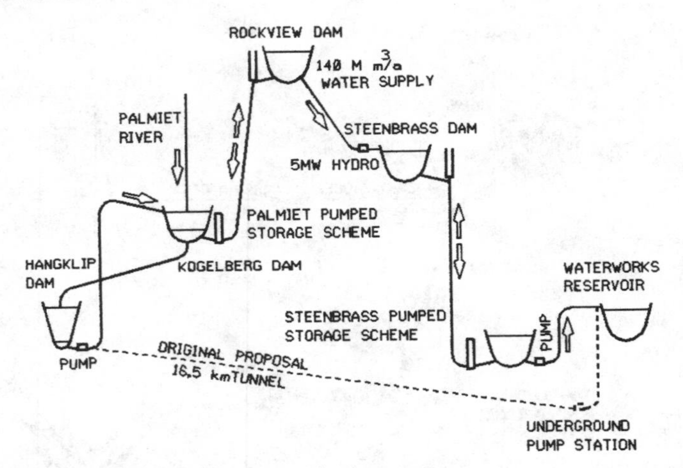

Fig. 3. Extended Palmiet scheme

set and secondly, by the existing 180 MW Steenbrass Pumped Storage Scheme. This latter scheme is owned and operated by the City of Cape Town.

5. CRACKING OF PIPEWORK EMBEDDED IN CONCRETE

Pipework is required to drain the space between the runner and guide vanes during pump starting and synchronous condesnor operation. At Drakensberg, this pipework which is solidly embedded in concrete cracked and could not be repaired. The remedial action for Drakensberg is discussed as well as a unique design developed for Palmiet which would permit repairs.

For Drakensberg, it was decided to solidly embedded the draft tube and bottom ring in concrete. This was done to maximise the compound structural integrity of the high-head pump-turbine and its concrete surround. Consequently, the drainage pipework was also embedded in concrete. Two years after commissioning a crack developed in this pipework, causing high-pressure leakage into the concrete surrounding the spiral casing. During standstill the leakage accumulated and was suddenly pressurised by the radial expansion of the spiral casing during pump start causing the surrounding concrete to crack in the horizontal plane. Before describing the repair solution, it is necessary to explain the pump start system for Drakensberg which is shown schematically on Figure 4 on the next page. The water pressure in the spiral casing is kept slightly above the tailrace pressure by sealing the guide vane faces in the closed position and by an auxiliary by-pass around the closed main inlet valve. After dewatering the runner space with compressed air, the water between the runner and guide vanes is drained during pump start-up by two separate pipework systems into the draft tube. The crack developed in the ring main of the outer pipework system marked (a) on

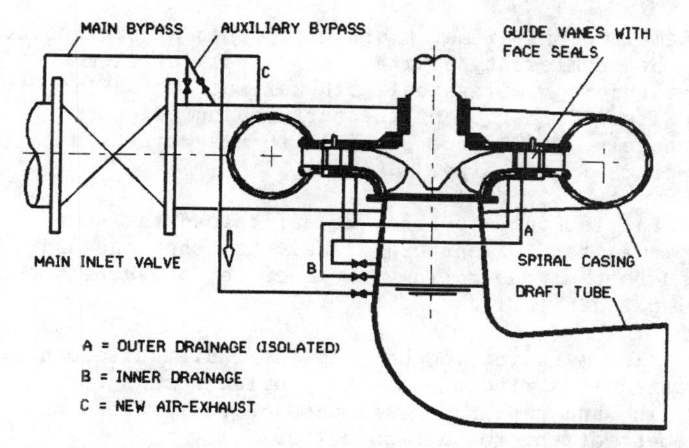

Fig. 4. Drakensberg drainage system

For Palmiet a different pump start system was adopted as shown on Figure 5. The guide vane faces are not sealed in the closed position and a water seal is created against the closed guide vanes by the centrifugal action of the rotating runner. The soffit of the spiral inlet is connected by pipework to the draft tube to exhaust air which may leak into the spiral casing. The main inlet valve is closed and no auxiliary by-pass is provided.

The eventual remedial action for Drakensberg was to change over to the start-up system used at Palmiet and to isolate the faulty pipework and rely on the single remaining pipework system which remained healthy. A new pipe for air-exhaust marked (c) on Figure 4 was installed and the seals of the guide vane faces removed. The modified system is working satisfactorily.

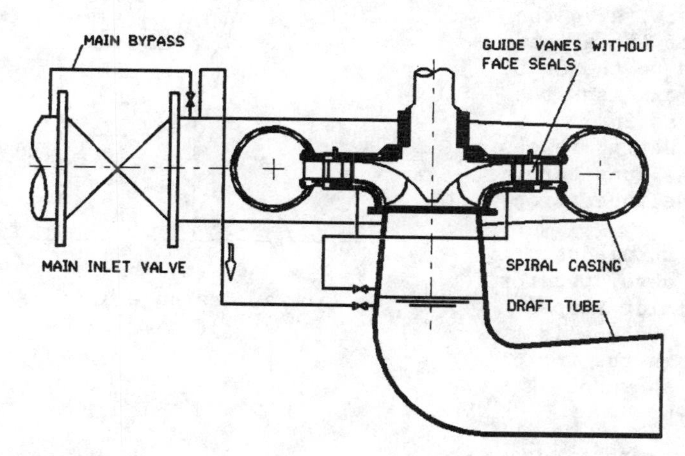

Fig. 5. Palmiet drainage system

The 115mm internal diameter drainage pipes at Drakensberg are of the seamless type with a thick wall of 11mm manufactured from mild steel with maximum ductability to accommodate deformation of the surrounding concrete. Unfortunately, it was not possible to determine the cause of the crack in the pipe.

For the Palmiet pump-turbines we maintained the same design principle as for Drakensberg, namely to embed the draft tube and bottom ring solidly in concrete, however, with the following modifications:

(a) The ring main interconnecting the individual down pipes are located in a ring-passage to provide access for dismantling and repairs. This passage facilitated also the arrangement of the foundation bolts.

(b) The pipes are of the thick-wall seamless type of mild steel (similar to Drakensberg) but all welded components are outside the concrete and accessible. Repair by a welding robot is possible, should a crack develop.

(c) The concrete foundation for the stayring is designed as a continuous circular structure incorporating the ring-passage. This integral circular structure, save the width of the access passage to the draft tube, maximises the stiffness for accommodating forces from the pump turbine into the foundation.

To our knowledge, this design developed for the Palmiet Pumped Storage Scheme is unique in the world and is shown on Figure 6.

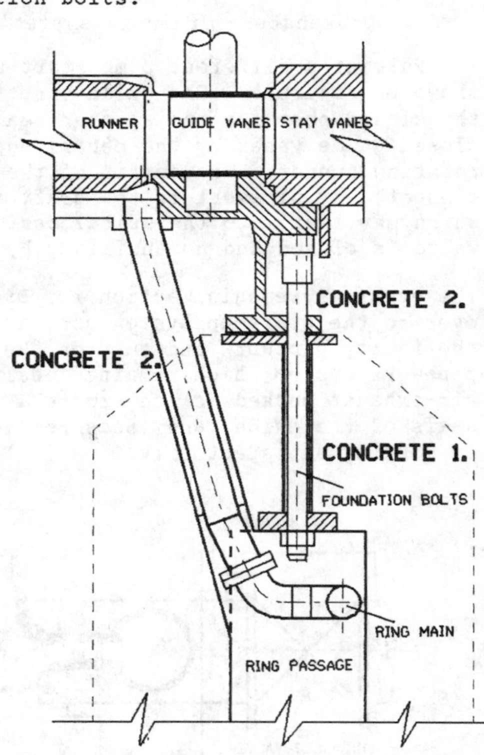

Fig. 6. Palmiet drainage arrangement

6. OPERATIONAL REGIME

Drakensberg Pumped Storage Scheme, since its inception has

been extremely well utilised. Three basic modes of operation are in use:
- Generating mode,
- Pumping mode, and
- Synchronous condenser mode (S.C.O).

The table indicates the utilisation of the plant in these modes.

YEAR MODE	1984	1985	1986	1987	1988
Gen Mode	33,86%	33,47%	29,10%	31,19%	23,70%
Pump Mode	52,41%	49,35%	48,73%	48,34%	43,75%
S.C.O. Mode	13,72%	17,18%	22,16%	20,46%	32,42%
TOTAL UTILISATION	70,14%	77,12%	73,17%	75,96%	58,42%

The table indicates a shift from the generation mode to the SCO mode since 1988. The reason is the excess capacity available in the Eskom system. The more expensive mid-merit and peaking thermal plants have been placed in storage and some plant was de-commissioned. As a consequence, the cost differential was reduced between cheap base load plant and the remaining mid-merit and peaking thermal plant. Consequently, the economics for operating the overall system required a reduction in using pumped storage for peak-generation and off-peak pumping cycle. At present, pumped storage is used more for emergencies to replace plant that has tripped off the system with little or not notice. This has proved to be invaluable during the winter of 1989 and has greatly contibuted to the satisfactory quality of supply by minimising load shedding due to under-frequency.

An advantage of the reversible pump-turbine sets is that the mode of operation can be altered within a short time period as shown on the table of mode changes below.

MODE CHANGE	DRAKENSBERG TIME	PALMIET TIME
Standstill-Generation (full load)	2 1/2 minutes	2 1/2 minutes
Standstill-Pump	8 minutes	5 1/2 minutes
Generation-Pump	13 minutes	8 1/4 minutes
S.C.O.-Generation	2 minutes	1 minute
S.C.O.-Pumping	-	4 1/2 minutes

This fact of operation has been used to the fullest extent by Eskom as indicated in the table of average daily mode changes recorded per machine for Drakensberg.

YEAR	NO. OF MODE CHANGES PER DAY PER UNIT
1984	5,25
1985	5,8
1986	5,2
1987	5,1
1988	4,2

It is expected that the growth in system demand will absorb the excess capacity towards the turn of the century and pump storage operation will revert to its original function.

7. PERFORMANCE

Drakensberg was the first pumped storage scheme in operation in Eskom and performance was initially measured in terms of the criteria applicable to thermal power stations. This, however, proved to be inadequate as pumping and S.C.O. modes were excluded from the performance calculations.

Consequently the plant performance measurement system was modified especially to suit conventional hydro and pumped storage schemes. The measurement system takes all modes of operation into account at these stations. This produced more realistic values for the actual performance of the stations.

A computer program was developed in-house to process the performance data automatically from information received from the process control computer. Owing to the operating regime of the plant, the evaluation criteria of the whole system, rather than being output related, is based on the successful achievement of mode status.

The following performance figures have been achieved at Drakensberg.

Mode Change Reliability = 97%
Availability = 80%

$$\text{Mode Change Reliability} = \frac{\text{mode changes achieved}}{\text{mode change attempts}}$$

$$\text{Availability} = \frac{\text{MCR x Period time} - \Sigma\underset{\text{loss}}{\overset{\text{forced}}{(\text{capacity}}} \times \text{Tf}) - \Sigma\underset{\text{loss}}{\overset{\text{planned}}{(\text{capacity}}} \times \text{Tp})}{\text{MCR x Period time}}$$

where: MCR = Maximum continuous rating
Tf = Time on forced outage
Tp = Time on planned outage

Although these figures are considered high, it is our belief that improvements could still be made through the elimination of program sequence failures.

A similar program has been implemented at Palmiet Power Station and performance figures achieved are:

Mode Change Reliability = 98,8%

Availability = 89,4%

8. PROBLEM AREAS

Certain problems have arisen since the commissioning of Drakensberg in 1982 and much of the experience gained has subsequently been incorporated into the design of Palmiet.

8.1 Generator-Motor Winding

The endwinding on the drive end of the stator are glassfibre base ropes impregnated with Aroldite Resin. The overhang of these end windings comprises a single roping for approximately 50% of their length. The dual cycle of the machine causes flux reversal in the end windings resulting in excessive vibration. The Contractor modified the end windings by counter binding the roping and thereby reducing the vibration.

Unit one is of particular concern at present, as it was subjected to the mandatory short circuit test on commissioning, as well as two inadvertent 400kV breaker closures while at standstill and the stator requires re-winding.

8.2 Pony Motors

A 16,5MW slipring pony motor is used for dynamic braking in both the pumping and generation directions and as the drive in the run-up sequence for pumping mode. The stator windings on the Unit 2 motor have failed twice, once in 1984. An extensive investigation was held to determine the cause as no other similar problems had been experienced. The results indicate that the surge protection damping which is augmented by the capacitive effect of the cable was inadequate thereby allowing surges of electricity to burn out the windings of the stator. This effect is particularly marked on Unit No. 2 due to the proximity of the motors to the vacuum supply breakers - 35 m compared to 70 m for the next nearest. The problem has been alleviated by fitting Zorc surge arrestors to the terminals of each pony motor.

8.3 Thrust Bearing

In 1985, both Unit 2 and Unit 4 thrust bearings were wiped. One of the causes of this problem was air in suspension in the thrust bearing oil even though anti-foaming agents had been added. A solution to this problem would have been the inclusion of a bubble releasing tank in the oil circuit.

On analysis it was found to be more viable to modify the oil groove, monitor the thrust bearing pads annually and re-metal them on a four yearly basis.

8.4 Other problems

Some of the other more significant problems that have been solved using local expertise are:

(1) The local manufacture of carbon segments for the turbine shaft seals. These were originally imported from Japan.

(2) The earthenware insulation pots for the liquid starters are no longer imported from the UK, but are manufactured out of glassfibre by a South African company.

(3) The thrust bearing temperature monitors have been replaced by locally made components.

9. CONDITION BASED MAINTENANCE

Palmiet Pumped Storage Scheme personnel make extensive use of plant monitoring techniques to assess the condition of plant and maintenance requirements. Presently 1 000 plant items are monitored making use of 220 procedures. Examples of where condition monitoring is employed are:

- pressure vessel leakage tests
- compressor efficiency
- brake lining
- blow down intervals
- drainage dewatering pumps
- sump inflow
- S.C.O. mode power consumption.

Information relating to parameters such as temperature, level, wear, vibration is gathered on a systematic basis and entered into the computerised maintenance management system. This information along with the initial commissioning values is then used for trend analyses and predictive maintenance.

10. MAINTENANCE MANAGEMENT SYSTEM

The choice of a suitable maintenance management system was the subject of an extensive investigation. Existing systems within Eskom exhibited certain drawbacks with respect to their applicability to Palmiet and Drakensberg. Eskom's computerised system PERMAC which is used on all fossil fired power stations and on Drakensberg until recently, can only be run on the central mainframe computer at Eskom's Head Office in Johannesburg. The distance between Johannesburg and the power stations resulted in poor line availabilities, reliabilities and response times. It was also found to be deficient in the accurate collation of past history, storage of general information and its adaptability/flexibility to cater for specific requirements.

In order to cater for the specific needs of Palmiet, a very powerful maintenance system, called POSAC (Palmiet On Line and Case History System) was developed, to control maintenance and collect plant history data making use of the plant location and identification code.

The main features of the system are:

(a) It is user-friendly, flexible and customised for hydro terminology.

(b) It provides comprehensive case history, asset description, specification and location data.

(c) It provides comprehensive scheduled heavy and light maintenance routines and listings.

(d) It provides a comprehensive and flexible reporting facility.

This system coupled to the plant condition monitoring system provides management with the information necessary to optimise maintenance and prevent recurring problems. The information will also be useful to the designers of new plant.

11. ESKOM'S CO-OPERATION

The experience Eskom has gained in the technology, operating and maintenance of Hydro plant, as well as the infrastructure that has been established to support hydro, has placed Eskom in a strong position to assist its neighbouring utilities. Co-operation exists with Malawi, Swaziland, Namibia, Mocambique and Transkei. It is Eskom's desire to build on this co-operation and set an example of what can be accomplished with co-operation, for the benefit of all.

8. Maintenance of pumped storage plants

B.E. SADDEN, BSc, MICE, Works International Consultancy, Singapore

Abstract

Pumped Storage plants are among the most sophisticated hydro plants that civil engineers are required to build, and the time required to plan and design a scheme ensures that relatively few design engineers are involved with the implementation of more than one such project in their professional career. This paper is a record of an investigation of existing pumped storage which was carried out to provide information to owners, operators and design engineers on the performance and maintenance record of plants so that future designs might avoid some of the pitfalls so far encountered.

Introduction

1. There are more than 20 pumped storage plants in the USA with an installed capacity of more than 200 Mw, and many more of similar size in various other developed nations. It appears that, for the foreseeable future, the capabilities of pumped storage plants are going to be desirable to most utilities to enhance the flexibility of the electrical system in which they are located. Because any utility manager or design engineer is unlikely to play a significant role in the implementation of more than one such plant, it was deemed to be useful to gather together highlights of operating and maintenance experience of existing plants for the use of planners and designers of future installations.

2. The Electric Power Research Institute of the USA funded the investigation and will be publishing the report.

Information Gathering

3. The scope of the study was limited in two ways:

o The study was initially limited to US plants

considered "normal", and operating daily in large electrical systems. The scope was later expanded to include some foreign installations which demonstrate heavy, or unusual, use which might be applicable to future US practice. Notable exceptions to the study were existing tidal plants at La Rance and the Bay of Fundy, both of which can enhance the tidal cycle by pumping and could strictly be classed as pumped storage. Although some utilities are considering pumped storage schemes using the sea as a lower reservoir, it was concluded that such exceptional operating conditions fell outside of this study.

- With few exceptions, plants included in the study were equipped with reversible Francis pumped turbines of such a size and number that the total plant capacity was in excess of 200 Mw. The study was also limited to those plants with significant operational experience.

4. These scope limitations resulted in a study of 24 US plants varying in size from Mount Elbert (200 Mw) to Bath County (2592 Mw), together with 11 foreign plants ranging in size from Le Truel (40 Mw) to Dinorwig (1550 Mw). Power plants studied were:-

- Bath County
- Bear Swamp
- Blenheim Gilboa
- Cabin Creek
- Carters
- Castaic
- Fairfield
- Helms
- Horse Mesa
- Jocassee
- Lewiston
- Ludington
- Mormon Flat
- Mount Elbert
- Muddy Run
- Northfield Mountain
- Salina
- San Luis
- Seneca
- Smith Mountain
- Taum Sauk
- Wallace

- o Yards Creek
- o Dinorwig
- o Ffestiniog
- o Turlough Hill
- o Le Truel
- o Montezic
- o Bajina Basta
- o Minghu
- o Numappara
- o Shintoyone
- o Masegawa
- o Okuyahagi

General information about the plants is given in Table 1A through 1K.

Study Method

5. For each plant, after basic information was compiled from published data, the station office was contacted to supplement and correct it from their own records. Later, a visit was made to each plant by a civil engineer, electrical engineer and mechanical engineer from the study team. Discussions were held with plant managers and maintenance managers, followed by a tour of the plant. Plant managers were encouraged to forward additional information about incidents occurring during the time during which the report was in preparation.

Review of Design philosophies

Although pumped storage plants are essentially similar to conventional hydro plants, several aspects of design have received extra attention during the development of the technology, because of the critical loads or operating criteria. In the areas of particular interest the plants reported the following:

- o Turbine materials - Almost every plant manager expressed a desire for all runners to be stainless steel. Early plants used mild steel runners with stainless overlay but the lowering of price of stainless steel has resulted in a trend to complete stainless steel runners. Normally 13 - 4 stainless steel is used. Only four of the US plants were originally equipped with stainless steel wicket gates although two others had replacement gates fabricated in this material. Cavitation damage to wicket gates was reported at many plants. Most foreign plants had stainless steel wicket gates.

- o Wicket gate restraint - About 65% of plants visited utilized only a breaking element to protect the gates, which was usually a shear

TABLE IA

NAME	RIVER OR WATER SOURCE	STATE	CONSTN TIME (YRS)	YR FIRST IN SERV.	CONSTN COSTS ($1000)	CAPACITY (MW)
BATH COUNTY	BACK CREEK	VIRGINIA	8	1985	1,700,000	2,682
BEAR SWAMP	DEERFIELD RIVER	MASSACHUSETTS	3	1974	105,899	600
BLENHEIM GILBOA	SCHOHARIE CREEK	NEW YORK	4.5	1973	153,293	1,000
CABIN CREEK	SOUTH CLEAR CREEK	COLORADO	3	1967	37,418	300
CARTERS	COOSAWATTE RIVER	GEORGIA	13	1977	107,000	750
CASTAIC	W BRANCH CA. AQUEDUCT	CALIFORNIA	–	1973	317,037	1,275
FAIRFIELD	BROAD RIVER	SOUTH CAROLINA	4	1978	201,549	511
HELMS	W FORK KINGS RIVER	CALIFORNIA	8	1984	980,000	1,053
HORSE MESA	SALT RIVER	ARIZONA	–	1972	22,952	366
JOCASSEE	KEOWEE RIVER	SOUTH CAROLINA	7.5	1973	107,000	612
LEWISTON	NIAGARA RIVER	NEW YORK	3	1961	632,485	240
LUDINGTON	LAKE MICHIGAN	MICHIGAN	4	1973	308,726	1,872
MORMON FLAT	SALT RIVER	ARIZONA	–	1971	16,205	86
MOUNT ELBERT	MT ELBERT CONDUIT	COLORADO	–	1985		200
MUDDY RUN	SUSQUEHANNA RIVER	PENNSYLVANIA	–	1967	83,834	800
NORTHFIELD MOUNTAIN	CONNECTICUT RIVER	MASSACHUSETTS	–	1972	126,549	846
RACCOON MOUNTAIN	TENNESSEE RIVER	TENNESSEE	8	1979	334,000	1,530
SALINA	GRAND RIVER	OKLAHOMA	–	1968	29,922	260
SAN LUIS	CA AQUEDUCT	CALIFORNIA	4	1968	61,692	388
SENECA	ALLEGHENY RIVER	PENNSYLVANIA	3	1969	66,193	422
SMITH MOUNTAIN	ROANOAKE RIVER	VIRGINIA	3.5	1965	88,684	565
TAUM SAUK	BLACK RIVER	MISSOURI	3	1963	45,854	408
WALLACE	OCONEE RIVER	GEORGIA	–	1979	204,007	321
YARDS CREEK	YARDS CREEK	NEW JERSEY	–	1965	32,879	330
FOREIGN PROJECTS						
DINORWIG	AFON PERIS	UK (WALES)	–	1982	N/A	1,550
FFESTINIOG	NONE	UK (WALES)	–	1961	19,500	360
TURLOUGH HILL	LOUGH NAHANAGAN	IRELAND	5.5	1974	30,424	292
LE TRUEL	TARN RIVER	FRANCE	–	1982	N/A	40
MONTEZIC	TRUYERE RIVER	FRANCE	–	1982	N/A	912
BAJINA BASTA	RIVER DRINA	YUGOSLAVIA	–	1979	N/A	348
MINGHU	SHUILI RIVER	TAIWAN	–	1984	N/A	1,000
NUMAPPARA	NAKA RIVER	JAPAN	4	1973	201,665	675
SHINTOYONE	OHNYU RIVER	JAPAN	4	1972	284,065	1,125
MASEGAWA		JAPAN	–	N/A	N/A	288
OKUYAHAGI PLANT 1	–	JAPAN	–	N/A	N/A	343
OKUYAHAGI PLANT 2	–	JAPAN	–	N/A	N/A	783

TABLE IB

NAME	COST PER KW ($) 1988	NUMBER OF REV. TURB	TOTAL GENERATION		PUMPING		CONDENSING	GENERATING TO PUMPING RATIO
			ENERGY MWHRS/YR	PLANT FACTOR	ENERGY MWHRS/YR	PUMP FACTOR	MWHRS/YR TOTAL	
BATH COUNTY	755	6	2,241,330	9.5	3,000,000	13.9	20,700	0.75
BEAR SWAMP	399	2	395,595	7.5	583,006	N/A	N/A	0.68
BLENHEIM GILBOA	367	4	1,953,328	22.3	2,917,700	30.6	706	0.67
CABIN CREEK	424	2	64,713	2.5	99,763	3.8	10,426	0.65
CARTERS	271	2	95,274	1.5	462,638	20.5	N/A	0.21
CASTAIC	596	6	N/A	N/A	N/A	N/A	N/A	0.67
FAIRFIELD	706	8	466,836	10.4	659,740	13.2	N/A	0.71
HELMS	1,175	3	595,137	5.6	958,498	10.6	N/A	0.62
HORSE MESA	159	1	N/A	N/A	N/A	N/A	N/A	0.5
JOCASSEE	419	4	1,162,000	21.7	1,479,000	23.8	N/A	0.79
LEWISTON	N/A	12	194,892	9.3	315,905	10.5	N/A	0.62
LUDINGTON	395	6	2,640,000	16.1	3,655,000	21.3	4,880	0.71
MORMON FLAT	507	1	104,170	13.8	124,662	29.6	N/A	0.84
MOUNT ELBERT	N/A	2	N/A	N/A	N/A	N/A	N/A	N/A
MUDDY RUN	356	8	N/A	N/A	1,500,000	21.0	N/A	0.71
NORTHFIELD MOUNTAIN	380	4	N/A	N/A	N/A	N/A	N/A	0.73
RACCOON MOUNTAIN	369	4	1,292,160	9.6	1,637,544	11.5	N/A	0.79
SALINA	369	6	N/A	N/A	N/A	N/A	N/A	N/A
SAN LUIS	510	8	N/A	N/A	N/A	N/A	N/A	0.67
SENECA	475	2	586,000	15.9	812,000	23.4	N/A	0.72
SMITH MOUNTAIN	600	3	435,200	8.8	450,000	20.8	5,080	0.89
TAUM SAUK	482	2	14,000	0.4	45,400	1.3	N/A	0.31
WALLACE	1,074	4	347,172	12.3	419,545	19.3	N/A	0.83
YARDS CREEK	381	3	571,041	19.8	.859,251	23.4	N/A	0.66
FOREIGN PROJECTS								
DINORWIG	N/A	6	1,823,000	13.4	2,664,000	16.9	96,000	0.68
FFESTINIOG	261	4	328,146	10.4	502,103	19.1	33,000	0.65
TURLOUGH HILL	236	4	369,764	14.5	567,265	23.7	N/A	0.65
LE TRUEL	N/A	1	N/A	N/A	N/A	N/A	N/A	N/A
MONTEZIC	N/A	4	N/A	N/A	N/A	N/A	N/A	0.75
BAJINA BASTA	N/A	2	N/A	N/A	N/A	N/A	N/A	N/A
MINGHU	N/A	4	N/A	N/A	N/A	N/A	N/A	N/A
NUMAPPARA	716	3	199,000	3.4	321,000	4.9	N/A	0.62
SHINTOYONE	641	5	518,667	5.3	618,305	5.4	N/A	0.84
MASEGAWA	N/A	2	N/A	N/A	N/A	N/A	N/A	N/A
OKUYAHAGI PLANT 1	N/A	3	N/A	N/A	N/A	N/A	N/A	N/A
OKUYAHAGI PLANT 2	N/A	3	N/A	N/A	N/A	N/A	N/A	N/A

TABLE IC

NAME	STORED ENERGY (MwH)	GOV SPEED DROOP (%)	NORMAL OPERATION DESCRIPTION	AV NO MODE CHANGES PER DAY	BRAKING SPEED %	PUMP TURBINE OUTAGE %	AVAILABILITY %
BATH COUNTY	23,700	BLOCK	DAILY CYCLE	8	12	-	90
BEAR SWAMP	3,690	N/A	DAILY CYCLE	N/A	9	10	90.3
BLENHEIM GILBOA	12,000	4	DAILY CYCLE	6	4	8	92.8
CABIN CREEK	1,450	BLOCK	DAILY CYCLE	6	3	-	91 to 96
CARTERS	6,245	5	DAILY CYCLE	6	13	1.4	95.6
CASTAIC	12,000	N/A	SYNC COND	6	12	-	-
FAIRFIELD	4,096	N/A	DAILY CYCLE	2	15	10	90
HELMS	N/A	N/A	DAILY CYCLE	N/A	4	-	91
HORSE MESA	620	BLOCK	DAILY CYCLE	N/A	N/A	-	-
JOCASSEE	32,000	5	DAILY CYCLE	N/A	15	-	89.6
LEWISTON	N/A	BLOCK	N/A	4/5	N/A	5.49	94.5
LUDINGTON	15,000	BLOCK	WEEKLY CYCLE	N/A	10	8	92
MORMON FLAT	287	BLOCK	DAILY CYCLE	N/A	36	-	-
MOUNT ELBERT	3,200	N/A	DAILY CYCLE	1	10	32.1	67.9
MUDDY RUN	11,800	5	DAILY CYCLE	4 TO 8	50	-	80
NORTHFIELD MOUNTAIN	8,500	BLOCK	N/A	N/A	N/A	-	-
RACCOON MOUNTAIN	N/A	5	DAILY CYCLE	9.7	2	-	88.4(74.5)
SALINA	N/A	BLOCK	N/A	N/A	20	-	-
SAN LUIS	465,000	BLOCK	SEASONAL	N/A	20	-	73.8
SENECA	4,200	5	DAILY CYCLE	4 TO 14	11	-	94
SMITH MOUNTAIN	26,656	5	DAILY CYCLE	UP TO 12	28	4	-
TAUM SAUK	2,700	BLOCK	N/A	N/A	15	less than	-
WALLACE	2,130,000	5	PEAKING	N/A	30	0.35	-
YARDS CREEK	2,850	BLOCK	DAILY CYCLE	6	12	10	90
FOREIGN PROJECTS							
DINORWIG	N/A	1	DAILY/SYNC C	40	DYNAMIC	-	-
FFESTINIOG	1,358	BLOCK	DAILY/SYNC C	27	25	-	89
TURLOUGH HILL	1,540	N/A	DAILY/SYNC C	6	N/A	-	88.75
LE TRUEL	N/A	5	N/A	N/A	5	-	-
MONTEZIC	30,000	5	WEEKLY/SYNC	6	5	-	-
BAJINA BASTA	N/A	4	N/A	N/A	3	-	-
MINGHU	N/A	N/A	N/A	N/A	N/A	-	-
NUMAPPARA	N/A	BLOCK	DAILY CYCLE	2	20	6.8	93.2
SHINTOYONE	N/A	5	DAILY CYCLE	6	8	9.45	90.55
MASEGAWA	N/A	BLOCK	N/A	N/A	N/A	-	-
OKUYAHAGI PLANT 1	N/A	BLOCK	DAILY CYCLE	N/A	N/A	-	-
OKUYAHAGI PLANT 2	N/A	BLOCK	DAILY CYCLE	N/A	N/A	-	-

TABLE ID

NAME	NUMBER OF GUIDE BEARINGS	RUNNER DIA (FT)	RATED TURBINE NET HEAD (FT)	UNIT TURBINE OUTPUT (MW)	RATED HEAD AS PUMP	RATED PUMP UNIT DISCHARGE (CFS)	TURBINE OPERATION SPEED (RPM)	OPERATING RANGE/ TURBINE %/%
BATH COUNTY	3	20.83	1262.0	350.0	1080.0	4500	257.0	85 TO 115
BEAR SWAMP	3	19.16	720.0	298.0	865.0	4430	225.0	50 TO 80
BLENHEIM GILBOA	3	19.90	1002.0	250.0	1085.0	2840	257.0	60 TO 120
CABIN CREEK	3	–	1190.0	160.0	1230.0	840	360.0	40 TO 106
CARTERS	2	20.66	345.0	129.0	347.0	4435	150.0	–
CASTAIC	2	19.25	1000.0	239.0	1060.0	1870	257.0	NOT 40 TO
FAIRFIELD	2	–	167.0	64.8	173/158	4615/4985	150.0	70 TO 100
HELMS	3	17.17	1625.0	350.0	1500.0	2400	360.0	15 TO 100
HORSE MESA	3	–	246.5	80.0	260.5	3800	150.0	–
JOCASSEE	2	24.00	294.0	170.0	294.0	6200	120.0	55 TO 111
LEWISTON	2	17.10	75.0	20.9	85.0	3400	112.5	165 TO 180
LUDINGTON	3	–	320.0	360.0	305/309	110/9100	112.5	–
MORMON FLAT	2	–	129.0	42.0	130.0	4070	138.5	–
MOUNT ELBERT	2	18.83	442.0	100.0	405.0	3200	180.0	–
MUDDY RUN	2	17.91	353.0	103.0	427.0	2610	180.0	40 TO 100
NORTHFIELD MOUNTAIN	3	–	745.0	260.0	740.0	3600	257.0	–
RACCOON MOUNTAIN	2	16.58	1020.0	391.0	1000.0	3850	300.0	9.7
SALINA	2	–	225.0	44.7	245.0	2100	171.4	–
SAN LUIS	3	17.50	197.0	52.2	290.0	2030	150.0	–
SENECA	2	18.67	646.0	162.0	700.0	3200	225.0	63 TO 100
SMITH MOUNTAIN	2	19.80	170.0	105.0	188.0	7080	90.0	40 TO 90
TAUM SAUK	2	21.50	790.0	220.0	764.0	2650	200.0	–
WALLACE	2	26.67	89.0	52.2	98.0	6500	85.7	–
YARDS CREEK	3	18.50	656.0	113.0	732.0	2145	240.0	80 TO 115
FOREIGN PROJECTS								
DINORWIG	3	12.45	1758.0	317.0	1787.0	1766	500.0	28 TO 104
FFESTINIOG	6	8.42	925.0	94.0	1000.0	745	429.0	55 TO 100
TURLOUGH HILL	3	5.12	941.0	73.0	943.0	781	500.0	–
LE TRUEL	4	5.49	1443.0	38.0	1427.0	244	750.0	10 TO 100
MONTEZIC	3	13.16	1367.0	228.0	1400.0	1512	429.0	44 TO 100
BAJINA BASTA	–	15.74	1819.0	294.0	2038/1794	1296/1744	429.0	–
MINGHU	3	16.70	1016.0	250.0	1070.0	2896	300.0	–
NUMAPPARA	3	16.40	1568.0	230.0	1503/1732	1766/989	375.0	60 TO 100
SHINTOYONE	3	–	666.0	230.0	590/804	4341/2910	VARIES	60 TO 100
MASEGAWA	3	16.40	327.0	149.0	360/180	4719/5916	180.0	
OKUYAHAGI PLANT 1	3	13.12	529.0	105.0	598.0	2402	300.0	–
OKUYAHAGI PLANT 2	3	16.08	1326.0	267.0	1463.0	1766	360.0	–

TABLE IE

NAME	PUMP SUBMERGENCE (FT)	GENERATOR OUTPUT (MVA)	GENERATOR POWER FACTOR	GENERATING VOLTAGE (KV)	GENERATOR ROTOR WEIGHT TOM)	GENERATOR MOTOR EXCITATION TYPE
BATH COUNTY	65.00	389.00	0.90	20.50	716	STATIC
BEAR SWAMP	60.00	333.00	0.90	13.80	583	ROTATING AMPLIFER
BLENHEIM GILBOA	50.00	278.00	0.90	17.00	497	DC GENERATOR
CABIN CREEK	–	167.00	0.90	13.80	–	–
CARTERS	21.00	132.00	0.95	13.80	285	ROTATING AMPLIDYNE
CASTAIC	–	250.00	0.85	18.00	550	STATIC
FAIRFIELD	30.00	71.00	0.90	13.80	175	STATIC
HELMS	–	390.00	0.90	18.00	520	STATIC
HORSE MESA	18.00	96.50	0.90	13.80	220	STATIC
JOCASSEE	12.00	170.00	0.90	14.40	400	STATIC
LEWISTON	13.00	25.00	0.80	13.80	110	DC GENERATOR
LUDINGTON	–	325.00	0.85	20.00	–	ROTATING AMPLIDYNE
MORMON FLAT	9.00	47.00	0.90	13.80	167	STATIC
MOUNT ELBERT	50.00	105.30	0.95	11.50	315	STATIC
MUDDY RUN	20.00	111.00	0.90	13.80	80	ROTATING
NORTHFIELD MOUNTAIN	106.00	235.00	0.90	13.80	–	ROTATING AMPLIDYNE
RACCOON MOUNTAIN	126.50	425.00	0.90	23.00	440	STATIC
SALINA	–	48.00	0.90	13.80	–	–
SAN LUIS	15.00	53.00	1.00	13.80	128	ROTATING AMPLIDYNE
SENECA	–	220.00	0.90	13.80	221	–
SMITH MOUNTAIN	3.16	141.18	0.95	13.80	414	STATIC
TAUM SAUK	30.00	204.00	1.00	13.80	138	–
WALLACE	7.20	52.20	0.90	14.40	250	STATIC
YARDS CREEK	25.00	125.00	0.90	14.40	327	ROTATING AMPLIDYNE
FOREIGN PROJECTS						
DINORWIG	198.11	313.50	0.95	18.00	–	STATIC
FFESTINIOG	49.00	95.00	0.95	16.00	600	ROTATING AMPLIDYNE
TURLOUGH HILL	81.33	87.50	0.85	10.50	–	–
LE TRUEL	–	41.50	0.90	5.65	–	STATIC
MONTEZIC	171.42	250.00	0.91	18.00	–	STATIC
BAJINA BASTA	177.12	315.00	0.95	11.00	–	–
MINGHU	125.30	280.00	0.90	16.50	440	STATIC
NUMAPPARA	151.30	250.00	0.95	16.50	488	STATIC
SHINTOYONE	46.00	250.00	0.95	16.50	550	STATIC
MASEGAWA		160.00	0.90	13.80	–	--
OKUYAHAGI PLANT 1	–	–	–	–	–	–
OKUYAHAGI PLANT 2	–	290.00	0.90	18.00	–	--

NAME	MOTOR OUTPUT (MW)	USABLE VOLUME UPP RES (AC FT)	TABLE IF UPPER DAM TYPE	DAM HEIGHT (FT)	DAM CREST LENGTH (FT)	UPPER DAM VOLUME (CU YD)
BATH COUNTY	410.00	22,500	ROCKFILL W/CORE	460	2,200	17,000,000
BEAR SWAMP	309.00	8,870	ROCKFILL W/CORE	155	–	123,713
BLENHEIM GILBOA	272.00	15,000	EARTH DIKE W/CORE	162	11,900	5,875,471
CABIN CREEK	136.00	1,457	ROCKFILL W/CONC FACE	210	1,458	1,194,000
CARTERS	123.00	134,900	ROCKFILL W/CORE	445	2,053	13,500,000
CASTAIC	231.00	179,000	ROCKFILL	386	1,080	6,952,000
FAIRFIELD	77.00	29,000	EARTH EMBANKMENT W/CORE	180	10,000	9,000,000
HELMS	320.00	119,200	EARTH W/CONC. FACE	304	862	123,300
HORSE MESA	84.00		CONC. THIN ARCH	305	660	162,000
JOCASSEE	174.00	215,700	ROCKFILL W/CORE	385	1,750	11,373,600
LEWISTON	28.00	60,000	ROCKFILL W/CORE	35	334,320	99,037,000
LUDINGTON	276.00	54,000	EARTHFILL W/ASPHALT FACE	–	–	132,777,000
MORMON FLAT	45.50		CONC. THIN ARCH	224	380	59,900
MOUNT ELBERT	127.00	7,126	EARTHFILL W/PVC LINER	90	1,500	845,700
MUDDY RUN	100.00	33,200	EARTH & ROCKFILL	250	4,800	5,185,600
NORTHFIELD MOUNTAIN	217.00		ROCKFILL W/CORE	160	–	–
RACCOON MOUNTAIN	345.00	36,340	EARTH & ROCKFILL	230	8,500	10,224,154
SALINA	48.00		EARTHFILL	200	2,300	3,700,000
SAN LUIS	53.3-47/25	2,000,000	EARTHFILL	385	18,600	75,000,000
SENECA	195.00	5,755	ROLLED FILL-SANDSTONE	115	7,800	14,693,000
SMITH MOUNTAIN	127.60	150,000	CONC. ARCH. DAM	235	815	175,000
TAUM SAUK	179.00	4,350	ROCKFILL DAM	–	–	3,750,000
WALLACE	61.90	150,000	EARTHFILL & CONCRETE	–	–	825,644
YARDS CREEK	140.00	4,650	ROCKFILL W/CORE	70	9,600	–
FOREIGN PROJECTS						
DINORWIG	313.50	5,189	EARTHFILL W/ASPHALT FACE	1	1,968	1,794,000
FFESTINIOG	78.90	1,012	CONC. GRAV. BUTTRESS	121	801	2,616,000
TURLOUGH HILL	–	1,705	ROCKFILL	112	4,740	1,705,565
LE TRUEL	42.00	–	–	–	–	–
MONTEZIC	234.00	24,321	ROCKFILL W/CORE	187	2,690	2,243,460
BAJINA BASTA	–	–	–	–	–	–
MINGHU	260.00	–	–	–	–	–
NUMAPPARA	250.00	3,421	EARTH W/ASPHALT FACE	125	5,420	1,643,043
SHINTOYONE	260.00	32,753	CONCRETE ARCH	382	1,020	69,975,358
MASEGAWA	160.00	–	–	–	–	–
OKUYAHAGI PLANT 1	–	–	–	–	–	–
OKUYAHAGI PLANT 2	283.00	–	–	–	–	–

TABLE 16

NAME	POWER POOL CHANGE (FT)	LOWER RES USEABLE VOLUME (AC FT)	LOWER DAM TYP	DAM HEIGHT (FT)	DAM CREST LENGTH (FT)	LOWER DAM VOLUME (CY)	LOWER POOL EL. CHANGE (FT)
BATH COUNTY	105.0	22,500	EARTH & ROCKFILL	135	2,400	3,600,000	60.0
BEAR SWAMP	50.0	4,900	EARTH & ROCKFILL	130	900	567,000	90.0
BLENHEIM GILBOA	48.0	12,700	EARTH & ROCKFILL	100	1,800	26,373,160	40.0
CABIN CREEK	90.0	1,905	EARTH & ROCKFILL	95	1,150	1,043,000	52.5
CARTERS	77.0	17,210	ROCKFILL DYKE	65	3,349	766,000	35.5
CASTAIC	-	33,000	ARTIFICIAL	170	2,000	5,903,000	65.0
FAIRFIELD	4.5	29,000	CONCRETE GRAVITY	37	200		10.0
HELMS	164.0	89,100	EXISTING	290	3,330	128,500	110.0
HORSE MESA	18.0	-	CONC. THIN ARCH	224	380	59,900	5.5
JOCASSEE	30.0	-	-	-	-	-	3.0
LEWISTON	35.0	-	-	-	-	-	17.0
LUDINGTON	67.0	-	-	-	-	-	-
MORMON FLAT	15.0	-	CONC. THIN ARCH	207	1,260	120,500	7.0
MOUNT ELBERT	57.0	68,000	EARTHFILL	55	3,140	624,000	32.0
MUDDY RUN	30.0	81,000	CONC GRAVITY	102	-	666,720	0.0
NORTHFIELD MOUNTAIN	84.5	-	-	-	-	-	9.0
RACCOON MOUNTAIN	142.0	36,340	CONCRETE	-	-	-	0.0
SALINA	-	-	-	-	-	-	-
SAN LUIS	217.0	-	-	-	-	-	-
SENECA	70.0	4,650	EARTHFILL	-	-	-	-
SMITH MOUNTAIN	1.8	35,000	CONC. GRAVITY	94	380	100,000	13.0
TAUM SAUK	92.0	4,460	-	-	-	22,000	16.0
WALLACE	7.0	27,600	CONC. GRAVITY	117	870	-	1.3
YARDS CREEK	49.0	4,650	EARTHERN	55	3,500	366,610	23.5
FOREIGN PROJECTS							
DINORWIG	108.0	5,189	ROCKFILL	-	-	-	19.7
FFESTINIOG	65.3	1,705	CONC. GRAV.	49	1,801	-	19.6
TURLOUGH HILL	65.0	1,705	NATURAL	-	-	-	32.5
LE TRUEL	-	-	-	-	-	-	-
MONTEZIC	75.4	24,321	THIN ARCH	199	892	69,700	50.9
BAJINA BASTA	-	-	-	13	-	-	46.7
MINGHU	69.8		CONCRETE GRAVITY	189	556	-	67.3
NUMAPPARA	131.3	16,944	ROCKFILL ASPHALT	248	1,095	2,572,755	105.0
SHINTOYONE	128.0	166,555	CONCRETE GRAVITY	510	963	427,534,336	131.0
MASEGAWA	-	-	-	-	-	-	-
OKUYAHAGI PLANT 1	-	-	-	-	-	-	-
OKUYAHAGI PLANT 2	-	-	-	-	-	-	-

TABLE IH

NAME	PENSTOCK NUMBER	PENSTOCK VELOCITY (FT/S)	PENSTOCK LENGTH (FT)	POWERHOUSE INLET VALVE DETAILS	TIME OPEN/CLOSE (SEC)	DRAFT TUBE GUARD VALVE NUMBER
BATH COUNTY	6	21.2-76.2	897-1257	114" SPHERICAL	40/60	NONE
BEAR SWAMP	2	5.5	350	132" SPHERICAL	30/30	NONE
BLENHEIM GILBOA	4	–	1,960	72" SPHERICAL	30/30	NONE
CABIN CREEK	1	60.0	1,563	NONE	120/120	2
CARTERS	4	21.2	838	NONE	–	NONE
CASTAIC	6	–	2,400	104" SPHERICAL	–	NONE
FAIRFIELD	4	22.6	800	NONE	–	NONE
HELMS	–	–	4,400	95" SPHERICAL	120/120	NONE
HORSE MESA	1	29.6	–	NONE	–	NONE
JOCASSEE	4	21.0	387	NONE	–	NONE
LEWISTON	12	10.0	180	NONE	–	NONE
LUDINGTON	–	–	–	NONE	–	NONE
MORMON FLAT	1	21.2	–	NONE	–	NONE
MOUNT ELBERT	2	10.0	3,000	NONE	–	NONE
MUDDY RUN	8	28.5	500	NONE	–	
NORTHFIELD MOUNTAIN	–	–	–	114" SPHERICAL	–	NONE
RACCOON MOUNTAIN	4	20.8	115	120" SPHERICAL	45	NONE
SALINA	–	–	–	BUTTERFLY	–	NONE
SAN LUIS	4	17.0	2,200	156" BUTTERFLY	183	NONE
SENECA	–	–	–	114" SPHERICAL	–	3
SMITH MOUNTAIN	5	15.1/19.1/	VARIES	NONE	–	NONE
TAUM SAUK	–	–	–	108" SPHERICAL	–	NONE
WALLACE	–	–	–	NONE	–	NONE
YARDS CREEK	1	29.0	2,375	84.5" SPHERICAL	180/180	NONE
FOREIGN PROJECTS						
DINORWIG	6	43.0	557	98.4" SPHERICAL	–	6
FFESTINIOG	4	23.0	754	70" STRAIGHTFLOW	48	NONE
TURLOUGH HILL	4	N/A	N/A	67" SPHERICAL	55/55	NONE
LE TRUEL	–	–	–	SPHERICAL	–	NONE
MONTEZIC	4	18.2/36.1/62.6		80.7" SPHERICAL	60/60	NONE
BAJINA BASTA	–	–	–	86.6" SPHERICAL	30	NONE
MINGHU	2	–	1705/1838	SPHERICAL	–	NONE
NUMAPPARA	3	41.4	2610/2643	94.5" SPHERICAL	150/150	NONE
SHINTOYONE	5	18.7	951/1033	137.8" SPHERICAL	70/70	NONE
MASEGAWA	–	–	–	NONE	–	NONE
OKUYAHAGI PLANT 1	–	–	–	118.1" BUTTERFLY	–	NONE
OKUYAHAGI PLANT 2	–	–	–	90.6" SPHERICAL	–	NONE

TABLE IJ

NAME	DRAFT TUBE GUARD VALVE DIA (FT)	POWERHOUSE TYPE	POWERHOUSE LENGTH (FT)	POWERHOUSE WIDTH (FT)	POWERHOUSE HEIGHT (FT)
BATH COUNTY	N/A	CONVENTIONAL SURFACE INDOOR	490	150	200
BEAR SWAMP	N/A	UNDERGROUND	227	79	152
BLENHEIM GILBOA	N/A	REINF. CONC. SEMI-OUTDOOR	366	175	132
CABIN CREEK	4.99	SEMI UNDERGROUND	143	100	107
CARTERS	N/A	CONVENTIONAL SURFACE INDOOR	362	114	159
CASTAIC	N/A	CONC. STRUC. 2/3 UNDERGROUND	600	187	190
FAIRFIELD	N/A	SEMI-OUTDOOR CONSTRUCTION	520	150	100
HELMS	N/A	UNDERGROUND	336	83	144
HORSE MESA	N/A	CONVENTIONAL INDOOR	83	75	93
JOCASSEE	8.99	-	-	-	-
LEWISTON	NONE	SEMI-OUTDOOR TYPE	974	233	160
LUDINGTON	NONE	-	576	171	106
MORMON FLAT	NONE	CONVENTIONAL INDOOR	94	72	98
MOUNT ELBERT	NONE	SEMI-OUTDOOR	146	104	182
MUDDY RUN		OUTDOOR TYPE W/ GEN. DECK	600	140	80
NORTHFIELD MOUNTAIN	NONE	UNDERGROUND CAVERN	328	70	155
RACCOON MOUNTAIN	NONE	UNDERGROUND	490	72	165
SALINA	NONE	-	-	-	-
SAN LUIS	NONE	INDOOR	483	97	49
SENECA	9.51	FULLY ENCLOSED	230	70	138
SMITH MOUNTAIN	NONE	SEMI-OUTDOOR	N/A	N/A	N/A
TAUM SAUK	NONE	REINF. CONC. SEMI-OUTDOOR	150	88	80
WALLACE	NONE	INDOOR	N/A	140	65
YARDS CREEK	NONE	REINF. CONC. SEMI-OUTDOOR	150	105	78
FOREIGN PROJECTS					
DINORWIG	12.3	UNDERGROUND	588	77	168
FFESTINIOG	NONE	SURFACE INDOOR	235	72	177
TURLOUGH HILL	NONE	UNDERGROUND	269	75	98
LE TRUEL	NONE	-	-	-	-
MONTEZIC	NONE	UNDERGROUND	476	82	157
BAJINA BASTA	NONE				
MINGHU	NONE	UNDERGROUND	417	70	132
NUMAPPARA	NONE	UNDERGROUND	423	66	131
SHINTOYONE	NONE	UNDERGROUND	451	70	69
MASEGAWA	NONE	-			
OKUYAHAGI PLANT 1	NONE	-	-	-	-
OKUYAHAGI PLANT 2	NONE	-	-	-	-

TABLE IK

NAME	MAIN CRANE CAPACITY (TON)	OTHER CRANE CAPACITY (TON)
BATH COUNTY	760	200
BEAR SWAMP	616	2 X 30
BLENHEIM GILBOA	510	-
CABIN CREEK	90	15
CARTERS	360	2 X 25
CASTAIC	375	30
FAIRFIELD	185	30
HELMS	2 X 270	-
HORSE MESA	N/A	N/A
JOCASSEE	_	-
LEWISTON	168	VARIOUS
LUDINGTON	360	-
MORMON FLAT	N/A	N/A
MOUNT ELBERT	450	10
MUDDY RUN		
NORTHFIELD MOUNTAIN	350	25
RACCOON MOUNTAIN	2 X 220	25
SALINA	_	-
SAN LUIS	2 X 175	50 & 10
SENECA	_	-
SMITH MOUNTAIN	350	30
TAUM SAUK	175	-
WALLACE	300	25
YARDS CREEK	327	25
FOREIGN PROJECTS		
DINORWIG	275	10
FFESTINIOG	2 X 122	2 X 30
TURLOUGH HILL	2 X 70	50
LE TRUEL	60	--
MONTEZIC	2 X 125	--
BAJINA BASTA	_	--
MINGHU	2 X 250	2 X 35
NUMAPPARA	2 X 190	--
SHINTOYONE	2 X 256	70
MASEGAWA	_	--
OKUYAHAGI PLANT 1	-	_
OKUYAHAGI PLANT 2	-	--

pin. About 27 % used a breaking element together with a friction restraining device, and about 8% used only a friction device. Some plants had experienced an unusually high incidence of shear pin failures and both Smith Mountain and Salina had suffered progressive failures . Friction devices were installed following these occurrences.

- Wicket gate Bearings - Most of the plants visited used grease lubricated wicket gate bearings, because there is a widely accepted belief that teflon bearings will not withstand the pounding occurring in a pump turbine. One plant in the US has started replacing grease lubricated bearings with self lubricating bearings but there is not yet sufficient operating experience to assess the success of this trial. The Japanese plants also have self lubricated bearings which have apparently operated with no problems. The most common problem reported concerned the seals, which occasionally failed and some plants have replaced the original seals with quad ring or chevron seals which are giving good service.

- Distributor seals - Apart from the French plants, all plants visited were equipped with seals at the top and bottom of the wicket gates. The majority of the seals are rubber or neoprene with some brass or bronze. Two plants however use nylon. The seals are normally confined by the facing plates, but at Turlough Hill where the seals were added after commissioning, they are installed in the wicket gates themselves. At Blenhaim Gilboa and Castaic, the original elastomeric seals were changed to have bronze contact surfaces.

- Turbine shutoff valves - Of the plants visited, 60% had spherical valves, 8% had butterfly valves and the rest had no valves. Spherical valves were used for turbine net heads between 646 feet(196 meters) and 1568 feet (478 meters). Butterfly valves were used for heads between 197 feet (60 meters) and 529 feet (161 meters)

- Starting methods - Plants visited included almost every starting method that has been devised. All the newer plants in the US and abroad tend to use static electronic equipment for synchronous starting. These have been trouble free, except for those at Mount Elbert which included a large number of field changes during commissioning.

Problems of equipment- Turbines

6. Problems in pump/turbines differed only in severity to those occurring in conventional hydro installations. Clearly the many mode changes, starts, stops and reversals

of rotation, put an increased strain on equipment, a fact recognised in the design criteria for the turbines for Dinorwig. It is not uncommon in the US for conventional turbines to operate for 30 or 40 years without a major rebuild. Pump/turbines, in contrast, need overhauling every ten to fifteen years. The following turbine problems were described by staff at the plants visited:

- Cavitation - Although all pump/turbines are subject to cavitation damage it was generally not considered to be a serious problem. Plants visited were almost equally divided between those with stainless steel runners and those with mild steel runners with stainless overlay. As might be expected, significant repair of the mild steel runners was required and the area of overlay increased each time the runner was repaired. The amount of repair decreased with each renewal of overlay. An extreme example was Horse Mesa, where the amount of weld material required per year decreased from an initial requirement of 1000 pounds to 10 pounds. Cavitation repairs to both runners and wicket gates have been successfully delayed for some time by the application of epoxy coatings.

- Vibration - At the plants inspected, many of the problems of the units were associated with excessive vibration, or resulted in bad vibration. Amongst those problems associated with vibration were fracturing of balance lines, excessive wear on linkages, failure of bearing spider assemblies, the near flooding of Racoon Mountain and other quoted examples. Although there are many reasons for heavy vibration, it is clear that this aspect of unit (and plant) design must be uppermost in the designer's mind and attention must be given to minimizing the forces driving the vibration and to increasing dampening. The same vibrations do occur in conventional plants but because of the reversals and the tendency to use pumped storage at varying unit flows, the machines spend more time in, or accelerating (or slowing) through the "rough zones".

- Runners - Although one or two plants actually lost parts of their runners or suffered cracks, the only significant problem of runners not associated with cavitation was the loosening of the cover plate on split runners. The usual cause was the failure of the bolts holding the cover plate in place. Loosening which occurred in four of the seven plants with this type of runner, caused serious damage and the problem was usually solved by welding the plates in place. Other solutions such as filling the cavity with epoxy, have been tried but abandoned.

- Wearing Rings - Wearing ring failures have included falling off and seizure. Seizure at Wallace, for instance, caused an outage of 16 months, while seizure at Bath County occurred because of a side load on a bearing caused by servo motor misadjustment.

- Bearings - Bearing failures have occurred in many plants and have been a result of misalignment, excessive downthrust, insufficient or uneven cooling, excessive side loads etc. Three plants with sprayed on babbitt failed when the babbitt came loose, and the bearings were changed to poured in babbitt which has given better service.

- Wicket Gate mechanism - The most common recurring problem reported was washing out of the bearing grease due to lack of, or failure of, the seals. Seals were of course replaced. Yards Creek suffered continuing shear pin failures while Smith Mountain and Salina suffered progressive failures. At Smith Mountain, during this incident three wicket gates failed, and gates also failed at Cabin Creek during a runner failure. Castaic, Ludington and Dinorwig suffered significant cavitation damage to gates.

- Balancing Lines - Many plants were subjected to damage to the balancing lines, resulting in lengthy and sometimes expensive repairs. Cavitation has occurred at the entrance, elbows and around valves in the line and the problem has been exacerbated and rendered more critical by vibration in the line. The most difficult repairs have been at the elbows embedded in concrete and plants have tried various alterations to the machines to try to avoid second failures in these lines.

- Shaft Seals - Incidents with the shaft seals have been relatively rare, but the most noticeable problem was at Mount Elbert where the holddown bolts loosened, resulting in turbine pit flooding. In general mechanical seals have performed better than packing boxes.

Problems of equipment- Generators

7. Generator/motors have performed well, and performance has inevitably improved in those plants where the stators were rewound with class H insulation to replace the original old class B insulation. The reported problems usually fell into one of the following groups:

- Stator windings - Several plants exhibited stator winding degradation mainly due to insulation deterioration due to corona, temperature cycling and age. The reversing and frequent starts tend to shorten the life of the stator and rewinding after 10 years could be said to be the norm. At some plants partial discharge analyzer equipment has been installed to monitor the condition of the windings.

- Corona - Corona damage has been limited, with only two plants exhibiting severe damage. Any corona problems have usually been improved by rewinding.

- Wedges - Loose wedges have been troublesome in almost all the plants, and in a few machines wedges dropped out of the slots. Most units are being checked at every scheduled maintenance. Many plants are now using corrugated (expandable) wedge backing strips, but it is too early to conclude that they have solved this particular problem completely.

- Alignment - This was not deemed to be a problem in most plants, but Castaic and Cabin Creek have both had bearing problems and have concentrated on near perfect alignment to minimize the bearing incidents. At Castaic, for instance, the shaft was aligned to within .002 inches using optical alignment. Several plants recommend the use of such equipment.

- Cooling - Although cooling was not highlighted as a major problem, leaks have occurred in the piping or coolers at several plants. Only one plant visited has water cooled stators and rotors and it has had periodic trouble with leakage, including a stator electrical fault directly as a result of leakage. The staff felt that the setting of the alarms was too high thus ensuring significant electrical damage before tripping an alarm. It was noted that a very small leak can quickly initiate a major electrical fault and the plant managers were, at the time of the visit considering resetting instrumentation to the minimum setting.

Problems of equipment- Ancillary equipment

8. As might be expected most ancillary equipment has been giving excellent service but there were a number of problems reported by operators that deserve noting:

- Unit circuit breakers - Whether at the generator/motor or in the switchyard, the unit circuit breaker has been reported as exhibiting continuing problems at the majority of plants. It is evident that at the design stage for the first generation of pumped storage plants, the use of this piece of equipment was severely underestimated. In the plants visited, a typical breaker would experience approximately 1000 close -open cycles per year, whereas the older designs of breaker have a rated life of 250 to 300 operations. In the light of this knowledge, their performance has actually been quite remarkable, exceeding the design life substantially. It should be noted that the early plants used the best breakers available at the time, but even in the newer US plants the equipment design life is only 3000 close - open operations, so the necessity for maintenance and replacement is clear. Some plants have been subjecting the starting and controlling breakers to take down inspections every month, while the majority are performing this exercise once every three months. At this time arc chutes, contacts and springs are usually replaced. Several plants reported difficulty in obtaining parts and many are fabricating their own to avoid long delivery. Where the older breakers have been replaced by SF_6 or vacuum breakers much better service has resulted.

- Control systems - The control systems have been giving excellent service and most of the designs in the plants visited have been superseded by more sophisticated type of programable systems. It is interesting to note that at some plants the original hardwired system included wiring for many modes of operation which have never been used, and repair is reported as rather confusing because of the extensive redundant wiring.

- Transformers - Transformers have given good service apart from several reports of gassing. However when serious trouble has occurred, it has been very time -consuming transporting the transformer to and from the manufacturer's shop. Every plant that has been obliged to send a transformer back to the manufacturer has recommended the storage of a spare transformer on site. The most extreme delay occurred at Blenhaim Gilboa, which reported a total of five years for a transformer repair because of transportation to and from the factory together with a second removal to the factory because of damage on the return trip from the initial repair.

- Valves - Of the 22 spherical and straightflow valve installations inspected, 36 % reported that the moveable seals had been wiped.

Problems of equipment- Civil and hydraulic

9. A major proportion of the investment in a pumped storage plant is spent on the large and often complex civil engineering works. Considering this fact and the cyclic or transient loading on much of the plant components, it is remarkable that the incidence of continuing problems is low. Most of the plants visited had experienced severe problems during the construction phase but the number of plants experiencing problems during operation was much less. Incidents include:

- Reservoir leakage - Plants reporting leakage from reservoirs have included Seneca, Ludington and Nummapara (which all have asphaltic liners), Muddy Run, Salina and Taum Sauk (which all utilize concrete liners) and Yards Creek, Jocassee, Carters and Mount Elbert (all of which use processed natural materials).

  The most extensive and time consuming repair was undertaken at Mount Elbert where an impermeable sheet of man-made fabric was spread over the floor of the whole of the upper reservoir. Also a significant modification was carried out at Seneca which had experienced leakage from the upper reservoir floor (which was not originally sealed). During the repairs a 10 acre PVC liner was installed and subsequently the whole floor was covered with asphalt.

  The failures of the lining at Ludington have been well documented in other conferences.

Muddy Run and Salina are both leaking from the upstream inlet channels which are lined with concrete. In both cases the leakage emanates from cracks and from the joints. At Salina the leakage is significant in the left abutment area and a slurry trench is being installed. Taum Sauk leakage has been determined to be through the asphalt covered floor, the toe of the embankment, and the joints in the concrete internal lining. Remedial measure have continued for some time and include grouting and cutting out and refilling joints. It should be noted that the embankment at Taum Sauk was uncompacted.

Of the Earth/Rockfill embankments, Yards Creek leaked, probably because of penetration of the filter into the core, Jocassee leaked through the abutments and large areas of natural material have been replaced by processed engineered fill. Most embankments have performed satisfactorily, but at San Luis there was an upstream slip because of a foundation failure.

o Hydraulic passages - No plant reported untoward transient pressure rises or problems or damage associated with that phenomenon. The most common problem of hydraulic passages were to do with trashracks. Incidents were reported at eight plants and included cracking and actual failure. At one plant a upstream trashrack failed during pumping and part of the failed rack were swept into the machines when generating. The reason was finally discovered to be a larger trashrack spacing on the lower trashracks than on the upstream trashracks. This detail allowed debris to be pumped up and onto the inside of the upper trashrack. At another plant with upstream trashrack failure during pumping, the upstream racks were modified to automatically raise during pumping.

One or two notable incidents of gate problems have occurred, including the lifting of a draft tube gate by blowdown air expelled through draft tube, causing the suspension cables to break as the gate dropped again. Upstream gates at Fairfield vibrated severely during pumping and eventually some fell. The gates were originally designed to hang just above the water passage but were modified to dog them as high as possible subsequent to the failure.

Two plants have reported serious problems with

the penstock. The regrouting of the penstock at Bath County is well documented , but the Taum Sauk experience is more interesting. The Penstock drains continuously become blocked with calcium carbonate. Upon draw down the penstock lining has buckled because of exterior pressure. The drains are now occasionally rota-rooted.

- Flooding - Although this occurrence cannot be described as a normal maintenance consideration, the substantial damage involved requires it to be considered. The worst case occurred at Northfield Mountain during the latter stages of commissioning. The valve in the drain line of the turbine shutoff valve seal control valve was accidentally left in the closed position, thus causing opening of the main valve seals. Because an access manhole to the scroll case was removed at the time, leakage water was able to enter the powerhouse. The upper reservoir was not full, but the penstock water completely flooded the powerhouse above all units. Yards Creek suffered two floodings, one from a fracture of the balancing lines and one as a result of human error. Raccoon Mountain suffered a near flooding when the bolts holding the valve bypass pipework became loose. Other plants had suffered less severe flooding which nonetheless could have had catastrophic results.

Maintenance time and costs

10. Required maintenance for pumped storage plants is significantly less than that normally needed at thermal plants but, because of the heavy use, rather more than can be expected for a conventional plant. Clearly the amount of work scheduled for maintenance at each plant depends on the age, complexity of design, required operation, history of outages, the time since the last major overhaul and many other factors.

11. Throughout the US, the plants were generally able to provide a complete maintenance history, and the philosophy tended to be the same. At discrete intervals , usually one year, preventative maintenance was carried out on each unit. Most plants carried out this work in one period, but one plant performed an inspection in the spring(before the summer Air conditioning peak) and as a result of this short (3 to 4 day) inspection, planned and ordered a longer maintenance period after the summer. Annual preventative maintenance usually consisted of cavitation damage repair, and inspection (and repair as

appropriate) of bearings, bearing clearances, valves, generator/motor windings and wedges, ancillary systems to the generator/motor and pump/turbine, cleaning of brushes, coolers, filters, circuit breakers, transformers etc.

12. Many of the plants visited had not yet performed a major equipment overhaul, but were planning on the first within the next year or two. Those who had performed this exercise reported widely varying intervals, from five years to sixteen years, but generally ten years would seem an appropriate time for planning. The time for performing the work varied from 6000 man hours (two and a half months) for a small unit, to nine months (projected) for large units. For the purposes of planning five weeks could be considered to be reasonable but will be influenced by the following factors:

- In house or contract labour
- Time of the year the work is done
- Laydown areas available
- Cranes available
- Unit size

13. Plants tended to plan a major overhaul because of increased forced outage, or awareness of a fault which would increase forced outages. Invariably the decision to perform a major overhaul was a function of the turbine condition. Some plants, particularly those outside the US were attempting to relate maintenance requirements to the need for major overhaul by plotting the downtime on each unit and observing the increase as the units aged, but only Yards Creek in the US was pursuing a similar analysis relating pumping/ generating ratio to the maintenance performed.

14. All plants generally tried to use units equally, but one plant was preparing to change its operating rule to use one unit as a "sacrificial unit". The chosen unit will always be first on and last off in both the pumping and generating modes. It is felt that this method of operation will encourage, in the chosen unit, such operational faults as are going to occur. Therefor the staff will be able to prepare for those same faults in the other units.

TABLE IIA

NAME	UNIT OVERHAUL			REG MAINTENFNCE		
	PERIOD (YEAR)	TIME REQUIRED (DAYS)	(MAN-HOURS)	PERIOD (MONTH)	TIME REQUIRED (DAYS)	(MAN-HOUR
BATH COUNTY	–	–	–	–	–	–
BEAR SWAMP	11	119	–	12	11	–
BLENHEIM GILBOA	10	90	36000	12	10	4000
CABIN CREEK	–	–	–	–	–	–
CARTERS	–	–	–	–	–	–
CASTAIC	–	126	–	–	–	–
FAIRFIELD	–	–	–	12	14	800
HELMS	–	–	–	12/36	5/35	500/7500
HORSE MESA	–	–	–	–	–	–
JOCASSEE	–	–	–	–	–	–
LEWISTON	–	75	–	–	8	3000
LUDINGTON	14	273(planne	–	–	–	–
MORMON FLAT	–	–	–	–	–	–
MOUNT ELBERT	–	–	–	12	30	12000
MUDDY RUN	–	30	240	–	–	–
NORTHFIELD MOUNTAIN	–	–	–	–	–	–
RACCOON MOUNTAIN	–	–	–	24	14	–
SALINA	–	–	–	–	–	–
SAN LUIS	15-20	60	2400	–	–	–
SENECA	–	–	–	–	–	–
SMITH MOUNTAIN	–	–	–	12	10	1000
TAUM SAUK	–	120	15000	–	12	–
WALLACE	–	48	9600	12	–	1200
YARDS CREEK	10	84	6720	12	10	2000
FOREIGN PROJECTS						
DINORWIG	–	–	–	–	–	–
FFESTINIOG	24	182	32000	12	24	700
TURLOUGH HILL	–	–	–	–	6	–
LE TRUEL	–	–	–	–	–	–
MONTEZIC	–	–	–	–	–	–
BAJINA BASTA	–	–	–	–	–	–
MINGHU	–	–	–	–	–	–
NUMAPPARA	10	60	28000	12	7	350
SHINTOYONE	10	60	28000	12	7	350
MASEGAWA	–	–	–	–	–	–
OKUYAHAGI PLANT 1	–	–	–	–	–	–
OKUYAHAGI PLANT 2	–	–	–	–	–	–

15. A summary of the maintenance data for the plants visited is given in Tables IIA and IIB.

General comments by operational staff

16. Plant managers indicated that plant performance was generally satisfactory and the versatility of pumped storage had been demonstrated repeatedly. However it was also mentioned that dispatchers, having discovered the merit of a pumped storage plant on the electrical system, tended to rely heavily on their characteristics. Such reliance, which may not have been foreseen by the utility managers originally, emphasizes the need for minimum planned and forced maintenance.
Aspects of the design of plants upon which the operating staff commented centred around the adequacy of the layout for easy and quick maintenance. The majority of managers considered that the total floor area in their plants was marginally sufficient for a complete overhaul and that designers should always err on the generous side. They

TABLE IIB

NAME	ANNUAL O&M COST($1000)			AVE O&M COST(Mil/Kmh)		
	1982	1983	1984	1982	1983	1984
BATH COUNTY	–	–	–	–	–	–
BEAR SWAMP	622	823	2,417	1.39	20.24	34.29
BLENHEIM GILBOA	7,219	6,910	3,530	–	–	–
CABIN CREEK	1,046	874	883	8.05	8.05	8.08
CARTERS	–	–	–	–	–	–
CASTAIC	6,449	7,821	10,178	–	–	–
FAIRFIELD	1,348	–	1,417	3.12	–	3.45
HELMS	–	–	–	–	–	–
HORSE MESA	876	997	1,932	22.41	5.19	31.73
JOCASSEE	3,786	2,322	1,467	5.68	2.23	3.55
LEWISTON	783	1,091	–	–	–	–
LUDINGTON	3,207	2,516	2,105	1.23	0.90	1.00
MORMON FLAT	1,856	367	527	66.42	3.89	8.16
MOUNT ELBERT	–	–	–	–	–	–
MUDDY RUN	4,518	5,595	5,651	4.01	5.71	4.58
NORTHFIELD MOUNTAIN	5,280	5,983	9,611	6.76	6.82	8.99
RACCOON MOUNTAIN	2,774	4,165	4,184	2.29	3.08	2.78
SALINA	360	1,162	411	4.83	11.19	8.46
SAN LUIS	–	–	–	–	–	–
SENECA	1,286	1,973	2,487	2.15	3.85	3.75
SMITH MOUNTAIN	1,463	1,363	1,431	3.71	2.60	3.52
TAUM SAUK	482	543	437	54.99	23.51	27.71
WALLACE	1,739	1,898	2,582	4.97	4.62	6.29
YARDS CREEK	902	1,710	1,536	2.18	3.58	2.60
FOREIGN PROJECTS						
DINORWIG	–	–	–	–	–	–
FFESTINIOG	–	–	–	–	–	–
TURLOUGH HILL	–	–	–	–	–	–
LE TRUEL	–	–	–	–	–	–
MONTEZIC	–	–	–	–	–	–
BAJINA BASTA	–	–	–	–	–	–
MINGHU	–	–	–	–	–	–
NUMAPPARA	–	–	–	–	–	–
SHINTOYONE	–	–	–	–	–	–
MASEGAWA	–	–	–	–	–	–
OKUYAHAGI PLANT 1	–	–	–	–	–	–
OKUYAHAGI PLANT 2	–	–	–	–	–	–

also tended to highlight their requirement for dismantling and workshop facilities to be located immediately adjacent to the powerhouse. Amongst other ideas discussed with them was the use of models at the design stage to finalize a unit bay layout. Many felt that this is the most realistic way of ensuring easy dismantling later.

17. The detailed criticism that was recorded most often concerned plant drainage. In many cases drainage channels were not sufficiently large, located dangerously, or easily blocked.

Conclusion

18. In general pumped storage plants visited were

operating with excellent records of availability and had overcome the inevitable setbacks during construction. There were no significant modifications or approaches that any plant was using that would drastically increase availability or efficiency. However there are a number of maintenance problems which continue to plague the operators :

- Wedge failures
- Balancing line failures
- Wearing ring failures
- Shutoff valve seal failures
- Trashrack failures
- Damaging vibration
- Power circuit breaker problems
- Lack of space during major overhauls

19. Aspects of plant design to minimise these difficulties should continue to be considered most carefully by those planning and implementing pumped storage projects.

Acknowledgements

20. The Author wishes to thank the Electric Power Research Institute for permission to publish the results of this study, and to thank all owners and operators of the plants visited, all of whom were exceptionally cooperative in discussing their problems. Discussions of plant deficiencies in this paper or in the report to be published, is not intended to reflect on the performance of the plant operators or on the designers or constructors, but rather to contribute to the general body of knowledge about pumped storage.

Engineering experience in the early years of operation of the Dinorwig generator-motor

E. BEEDHAM, BSc (Eng), MICE, and I.E. McSHANE, BSc, GEC Alsthom Large Machines, Rugby, UK

SYNOPSIS The authors describe two engineering investigations carried out on the Dinorwig machines. The first case arose from the discovery that stator core laminations near the bottom end of the machine had moved radially inwards; the development of a method to prevent further movement, by the use of ceramic dowels, is described. In the second case, the fracture of rotor field coil interpole connections in two machines was found to be due to fatigue. Modifications were developed to increase the fatigue life of all field connectors. The modifications are being incorporated in the machines as they become available.

INTRODUCTION

1. Pumped storage has become an increasingly important component of the modern electricity supply system as the demand for power has increased and the need to maximise system efficiency and reliability has become paramount. To assist in meeting these requirements, pumped storage can be used for such duties as frequency control, peak lopping or emergency stand-by, whichever mode of operation is appropriate at any given time. Inherently, individual units tend to be large in rating and hydraulic conditions are such as to require high speeds for the ratings. This combination, together with the highly variable and cyclic type of loading, impose a total duty on pumped storage equipment which is far more severe than that on conventional hydro stations. The duty will influence all equipment in the station but will particularly affect the major items such as the generator-motors.

2. An outstanding example of a large pumped storage installation is at Dinorwig, North Wales, which was originally the responsibility of the Central Electricity Generating Board (CEGB) but, with privatisation, is likely to form part of the National Grid Company. In the generating mode the output of this station is 1880 MW; this is obtained from six units with a generator rating of 330 MVA, 0.95 p.f., 18 kV. The generator-motors have 12 poles and run at 500 r.p.m.; they are each coupled to a pump - turbine which necessitates reversal of rotation to change between the

generating and pumping modes. A number of articles have been written describing the machines (refs. 1 to 3) including changes that were made during the erection and commissioning period (ref. 3). A vertical cross-section of the machine showing the main features is given in Fig. 1. The stator core has an outside diameter of 6.2 m and a bore of 4.6 m, the axial length being 3.6 m.

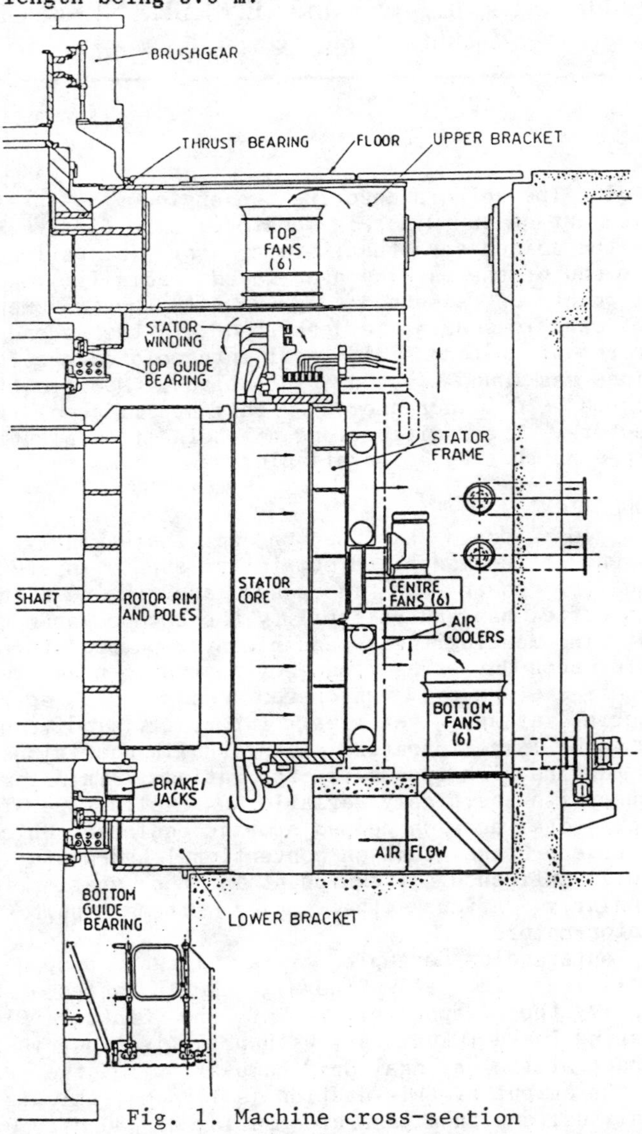

Fig. 1. Machine cross-section

3. The Dinorwig machines were commissioned over the period 1982 to 1984, Unit 3 being the first and Unit 5 the last to go into service. During the period 1985 to 1987 a number of

situations arose which led to engineering investigation, as a result of which either modifications have been carried out or plans laid to permit implementation of modifications should they ever be required. Two generator components of particular concern were the rotor field connections and the stator core. This article describes the situation which developed and the engineering experience in each of these cases. Other items of interest may be the subject of articles published at a later date.

STATOR CORE MOVEMENT

Core construction

4. Core construction is conventional, using layers of silicon steel laminations to form packets interspersed with radial ducts. The core is located at its outer diameter by a series of dovetail keybars secured to the stator frame. Axial clamping of the whole core is effected by the use of core studs spaced round the core periphery. There are 135 slots and 22½ segments per circle. Assembly of the core was carried out at site, the segments being laid spirally so that no vertical splits occur in the core.

5. As further security for the core, the laminations used in the two packets at each end of the core were given a coating of an impact adhesive; this was intended to produce bonded end packets due to the combined influence of core pressure and temperature during commissioning. The treatment was not applied to the tooth support plates.

Movement of tooth supports

6. Reference is made in an earlier article (ref. 3) to the occurrence of radially inward movement of the tooth support assemblies at the bottom end of the machine. This happened early in the life of the machines. The phenomenon is known to occur in vertical machines which are subjected to frequent thermal cycling and arises from radial expansion at the ends of the core relative to the clamping structure, in this case the bottom frame plate. Forces causing movement arise from axial clamping pressure combined with friction; a small ratcheting effect, due to differences in friction and constraint in the two radial directions, results in a net movement, usually radially inwards. The force may be sufficient to pull laminations off the dovetails. In the Dinorwig machines it was possible to pull the tooth support assemblies back into position and secure them with anchor blocks and bolts to plates behind the core studs. This arrangement has been successful in preventing further movement of the tooth support plates.

Movement of adjacent laminations

7. During 1986, inspection on two machines revealed that radially-inward movement had occurred of a few layers of laminations immediately above the bottom tooth support plates. The movement ranged in value from 0 to 2.5 mm and

only occurred over part of the periphery. It was not certain if the movement would continue to increase in value or extend to other parts of the periphery. 2.5 mm corresponds to the thickness of the winding packer at the bottom of the slot so further movement was undesirable. Unlike the tooth support assemblies, the disturbed laminations could not be pulled back in position, so further movement had to be prevented. As explained earlier, the end packets should have been bonded, so either the bond had not been formed or the shear strength of the bond was not sufficient to prevent movement. It was therefore decided to develop a mechanical means of restraint and the method chosen was the use of insulating dowels, to be inserted in an axial direction.

8. Dowelling arrangement. The completed dowelling arrangement is shown in Figs. 2 and 3. It is proposed to use 90 dowels, 16 mm diameter, of moulded ceramic, located near the back of the core and in pairs adjacent to each keybar. Ceramic material was chosen as it is a robust insulator capable of withstanding the load imposed by laminations on edge. Access to the bottom end of the core can only be gained through the bottom frame plate, as shown in Fig. 2. The dowels would be screwed to stainless steel rods, the assembly being held in position by spring-loaded clamping plates bolted to the underside of the bottom frame plate. The arrangement allows easy insertion of the dowel, it holds the dowel in position and permits checking of dowel tightness if required. Relative movement of the core and bottom frame plate is accommodated.

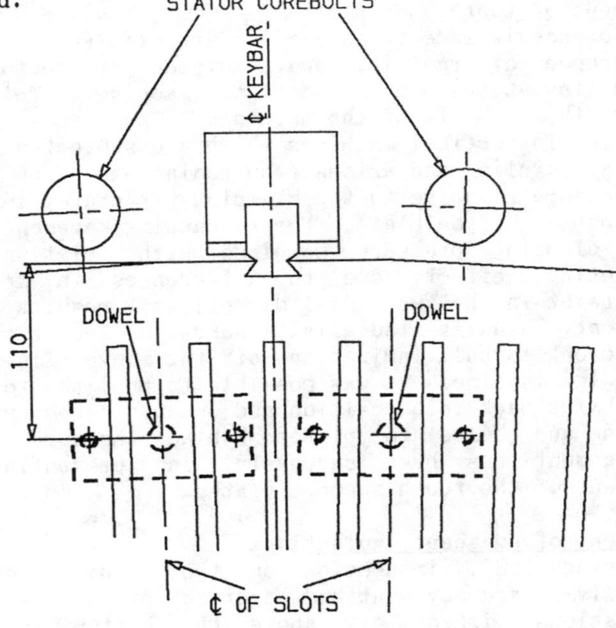

Fig. 2. Plan of core

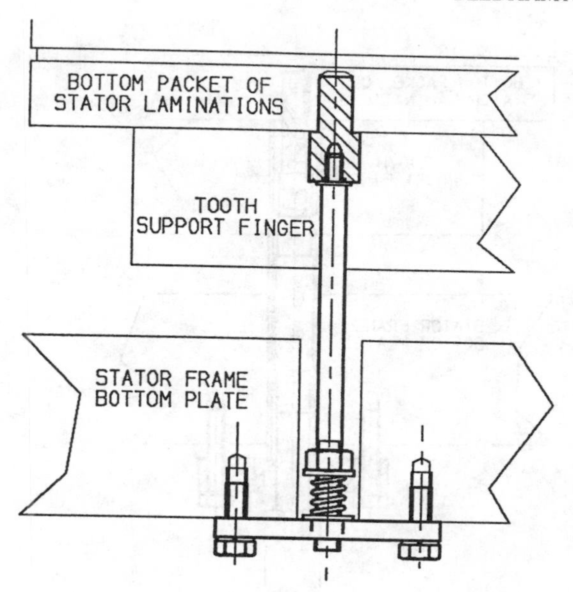

Fig. 3. Arrangement of dowel

9. Prior to placing the dowel assembly securely in position, a sequence of three operations would be required:-

Machining the bottom frame plate.
Machining the stator core dowel hole.
Etching the dowel hole.

The techniques adopted to carry out these operations were developed in response partly to the requirements and partly to the difficulty of access. Machining and etching arrangements are shown in Figs. 4 and 5.

10. Machining of bottom frame plate In order to carry out work on the core itself it is necessary to pass a tube with as large a diameter as possible up through the bottom frame plate and to press it against the underside of the core. This tube acts as a guide for tools and prevents deposit of unwanted material in the region above the frame plate.

11. The nominal size of the tube and the hole in the frame plate is 30 mm, the tube being a sliding fit in the hole. Machining of the hole is carried out using a drill head magnetically clamped to the underside of the frame plate and fitted with a Rotabroach cutter; the latter could best be described as a hollow end-mill. For this application the bit has a concave head that ensures the outer diameter leads the cutting process. Thus, with care, it is possible just to break through the bottom frame plate with a clean hole but leaving very little, if any, swarf on the top of the plate.

12. Machining of dowel hole. Having machined the bottom frame plate including holes for the clamp bolts, the 30 mm tube is inserted and clamped in position. There then follows the machining of the dowel hole itself; the arrangement is shown

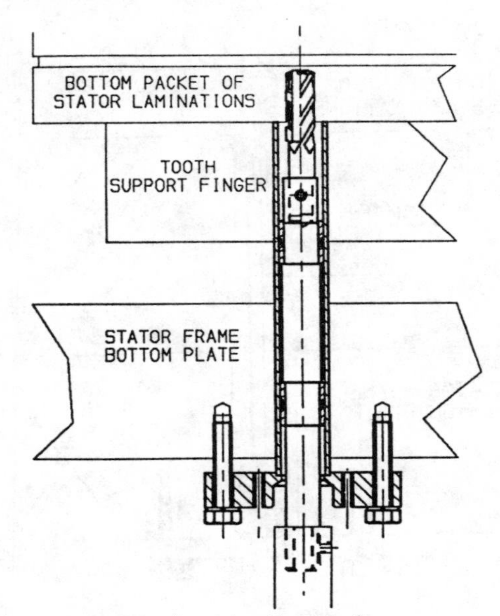

Fig. 4. Machining

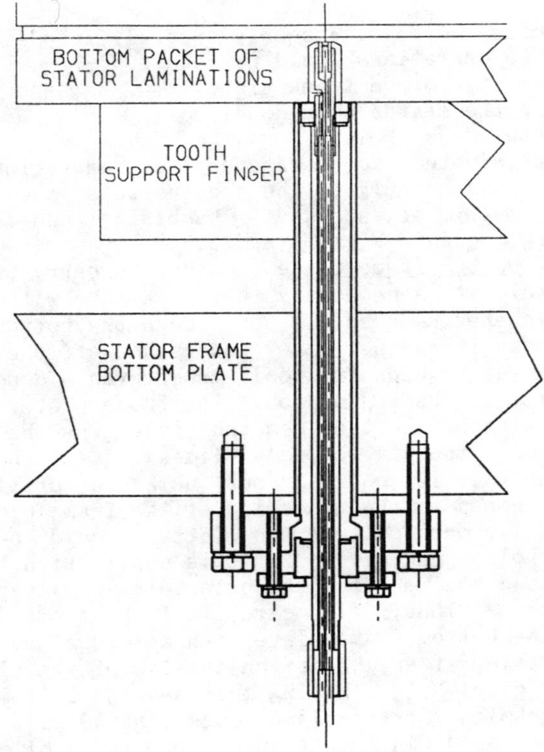

Fig. 5. Etching

in Fig. 4. The cutter is connected to the drill head through a shaft extension which has lubricated bushes for guidance in the tube. Once again the bit is a Rotabroach, but since the material is laminated the head of the bit is convex. A hole is produced which does not break through the bottom core packet and is 30 mm deep. This will allow an engaged length of the dowel of 25 mm.

13. Etching of dowel hole. As a mechanical surface the bore of the hole produced is very good, being smooth and uniform. However, it cannot be left in that condition since the laminations are shorted and may give rise to local heating. Burring of the laminations would be removed using etching by electrolysis with dilute Phosphoric Acid, similar to the well known technique for recovery of damaged cores. The arrangement is shown in Fig. 5. A tubular electrode is insulated at the bottom end by a bush around the stem and at the top by an extended insulation tube in the head plus sponge rubber bushes round the stem. With the electrode pressed home and clamped the rubber bushes also seal the mouth of the hole. A continuous flow of acid is required to remove the gasses formed and to obtain uniform etching. Tubes run up the inside of the electrode, the inlet discharging near the mouth of the hole and the return being through a slot in the top insulation tube. The back of the core is cleaned and a strip of solder pressed on to make contact with the complete end packet. Current is supplied and controlled from a battery and rheostat connected between the core and the electrode.

14. Testing the techniques. Having determined the method of hole preparation, the sequence of operations was demonstrated in the following manner. Using a mock-up of part of the bottom frame plate, plus tooth supports, the marking-off routine and frame plate machining were successfully carried out at the manufacturers Rugby works. Machining and etching of the stator core was demonstrated at CERL, Leatherhead who provided for the purpose a bonded section of core stack from a turbogenerator approximately 2.5 m outside diameter, 1.25 m bore and 100 mm thick. The stack was laid flat on a circular support and a steel platform was bolted to the underside at suitable places to represent the bottom frame plate of the Dinorwig machines. The platform was used to set up the machining arrangement and to produce a number of dowel holes. In so doing it was possible to optimise the general technique. This was followed up by the etching process, for which it was possible to determine a suitable current and time to etch satisfactorily. The gravitational head required to give a suitable rate of acid flow was also established. The state of each hole was considered satisfactory if, on examination with a boroscope, the surface was a uniform matt grey in colour and individual laminations were discernible.

15. Having produced physically acceptable holes, the final test was to apply a core flux test at rated flux density.

During the test the core pack was shorted at the outside diameter near each hole. When the temperature was steady the value in the holes was found to be less than 2°C above the surrounding core. As a control check, on a second run the laminations inside four holes were shorted out. In this case the temperature in the holes rose by up to 50°C above the unshorted values. The tests thus demonstrated that the proposed technique would produce satisfactory holes in a laminated core.

16. Present situation. Having successfully demonstrated the sequence of operations to the CEGB, the proposed method of fitting dowels into the core was accepted. Material and equipment are now on site for modifying the machines. However, no lamination movement has been detected since the initial occurrences and as the insertion of dowels is a major operation, implementation has been postponed. The situation will be reassessed if further movement is detected.

FIELD COIL CONNECTIONS

Interpole connections

17. The field winding is connected in series by an arrangement of soldered copper laminates spanning the interpolar gap. Poles are connected in an alternating pattern of top to top and bottom to bottom to give alternate North and South poles. The top is defined as the connection nearest the air gap. The arrangement is shown in Figure 6. This scheme has been successfully used for many years on conventional hydro-generators and pumped storage machines. No problems were encountered in the early operation of the units until May 1987 on Unit No.3 when an open circuit and earth fault occurred on the field winding during a starting sequence. It was found on examination that a length of about 50 mm of a top to top pole connection between a coil and the pair of support studs was absent. The loss of copper and of some associated insulation and supporting metal was consistent with destruction by heavy current arcing. Arcing had also occurred between the connection and the adjacent aluminium alloy coil support which would be earthed, causing local erosion of the support.

18. Within a space of a few days a similar fault occurred on Unit 2. A close "in situ" examination of all interpole connections had already been initiated. This examination proved to be difficult and the usual non-destructive techniques did not prove very practicable. In the event it was found that careful controlled visual examination by a skilled technician was the most satisfactory. This showed that cracking was occurring on the top to top connection in the copper laminates to some degree on every unit. The bottom to bottom connections, which incorporated a long radially-disposed section were found to be free of such problems. The cracking appeared to start on the inner laminate having the least radius, at its apex, and then later

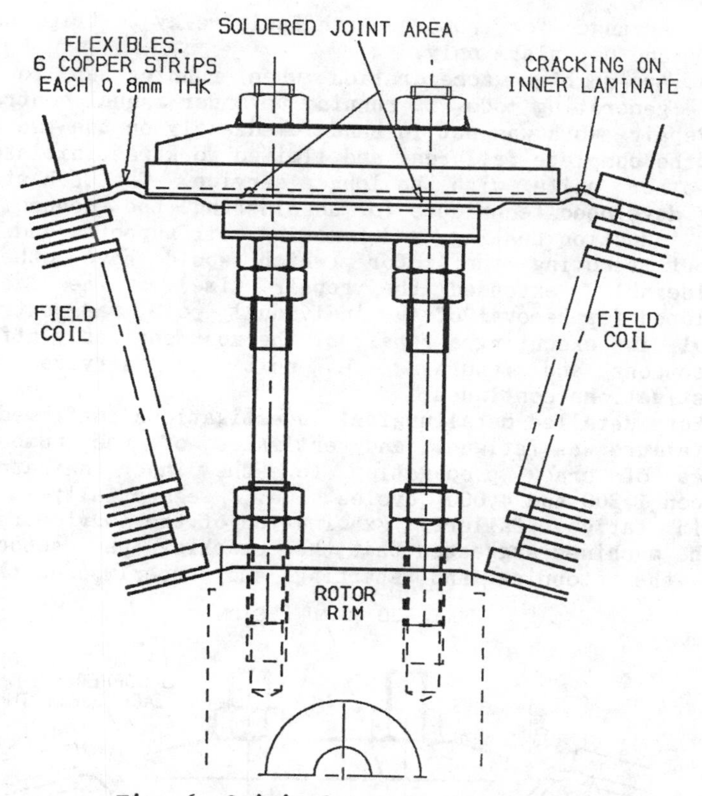

Fig. 6. Original top-top connection

occurred in the adjacent laminations with greater radius.

19. In view of the relatively serious potential loss of an important power system facility, the CEGB set up a panel of inquiry to report on cause of failures and look in particular at the longer term. GEC were members and worked with CEGB on both aspects.

20. There were three problems:-

(i) identification of the likely cause, to enable a viable repair to be executed,

(ii) making repairs with the minimum loss of station availability,

(iii) maintaining operation on the remaining sets if safe to do so.

21. The initial examination of other connections removed from Unit 2 showed fatigue mechanisms were present. A judgement was made that units showing the least extensive cracking could continue in service with a small restriction on their operation, in order to extend their life and minimise damage in the event of a fault. The operational changes were:-

(a) minimise the number of shut downs.

(b) Limit the field current to 1050A, compared with the permitted maximum of 1500A.

(c) Arrange for rotor earth fault relay to trip the set and not alarm only.

(d) Avoid high acceleration when running up to spin generating mode, by running up under manual control.

22. Repair work was put in hand immediately on the two units with the complete failures, and limited work was initiated on two other units with the longer service. The CEGB station staff developed techniques for unsoldering the connections, using induction heating, and removing coil supports and poles without removing the rotor (which would have otherwise considerably extended the repair time). The facility developed for removal of an individual pole was extremely useful in executing a repair of the most seriously affected connections and returning a unit to service while investigations continued.

23. More detailed metallurgical investigations confirmed that the failure was fatigue, and estimates of the number of cycles of crack propagation for the inner laminate were between 1,200 and 4,000 cycles i.e. essentially a high strain fatigue failure. Examination of the service records of the machines suggested that the cracking was associated with the stopping and starting and superimposed thermal

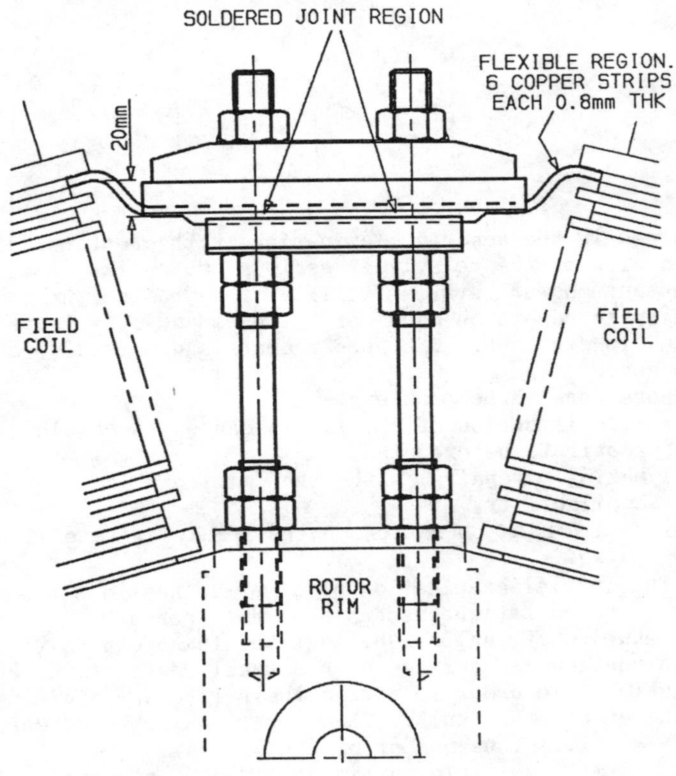

Fig. 7. Modified top-top connection

cycles, rather than with mode changes which numbered between 19,000 and 36,000. Unit 3 had the highest number of all types of cycles; about 60% more than the unit with the least.

24. Calculations using simple finite element models including plasticity suggested that cyclic circumferential movements of less than 0.5 mm could produce failure at the observed position. The structure is a complex one of copper, insulation and laminated steel, for which only estimates can be made of relative movements due to differential thermal expansion and of the effects of rotation. There was considerable debate as to the predominant mechanisms. It was essential at an early stage of investigation to put in hand repairs and a modified design was prepared (Fig. 7). The changes made were:-

(i) a 20 mm offset was introduced to substantially increase the flexibility in the circumferential direction, and improve it slightly for radial movement,

(ii) steps were taken on installation to control the flexible shape closely by using formers, and to minimise solder penetration into the flexible area.

25. Repairs proceeded on this design basis while investigations continued. A simple fatigue rig was set up to compare the original and repair designs. A constant amplitude displacement was applied in the circumferential direction and fatigue curves obtained for the two designs (Fig 8). There was clearly a substantial improvement in the performance, giving confirmation of the integrity of the repairs. Meanwhile CEGB Scientific Services set up tests to to assess the level of strains present in service. It was not practicable to measure relative movement other than in unrepresentative conditions (e.g. at standstill). However

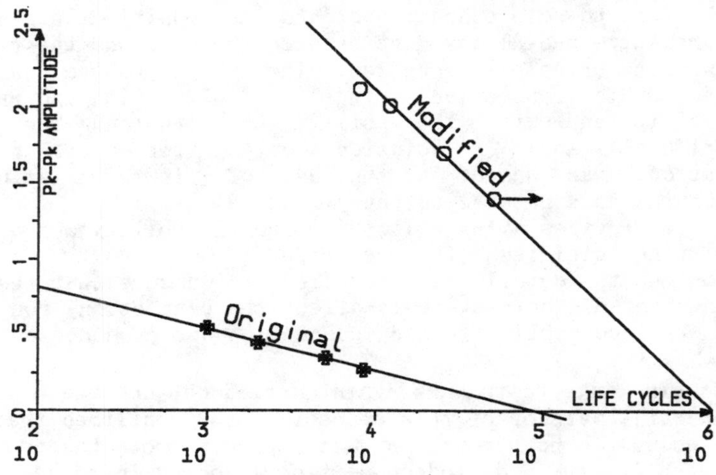

Fig 8 Fatigue test results for field connections

strain gauges were fixed at points of expected maximum strain on both designs, and measurements made at standstill (hot and cold), running (hot but unexcited, and cold). Measurements could not be obtained running excited due to electrical interference. However the results obtained represent the likely extreme conditions and give a valuable guide. The results showed:-

(i) that as expected from theory, the thermally induced strains opposed the rotationally induced strains. The maximum strain cycle is therefore obtained when a unit starts from cold, reaches a maximum temperature in service and is then shut down while hot.

(ii) there was a substantial reduction in the strain range on the new connection compared with the original one. On the original connection strains of the order of 6000 μ strain were identified at 500 r.p.m. and similar levels of strain change were observed for rotor temperature changes in the order of 40°C. On the new connections reductions of about an order of magnitude were obtained.

26. The results were very encouraging. Even allowing for the possibility that more adverse thermal cycles could arise and the fact that strain gauges are not necessarily measuring the highest actual strain, a life of the order of 10^6 cycles was expected. This would be acceptable. All units were returned to commercial operation for the 1987/88 winter.

<u>Slipring Connections</u>

27. During the course of work on the pole to pole connections it was discovered that similar, but less advanced cracking had occurred in some of the similar connections between the first or last field coils and the rods to the sliprings. These connections were situated at the top end of the rotor. The cracking appeared to be essentially of the same nature as on the pole to pole connections, and its position suggested that relative radial movement between the coil and the radial rod was the principal mechanism. The design problem was to replace this connection with one of high axial and radial flexibility and with the ability to withstand the high centrifugal loads. The solution adopted after several design iterations was based on the use of flexible laminates conforming to a natural catenary (Fig. 9).

28. This provides a large loop for the flexibility and at the same time minimises the centrifugal loadings. Careful attention to detail was required to ensure that the new arrangement did not adversely affect the connection rod and that the correctly formed connection was produced on site assembly.

29. Strain gauge tests were again carried out on site by CEGB on an early version of this connection and confirmed that the strains due to both rotation and temperature changes were acceptable. The modified connection is being introduced on a routine maintenance basis.

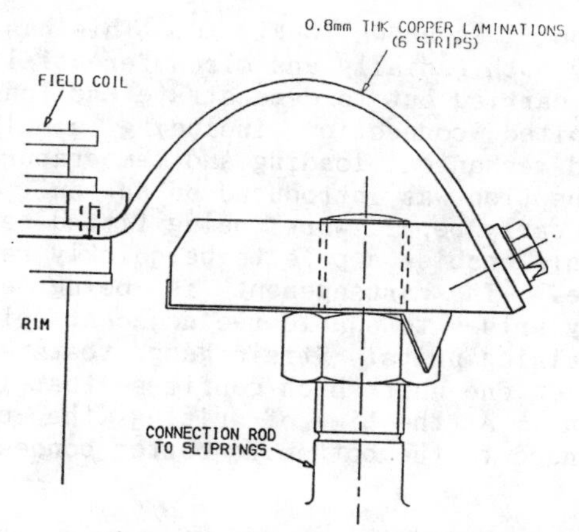

Fig. 9. Slipring lead

Bolted connections

30. The development of the methods for removal of poles focused attention on the possible long term advantages of such a facility for future maintenance. The removal of say one or two poles would enable detailed examination to be carried out on the stator bore and winding with the minimum dismantling work. There was therefore a powerful incentive to develop the method to capitalise on the experience. In particular the most difficult aspect was making and breaking the coil-to-coil soldered connections in situ. The idea of a bolted connection was therefore attractive. The sections of copper on the field coil were such that they would not carry the bending loads associated with a joint carried on the coil itself. The design had therefore to use flexible leads from the coils to a central connection block. The design which

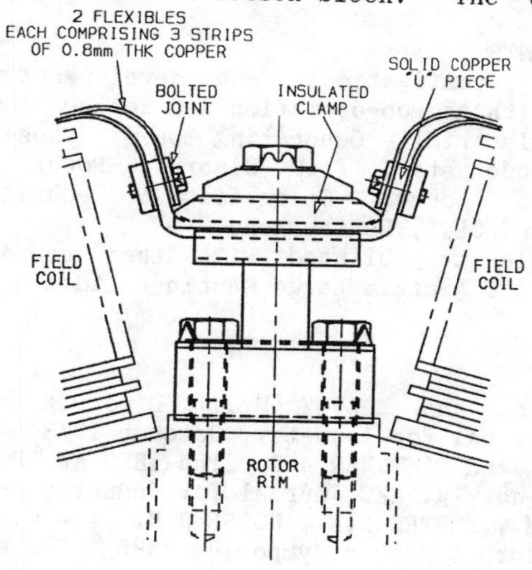

Fig. 10. Bolted top-top connection

was developed is shown in Fig 10. This has a high inherent flexibility both radially and circumferentially. Extensive work was carried out to demonstrate the long term integrity of the bolted connection including simultaneous cyclic current and mechanical loading and temperature rise.

31. The connection was introduced on one unit on a trial basis during the early repair work, being fitted between poles 11 and 12. This enables a pole to be quickly removed for stator maintenance. The arrangement is being extended as the opportunity arises to enable two adjacent poles to be removed on the remaining units. Strain gauge tests were conducted by the CEGB on one unit which confirmed that low strain levels were present. At the time of writing the concept is also being extended to the bottom to bottom connections.

CONCLUSIONS

32. The experience described in this article is associated with details which are not amenable to rigorous analysis at the design stage. They depend significantly on the evolved successful product experience on similar machines. The frequency and modes of operation of Dinorwig have imposed severe conditions on many items of plant, so experience gained is of value for future installations.

33. With regard to the interpole connections, machines are being modified in turn, in a sequence to minimise the effect on availability. To date, units 1, 3 and 5 have been fully modified and returned to service. The performance of these machines has so far been entirely satisfactory.

34. As stated earlier in the article, since the first signs of movement the stator cores appear to have stabilised. Therefore it has not so far been considered necessary to dowel the ends of the cores using the technique developed.

ACKNOWLEDGEMENTS

35. All the investigational and development work has been carried out with the co-operation and active involvement of the Central Electricity Generating Board. Those specifically involved include staff from Dinorwig Power Station, APB (North)-Europa House, Scientific & Technical Services, Wythenshawe and CERL, Leatherhead.

36. This article is published with the permission of the Directors of GEC Alsthom Large Machines Ltd.

REFERENCES

1. RIDLEY G.K. and MALTBY D.J. Dinorwig generator/motor units. GEC Journal for Industry, October 1978.

2. FOSTER E.N. and FLETCHER A.P. The 1880 MW Pumped Storage Station at Dinorwig. GEC Journal for Industry June 1983.

3. BUTTREY M., FOSTER E.N., HODGE J.M. The Dinorwig 330 MVA Generator Motors I Mech E Symposium 1985 - The Dinorwig Power Station

Twenty years operating experience with reversible unit pumped storage stations

A. SIDEBOTHAM and A.S. KENNEDY, Hydro-Electric

SYNOPSIS

The pumped storage stations at Cruachan and Foyers incorporated design philosophies which were at the time significant extensions of existing experience and design and were of a prototype nature. The operating regime, particularly the number of mode changes, has changed significantly from that envisaged when the stations were designed and constructed. This rigorous operating regime imposed frequent thermal cycling on the generators and also very frequent operation of other equipment. Research carried out elsewhere in Europe has indicated that in terms of equipment deterioration each mode change is equivalent to 10 hours normal running. Experience at Cruachan and Foyers has also clearly indicated reduced operating life as a result of the high number of mode changes and highlighted the requirement to design reversible machines to a much higher standard than conventional hydro machines. The paper reviews the problems encountered and the solutions developed initially and further improvements made during refurbishment. It is believed that this will be of assistance to other utilities when planning stations for similar arduous duties.

INTRODUCTION

1 The North of Scotland Hydro-Electric Board is responsible for the generation and supply of electricity to customers north of a line running roughly south of Perth in the east to Loch Long in the west, including the Argyll peninsula and the cities of Aberdeen, Dundee and Perth. This area covers two thirds of the area of Scotland and one quarter the area of the UK.

2 The Board has operated two pumped storage stations at Cruachan and Foyers for a number of years, the 4 x 100 MW units at Cruachan being commissioned in 1966 and 1967 and the 2 x 150 MW units at Foyers being commissioned in 1974. During this time, until April 1989, the generating

Pumped storage. Thomas Telford, London, 1990.

plant of the two Scottish Boards was operated on a joint basis.

BACKGROUND

3 Before considering the station designs for Cruachan and Foyers and their operating history, it is necessary to appreciate the original reasons for their construction.

4 When Cruachan was planned, the construction of larger thermal plants and of Hunterston 'A' nuclear station was proposed. These stations, although much more efficient than the older plants, would not be so responsive to load variations and would be best suited for base load operation. Pumped storage was attractive on two counts, in that it could be brought on to the system very quickly to generate during peak load periods and also it would enable more of the new high efficiency thermal and nuclear plant to operate on base load.

5 Studies showed that the storage required was about 6×10^6 kWh of generation (15 hours at full load) and this was increased to 7.5×10^6 kWh in order to accommodate the natural run-off from the catchment. The study showed that when the whole storage was being used, the predicted annual generated load factor would be around 15%.

6 It was considered that a weekly operating cycle would be the most economic as it would permit the larger weekend surplus of pumping energy to be stored and returned to the system during the following week, supported by a smaller amount of pumping during week-nights. (See Figure 1).

7 The need for a second pumped storage scheme was also identified and the decision was taken at the end of 1967, just after Cruachan was commissioned, to proceed with the Foyers station.

8 The main factor influencing the location of the second pumped storage station was to attain a better balance between the north and south of the Scottish system with Cruachan operating in conjunction with the thermal and nuclear plant in the south and Foyers operating in conjunction with planned nuclear stations in the north east of Scotland at Stakeness and Dounreay. The Stakeness nuclear station did not go ahead but instead, a 1,200 MW gas/oil fired station at Peterhead was constructed.

9 The Foyers site had an economic advantage as the upper reservoir of Loch Mhor already existed and had been used by British Aluminium Co for generation in their Foyers smelter from 1896 until 1967. The catchment area was increased to provide more generating capacity from run-

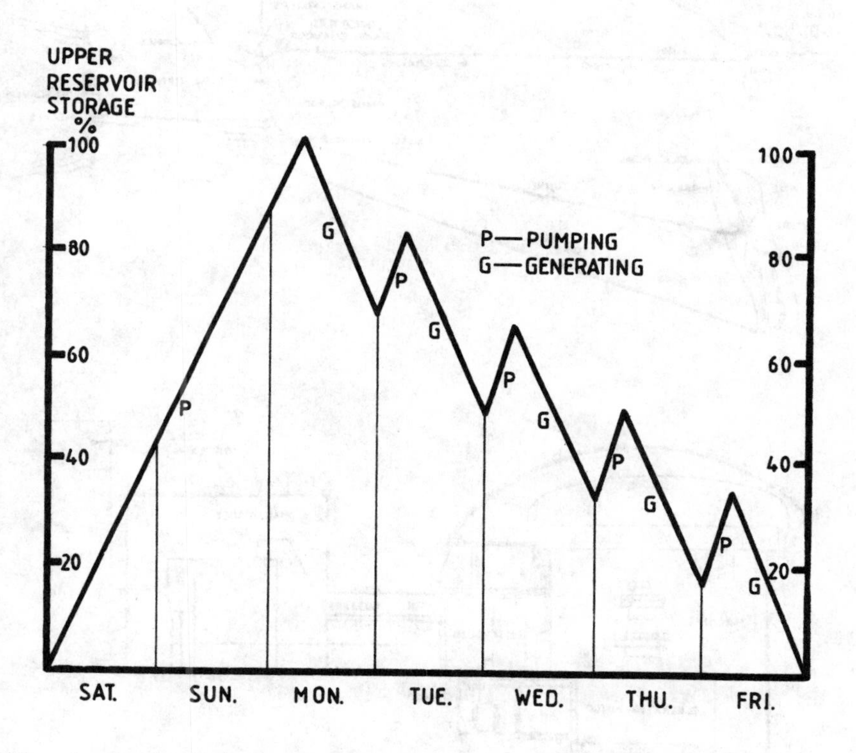

Fig. 1. Variation of upper reservoir level during original Cruachan power station weekly pumping/generating cycle

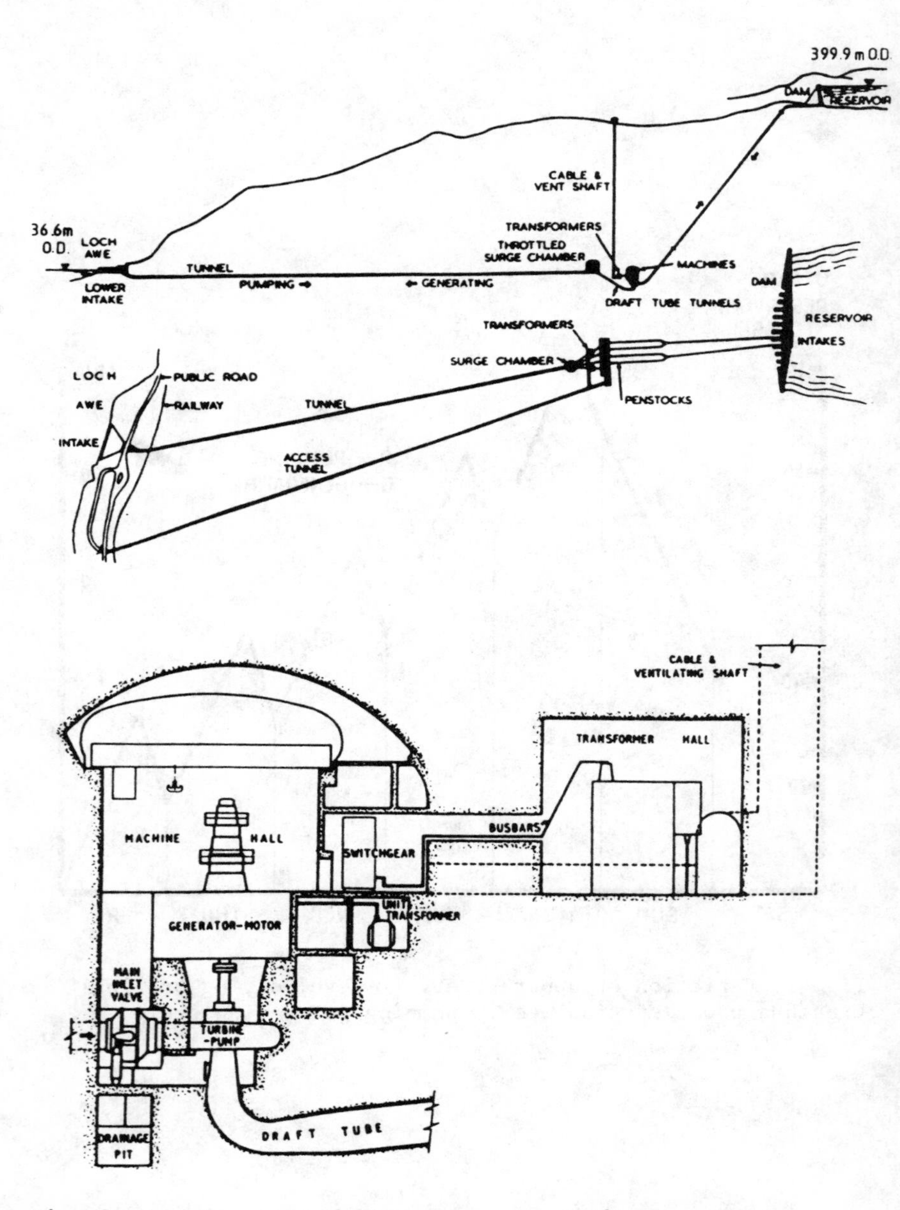

Fig. 2. Cruachan pumped storage power station arrangement and cross-section

off, but the reservoir was of sufficient capacity to operate in a manner very similar to the original Cruachan concept. (Approx 19.5 hours at full load.) The expected operating regime was 10 hours generating and 8 hours pumping each weekday and weekend pumping to re-establish the upper reservoir level for the following week.

CRUACHAN DESIGN

10 The site selected at Cruachan permitted the station to be located directly below the upper reservoir with minimum environmental impact. The head available was 365 metres with the machines situated 46 metres below the level of Loch Awe, the lower reservoir, to prevent cavitation damage to the machine impellers when pumping. As the station was to be located in an underground cavern, consideration had to be given not only to plant costs but more particularly to Civil costs.

11 Several arrangements of pumping and generating plant were considered for Cruachan:

a) Separate pumps and turbine generators.

b) Separate pump and turbine on same motor/generator shaft.

c) Reversible pump/turbine and motor generator.

12 Of the three plant arrangements considered, the most cost effective in terms of both plant and Civil works was a reversible machine, but at that time, Francis type pump/turbines had not been developed for such a high pumping head.

13 The Board therefore gave encouragement to two separate Companies, and their European associates, to develop suitable single stage pump/turbines. As a result of this initiative, considerable research and development was carried out at home and abroad and two acceptable designs of machine were obtained, one operating at 500 RPM and the other at 600 RPM. The high speeds permitted relatively small turbines and generators for the nominal rating of 100 MW per machine. Two machines of each of these designs were ordered.

14 For pump starting, the pony motor start method, using 7 MW motors mounted on the main shaft above the motor/generator was selected as being the simplest and best solution to the pump start problem.

15 It was decided that the most economical location for the generator transformers was in the cavern near the generators, and transport restrictions dictated that the

transformers be site erected. To save space, and reduce cost, it was decided that instead of separate transformers for each machine, two 3 phase 230 MVA, 3 winding 16/16/275 kV transformers would be installed with each transformer serving two machines.

16 All control systems at Cruachan, including the machine control sequences, were of the 'hard wired' electro-mechanical type.

17 The arrangement of the Cruachan station is shown in Fig 2.

FOYERS DESIGN

18 Because Loch Mhor already existed, a number of options for the station arrangement at Foyers were considered, both surface and underground. Economic appraisals resulted in the arrangement shown in Fig 3 being adopted with the machines being located in 47 m deep shafts adjacent to the lower reservoir of Loch Ness.

19 The Cruachan experience had proved that there was no case for employing anything other than reversible machines at Foyers. The system requirement was for a 300 MW station and a 2 x 150 MW generator arrangement was adopted. This provides greater security and could more closely match the system requirement for generating and pumping than a single 300 MW generator.

20 Because the lower head, 192 m, and the larger capacity at the lower running speed of 273 RPM, the Foyers pump/turbines were, at that time, physically the largest ordered in Europe.

21 Since Cruachan had been designed, advances in technology had made other possible arrangements of pump starting feasible.

22 After considering all methods, the choice was narrowed to well established back-to-back synchronous starting combined with auto-transformer or auxiliary generator starting for the last machine, and a new proposal employing a synchronous starting method with the machine supplied from the main system via static variable frequency convertors. The static variable-frequency convertor system was adopted.

23 The complete starting operation takes about 13 minutes and the input power to the machine is 4.5 MW at full speed. The starting time and the input power, together with the number of starts, specified as four at 15 min intervals, determined the rating of the convertor equipment.

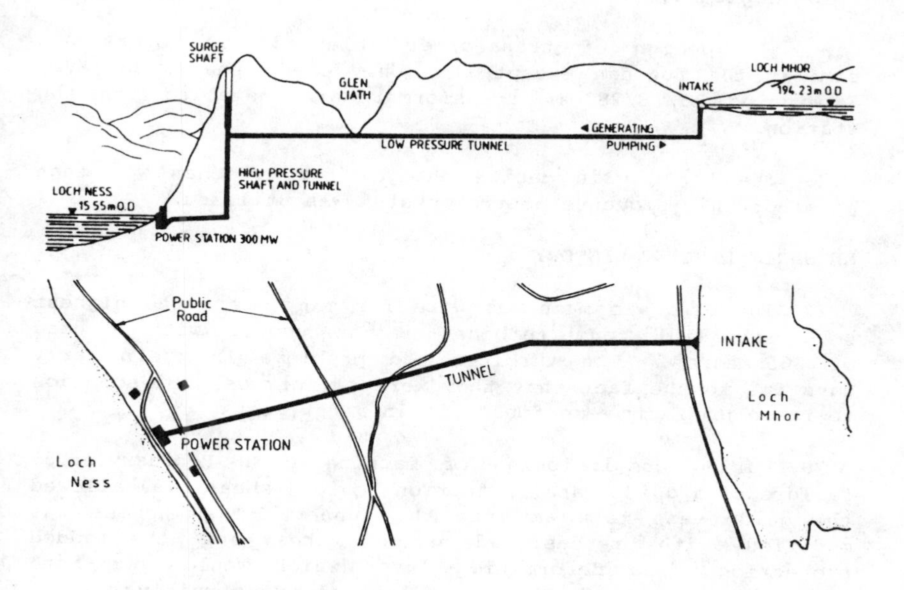

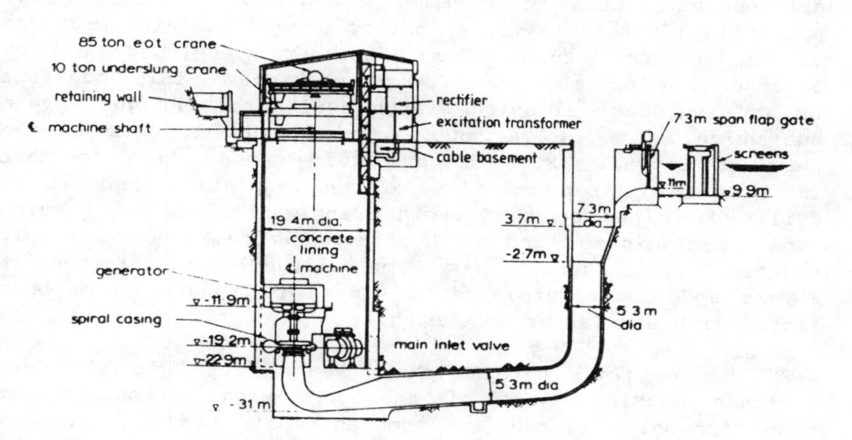

Fig. 3. Foyers pumped storage power station arrangement and cross-section

24 Common starting equipment is provided for the two machines, and the busbar inter-connections provided for the convertor also allows the back-to-back starting method to be employed.

25 The design of transformer selected for Foyers was similar to Cruachan except that only one 3 phase, 330 MVA, 3 winding 18/18/283 kV transformer was installed for the station.

26 The other main design change from Cruachan was that an electronic sequence control system was utilised.

MECHANICAL PLANT HISTORY

27 The Cruachan machines were for many years the highest head reversible pump/turbines in the world with a head of 365 metres. The turbines had problems in their early days due to the fact that they were prototypes, but solution to these problems were found within a few years.

28 During commissioning of machine 1 in February 1966 failure of a guide vane protection device (shear pin) allowed the guide vane to slam into the runner. The machine was modified with re-designed guide vanes and the added protection of a deformation lever which would come into action if a shear pin failed and act as a restraining device to prevent excessive movement of the guide vane. This arrangement fitted to machines 1 and 2 was successful in preventing further failures, but as a result of the Cruachan experience the turbine manufacturer developed a new arrangement for the two Foyers machines. This consists of a "Ringfeder' slipping clutch built into the guide vane hub which is set by a double action hydraulic ram to the required torque. Even if one guide vane slips so that is it out of step relative to the regulating gear it is still restrained so that rapid transient motion of a guide vane is eliminated and risk of cascade failure is greatly reduced. This system has been virtually troublefree at Foyers and the refurbished machines at Cruachan have been fitted with similar arrangements.

29 The original turbine shaft seals fitted to all four Cruachan machines were of the multi-ring carbon segment type commonly fitted to conventional hydro machines. Problems were experienced due to rapid wear, especially during the spin modes when the seals were cooled/lubricated by an external water system. The frequent maintenance required fairly lengthy outages because the turbine bearing had to be dismantled to gain access to the seals. Excessive seal wear also caused operational problems because of the high air loss when the machines were in spin modes. The blow down air system rapidly became depleted and the spinning

mode had to be shut down until the system pressure could be re-established.

30 A mechanical seal consisting of a stainless steel cone fitted to the turbine shaft and bearing on a composite seal in a retaining ring was fitted to turbine 1 in 1969. This arrangement improved the seal wear life from 3,000 to over 20,000 hours and was then fitted to the other Cruachan machines. A modified version was fitted to the Foyers machines as original equipment.

31 The Cruachan machines were originally designed to discharge the blow down air through the head cover into the station during change of mode from spin generate to generate and from spin pump to pump. The original operating regime was changed in 1970 to include much more use of the spinning modes. Changing from spin generate to generate took almost 2 minutes, almost as long as a start from standstill, because of the time required to discharge the air. Tests were carried out in 1971 which proved that priming could be improved by not venting the air through the headcover, but instead displacing it downwards by opening the guide vanes to admit water and letting the air vent through the tailrace surge shaft. The mode change speed was thereby reduced to 35 seconds and the priming period, which is the worst condition for vibration, was greatly reduced. All Cruachan machines have operated in this way since 1971, and the Foyers machines since commissioning.

32 Cruachan machines 3 and 4 have always been subject to high vibration problems which was proved to be because the normal operating speed was too close to and just above a shaft critical speed. The machines were kept available by operating with wide clearances in the upper guide bearing thus changing the rotor dynamics and reducing the vibration level.

33 Also these machines were originally constructed with labyrinth seals top and bottom, but the lower seals were changed to wear ring type parallel seals in order to introduce some hydraulic damping at the lower end of the machine.

34 In 1972 damage to turbine 3 was caused when the head of the by-pass valve, which is integral with the main inlet valve, became detached and passed through the turbine. Some guide vanes had to be replaced and major weld repairs had to be carried out on the runner resulting in a lengthy outage. The by-pass valve was operated by a hydraulic cylinder which opened the valve inwards into a port in the body of the main inlet valve. The mushroom head retaining nut became detached and the nut and head passed into the turbine. The arrangement was modified to improve

the head attachment and stops were fitted in the port to arrest the valve head in case of a repeat failure. This was not a complete solution as the valve nuts could still pass into the turbine, so that during the recent refurbishment of the machine the by-pass valve was re-designed using an hydraulically operated spear type valve which opens outwards and has no components which can pass into the machine.

35 Considering that the Cruachan pump/turbines were prototypes the amount of cavitation damage that has occurred over the years has been minimal and rectification has required very little welding, the normal repair being localised grinding to re-establish the original profile. The Foyers turbines have never required weld repairs, cosmetic grinding has been adequate.

36 After Cruachan machine 3 was re-commissioned in 1988 it was found that it could not operate in the pump mode because during the watering-up period to change from spin-pump to pump, the hydraulic uplift on the runner was greater than the combined weight of the turbine, generator and pony motor. This was due to the fact that the new water cooled generator rotor weighed significantly less than the original air cooled machine.

37 Tests were carried out varying the machine sequence. In the original control sequence a change from spin-pump to pump initiated opening of the top cover air release valve, and when the water level rose within the turbine sufficiently high to generate pressure at the turbine periphery the peripheral drain valve was closed and guide vane opening initiated. The sequence was changed so that closing of the peripheral drain was inhibited until the guide vanes were partially open. This reduced the upthrust condition and permitted machine operation in all modes once again.

38 Before any of the reversible pump/turbines are run in a spin mode the water in the runner is displaced by blowdown air. If air is lost due to seal leakage etc additional air is introduced to ensure that the runner continues to spin in air. On the Cruachan machines this was controlled by means of level switches connected to the machine suction cone, but the switches required frequent maintenance:

39 The Foyers machines were equipped with an external gauge tube connected to the suction cone and the level detection devices were fitted to this tube. This arrangement was much more satisfactory and was also adopted for the Cruachan machines.

40 With the increase in spinning mode operation at Cruachan the blowdown air requirements increased and additional air receivers were fitted.

41 At Foyers the air blowdown system was also modified, an additional air compressor being installed and additional air storage provided.

ELECTRICAL PLANT HISTORY

Cruachan

42 Motor/generator problems have resulted in reduced availability throughout the operating life of the Cruachan station and these were principally attributable to the frequent mode changes and the resulting thermal cycling effects.

43 Problems were encountered because of the radially inward movement of laminations at the bottom sections of the stator core, something that had not been previously encountered with conventional hydro machines. Generator stator winding problems were also encountered as they deteriorated much more quickly than on conventional plant. Both stator core and winding problems occur due to rapid thermal cycling associated with frequent mode changes.

44 To prevent the lamination movement problem, stator cores have been modified by application of adhesive to laminations in the end packs when machines have been re-built, but prior to this the movement was arrested by external application of a liquid adhesive which penetrated the laminations by capillary action.

45 Generator winding life has been improved by changing the insulation to high temperature epoxy type during repairs and rewinds.

46 Stator bar end caps also failed on Cruachan No 2 due to delamination of the insulation allowing ingress of oil and water. This problem has been overcome by developing a more effective air baffle system to improve end winding cooling. A modified arrangement was fitted to No 1 generator during the machine re-core and rewind in 1982/84. This proved very successful and was incorporated in machine 2 during its refurbishment in 1986/87.

47 All Cruachan generators have suffered problems caused by contamination by oil mist, brake dust, carbon/copper dust from slip rings and moisture.

48 The Cruachan machines were fitted with mechanical brakes, and all the machines suffered generator brake dust

contamination. In 1973 the machines were modified to incorporate "pony motor braking" whereby the pony motor is switched in, when stopping from the generate modes, to apply a dynamic braking torque. This system has been very satisfactory with the mechanical brakes then being applied at low speed significantly reducing brake dust. All brake pads have now been changed to non-asbestos material.

49 The greater part of the carbon/copper dust contamination in the machine came from the pony motor brushes and sliprings. The brushes were permanently applied although they were only required for pump mode starting and generate braking. A brush lifting arrangement was designed by the Board and fitted to machine 3 so that the brushes were lifted off the sliprings when not in actual use. This proved very successful in reducing contamination and brush and slipring wear. During refurbishment brush lifting arrangements have been incorporated in all the other machines in the station.

50 As previously described in 1980 the decision was taken to change unit 3 to a water cooled machine as part of the development programme for the then proposed station at Craigroyston. The change in design eliminated the critical speed/vibration problem. Motor/generator 4 is presently being replaced by a re-designed, air cooled, machine which will also eliminate this problem.

51 The water cooled machine 3 suffered a major failure in 1984 when the rotor smoothing plate became detached causing severe damage to stator and rotor which necessitated a complete re-build.

<u>Foyers</u>

52 The Foyers machines have also been subject to periods of unavailability due to motor/generator problems.

53 The Foyers machines were among the first to utilise a core clamping arrangement using the stator frame and pressure screws, instead of the conventional system of through bolts.

54 This system provided inadequate adjustment to follow core movements and keep the core properly compressed. The lack of proper core compression has caused additional core vibration and caused failure of pressure finger weld fixings to top core laminations. This has resulted in the inward radial migration of pressure fingers into the air gap, necessitating regular fortnightly inspections to check for this condition.

55 The core duct spacers are made from 2 MM x 6 MM section straight steel strip. These were manufactured by guillotine from steel plate which left a shear edge. Core compression has caused spacers to tilt onto the shear edge causing core collapse and cutting through adjacent laminations. The problem has been halted by the insertion of epoxy spacers into the ducts to prevent further collapse.

56 Signs of slot discharge were detected early in the life of the machines and all slots were rewedged using parallel wedges and packing in preference to the opposed taper wedge system originally installed. Despite this modification progressive slot discharge still gives cause for concern, although the windings have now been in service for 15 years without failure due to this problem.

57 Problems have been experienced with the continuously interconnected damper windings. Even after a redesign of the interconnections a failure of an interconnection support occurred causing damage to the stator winding of one generator.

58 After appraisal of the system it was decided to remove the damper winding interconnections completely, and no problems have been experienced since that date.

59 In 1981 machine 1 LV winding failed in the Foyers generator transformer. A rewind of the transformer would be very costly because of the high transport charges over and above the repair costs, and both machines would be out of service for a considerable period.

60 In order to keep one machine available a new 165 MVA, 18 kV/275 kV transformer was purchased and connected to machine 2. Machine 1 was connected to the machine 2 connections of the original transformer.

61 The switchgear installed in both stations was of a similar air blast design but 16 kV at Cruachan and 18 kV at Foyers. Problems with breakers requiring increased preventative maintenance were experienced at Cruachan, but at Foyers the problems were considerably worse. To contain this inherent problem inspection/maintenance of the Foyers breakers is carried out on a fortnightly routine.

CRUACHAN REFURBISHMENT

62 Cruachan station mode changes from 1971/72 until commencement of refurbishment is shown in Table 1.

63 Nett availability at Cruachan was below target over this period, mainly due to generator electrical problems.

Table 1. Cruachan power station mode changes 1971-72 to 1984-85

PERIOD	TOTAL MODE CHANGES	AVERAGE MODE CHANGES PER MACHINE	AVERAGE MODE CHANGES PER MACHINE PER DAY	FAILS TO CHANGE MODE AND FAILS TO START*	FAILS AS % OF TOTAL MODE CHANGES
1971/72	11,321	4,126	11.3	192	3.06
1972/73	10,837	5,083	13.9	112	1.89
1973/74	7,782	6,080	16.7	42	1.09
1974/75	8,077	5,177	14.2	169	4.91
1975/76	12,096	3,969	10.9	205	3.52
1976/77	11,237	3,736	10.2	148	2.79
1977/78	11,962	4,919	13.5	104	1.95
1978/79	12,617	4,399	12.0	108	1.71
1979/80	15,027	4,885	13.4	25	0.17
1980/81	14,051	5,675	15.5	26	0.18
1981/82	12,614	7,806	21.4	15	0.12
1982/83	8,402	5,136	14.0	12	0.14
1983/84	7,383	4,253	8.6	33	0.45
1984/85	7,434	3,134	8.6	17	0.23

Average Mode Changes is based on machine availability.

*Failures to change mode or start occur when the system control instruction is not achieved within one minute. Target is 1%.

64 In the early years there were some turbine problems due to the fact that the machines were prototypes, and generators 3 and 4 had generator rotor problems.

65 In 1970 the System Control staff began to take advantage of the flexibility which could be achieved by using spin modes to permit rapid loading in both generate and pump modes.

66 The nett availability, which had been improving up until that time as the original teething troubles were overcome, deteriorated over the next few years due to the change in operating regime which resulted in the core and insulation problems previously detailed.

67 It was decided that sustained improvement in availability could only be achieved by major refurbishment of all the main plant in the station and it was essential that high availability be achieved when the Torness Nuclear Station came on load. The decision to undertake complete refurbishment of the pumped storage stations, starting with Cruachan, was agreed by the two Scottish Electricity Boards in 1985.

68 At that time the water cooled generator 3 had suffered major accidental damage, and inspection had revealed severe cracking in the turbine runner. The first contracts placed were therefore for refurbishing generator 3, still as a water cooled machine, and refurbishing turbine 3 including a replacement runner.

69 The station system, Fig 4, has duplicate systems, ie 2 machines supplied from a common penstock and supplying a common transformer.

70 Both penstocks required inspection, shot blasting and coating, so the east penstock was dewatered first to carry out this work as this would result in only one additional machine availability being lost. In parallel with this penstock work main inlet valves 3 and 4 were removed to contractors works and completely refurbished.

71 Surveys were then carried out to ascertain the extent of work required to re-establish long term reliability and availability of the station.

72 The major common items identified were:

<u>(i)</u> <u>Static Excitation</u>

73 Considerable maintenance problems had been experienced with the four machines exciters resulting from oil and brush dust contamination. A contract was therefore placed

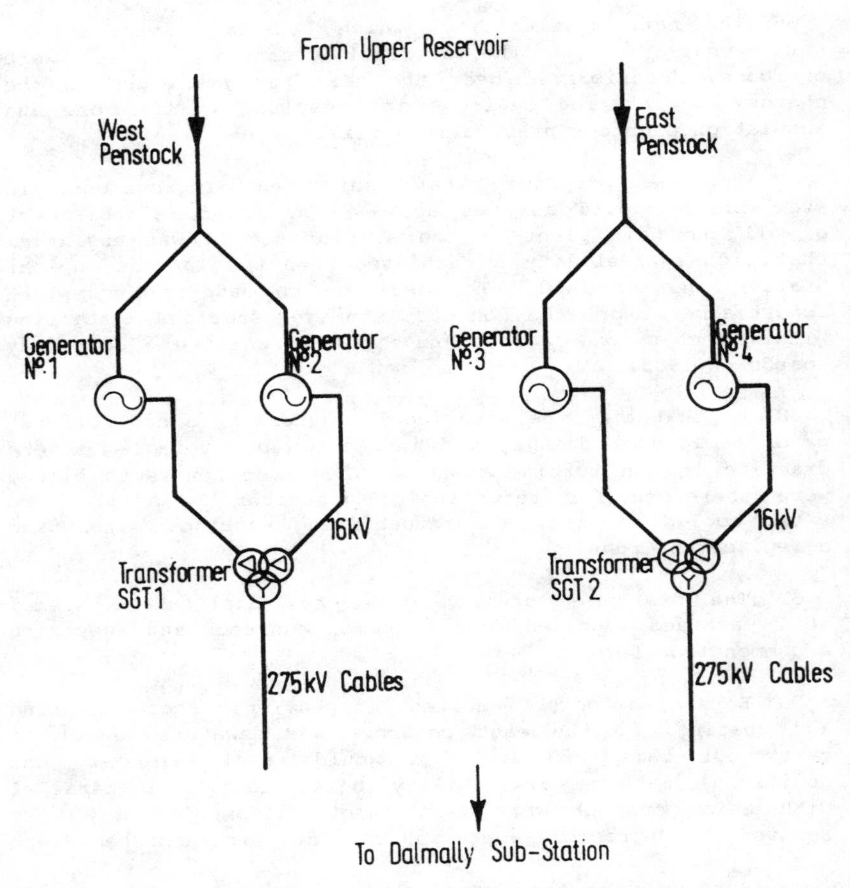

Fig. 4. Crucachan generating station

to provide static excitation on all generators thus eliminating this maintenance problem.

(ii) 16 kV Switchgear

74 The 16 kV air blast circuit breakers had been unreliable in service and at times had failed to open. Inspection/maintenance was required every five weeks.

75 A contract was placed for replacement by simpler and more robust SF6 type breakers.

(iii) Sequence Control and Alarm Systems

76 The sequence control system had been fairly reliable, but the onerous service conditions with numerous mode changes under which it was operating resulted in deterioration of relays and contactors which were by then obsolete.

77 The work to be carried on other plant required a complete new sequence control and alarm system as the number of inputs was being greatly increased. For example the number of alarms per machine was increased from 30 to approximately 200.

78 The system adopted to replace the relay/contactor system was a solid state system using programmable logic based on the GEC GEM80 range of controllers having control room displays for all systems. The installation also has a 'hard wired' emergency control system for circuit breakers and emergency back-up supplies to enable safe operation in the event of failure of the GEM80 equipment.

(iv) Cabling

79 A cabling contract was placed to carry out all the cabling required for the control and alarm system and to interface all the other electrical and control equipment, and to replace the existing LV power cables with PVC insulated and sheathed cable as the MICC cable and copper sheaths of the original had seriously deteriorated in the damp cavern atmosphere.

(v) Fire Detection and Alarm

80 The existing equipment was obsolete and its cabling had also deteriorated. A modern detection and alarm system was fitted and the CO2 protection system on various items of plant was changed to Halon.

(vi) Metering and Protection

81 The protection and metering equipment on the 16 kV

and 3.3 kV systems was obsolescent and required complete replacement with modern equipment.

(vii) 415 V Switchboards

82 The 415 V switchboards were obsolete and spares were unobtainable so they were all replaced.

(viii) Turbines and Generators

83 Complete overhaul of turbines has been carried out with fitting of 'ringfeeder' slipping clutches to guide vane levers. Generators had new stator windings installed and improvement to oil mist control and other detail improvements.

84 In the case of No 4 generator this is being replaced with a new generator to eliminate the critical speed problem.

85 The electric governors were obsolete and spares were no longer available. New digital electronic governors have therefore been installed.

(ix) Pipework

86 Investigation showed that all the original steel pipework in the station was badly corroded and smaller diameter pipework was partially blocked due to corrosion products and peat deposits.

87 All common cooling water system, dewatering and drain pipework, and transformer cooling systems were renewed using stainless steel pipe for smaller diameters and carbon steel pipe coated internally and externally with an epoxy material for larger diameters.

88 The various system valves were renewed or overhauled and all electric actuators were renewed.

(x) Main Inlet Valves (MIV)

89 All main inlet valves were fully overhauled in contractors works, with improvements to seals and to control systems being incorporated.

(xi) Transformers

90 A Generator transformer failed during the station refurbishment period and was rewound with redesigned windings to prevent recurrance of the LV winding fault. Following this failure, similar work was carried out on the other generator transformer.

Table 2. Foyers power station mode changes 1978-79 to 1987-88

PERIOD	TOTAL MODE CHANGES	AVERAGE MODE CHANGES PER MACHINE	AVERAGE MODE CHANGES PER MACHINE PER DAY	FAILS TO CHANGE MODE AND FAILS TO START*	FAILS AS % OF TOTAL MODE CHANGES
1978/79	4,285	2,715	7.4	89	2.07
1979/80	4,554	3,149	8.6	67	1.47
1980/81	5,150	3,781	10.4	81	1.57
1981/82	3,812	4,433	12.1	22	0.58
1982/83	6,573	8,581	23.5	55	0.84
1983/84	6,786	3,978	10.9	64	0.94
1984/85	3,910	2,426	6.6	23	0.59
1985/86	4,445	3,941	10.8	25	0.56
1986/87	5,231	3,432	9.4	35	0.67
1987/88	4,982	2,850	7.8	25	0.50

Average Mode Changes is based on machine availability.

*Failures to change mode or start occur when the system control instruction is not achieved within one minute. Target is 1%.

FOYERS REFURBISHMENT

91 Foyers station mode changes from 1978/79 until commencement of refurbishment is shown in Table 2.

92 Although availability is higher than Cruachan, it has been achieved at the expense of fortnightly daily outages of each machine for inspection and rectification of the inherent core loosening problem, switchgear and starting/excitation equipment. This close monitoring has resulted in containment of failures to relatively short outages.

93 The decision to refurbish Foyers was taken after plant surveys and operating history/experience had highlighted the requirement major work to be undertaken in order to achieve long term availability and reliability. The principle areas of work are:

(i) Motor/Generators

94 As detailed earlier considerable problems have been experienced with the generator stator cores, particularly on machine 1. A contract has been placed to completely rebuild the machines with new stator cores utilising core clamping bolts and also to instal new stator windings. The rotor poles will be retained and refurbished and the field winding pole-to-pole connections will be replaced to an improved design.

(ii) Generator Transformer

95 The original three winding generator transformer which had an LV winding failure in 1981 and is now only connected to generator 1, is constantly monitored for dissolved gases and gives cause for concern. It will be replaced with a single transformer of the same 165 MVA, 18 kV/275 kV rating as the generator 2 transformer.

(iii) 18 kV Switchgear

96 The 18 kV air blast circuit breakers will be replaced with the simpler more robust SF6 type of breaker recently installed at Cruachan.

(iv) Turbines

97 The turbines are in remarkably good condition and the work will be limited to redesigning guide vane axial thrust restraints, replacement of ancilliary equipment, improved control equipment and replacement of pipework in stainless steel.

(v) Governors

98 The existing governors are to modern standards so, because of their continuing reliability, no major work is intended.

(vi) Main Inlet Valves

99 The main inlet valves will be removed to a contractors works and both seals renewed and the trunnion thrust will be renewed with a redesigned arrangement and the trunnion bushes will be renewed.

(vii) Sequence Control Equipment

100 The solid state sequence control system has been less reliable than at Cruachan probably due to the fact it was new technology when installed.

101 As the scope of control is to be extended, and to improve reliability and fault diagnosis, a new sequence control system will be installed based on the GEC GEM80 range of controllers used at Cruachan.

(viii) Starting/Excitation System

102 The system employing static variable - frequency converters was new technology when installed and a number of failures occurred during the early years.

103 This system has also been maintained at the expense of fortnightly inspections and testing of thyristors. The existing system does not lend itself to thyristor monitoring and therefore thyristors have to be rejected if they show any sign of deterioration. A new arrangement using modern technology will be installed which will not only radically reduce the number of thyristors in the system but also incorporate thyristor monitoring.

(ix) Pipework

104 Deterioration of the pipework systems has taken place and replacement would be necessary in the near future. As this would require machine outages the pipework will be replaced with stainless steel pipework during the machine refurbishments to ensure that all plant has a similar life expectancy.

CONCLUSIONS

105 The stations at Cruachan and Foyers were designed principally for long periods of pumping and generating with few mode changes. The system requirements changed

in the early years of operation to a more dynamic operation resulting in large numbers of mode changes, and with more frequent but shorter pump and generate runs. This resulted in problems which could be attributed to thermal cycling effects on generators and also to significantly increased frequency of operation of such components as main inlet valves, guide vanes, switchgear and sequence control equipment.

106 In order to retain availability to the system, regular inspections of equipment where problems had been identified, were carried out during short outages of a few hours. As major refurbishment was carried out, some equipment was changed and design of other equipment was modified. In this way the need for such regular inspections was eliminated.

107 As a result of 25 years and 15 years of very arduous operation of the plant in Cruachan and Foyers respectively, invaluable experience has been gained and development in the design and operation of the machines has led to steadily improved performance and confidence in the future high availability of the machines following refurbishment.

108 When planning new pumped storage stations where the anticipated operation indicates limited mode changes, consideration should be given to the possibility, during the station life, of the operation changing to be more dynamic and taking account of this in the initial design.

Discussion

A. FERREIRA, ***Northwest Utilities, USA***

What has been the experience at Cruachan and at the South African pumped storage plants with respect to the equalizer-balancing lines? What have been the corrective measures for repairing or replacement of eroded/corroded embedded steel piping?

B.W. GRABER, *Author*

In reply to A. Ferreira, to date, we have experienced no problems with the equalizer-balancing lines on South African installations. At Drakensberg, regulating valves were installed, but these were omitted at Palmiet with no adverse results.

A. SIDEBOTHAM, *Author*

In reply to A. Ferreira, at Cruachan there was no undue erosion or corrosion in either the main inlet valve bypass lines, or in the turbine casing peripheral drain lines.

In general the main embedded pipework which had been part of the civil construction contract had been coated internally and for some of the small lines had been in alloy steel. These embedded lines were cleaned and re-coated during the refurbishment.

The main cooling water and drainage lines were not embedded and had been constructed from carbon steel without internal coating. It was originally intended to repair possibly to replace some bends during refurbishment.

However, when the pipework was dismantled, corrosion was very severe, and in places almost through the pipe thickness; all the cooling water and drain piping was replaced.

Performance of civil engineering structures on pumped storage schemes

F.G. JOHNSON and C.K. JOHNSTON, Scottish Hydro-Electric
(formerly North of Scotland Hydro-Electric Board)

SYNOPSIS. The design, construction, and maintenance of the civil engineering structures associated with the Board's Pumped Storage Schemes at Cruachan and Foyers are described in outline. The performance of the Civil Engineering works are reviewed and evaluated over the period they have operated extending up to 23 years. The lessons to be drawn from the Projects are interesting since the Cruachan Scheme is based on an underground station and the Foyers Station on a shaft layout of station.

DESIGN OF SCHEMES

1. The Board have operated two pumped storage schemes, Cruachan which was constructed in the early 1960 s and commissioned in 1966/67 and Foyers which was built in the period 1969-75 and commissioned in 1974/75 - FIG 1. In addition to the major pumped storage facility, both schemes have natural run-off and diverted catchments to supplement their output.

Cruachan

2. The Cruachan scheme ref 1 which is part of the Awe project, comprises an underground power station immediately below the upper reservoir and drawing water from Loch Awe, the lower reservoir. The Station accommodates four reversible 100 MW Francis pump turbines which were of a world prototype design.

3. The layout of the Cruachan Project is illustrated in FIG 2. A gated barrage was constructed at the outlet from Loch Awe, the lower reservoir, which, in addition to regulating the level of Loch Awe forms the reservoir for Inverawe Power Station, a conventional hydro station of 22 MW capacity, which is located approximately 4 km downstream of the Barrage. Cruachan upper reservoir was formed by constructing a multiple buttress dam on good rock foundations utilising a small corrie on Ben Cruachan.

CRUACHAN AND FOYERS CATCHMENTS

FIG. 1

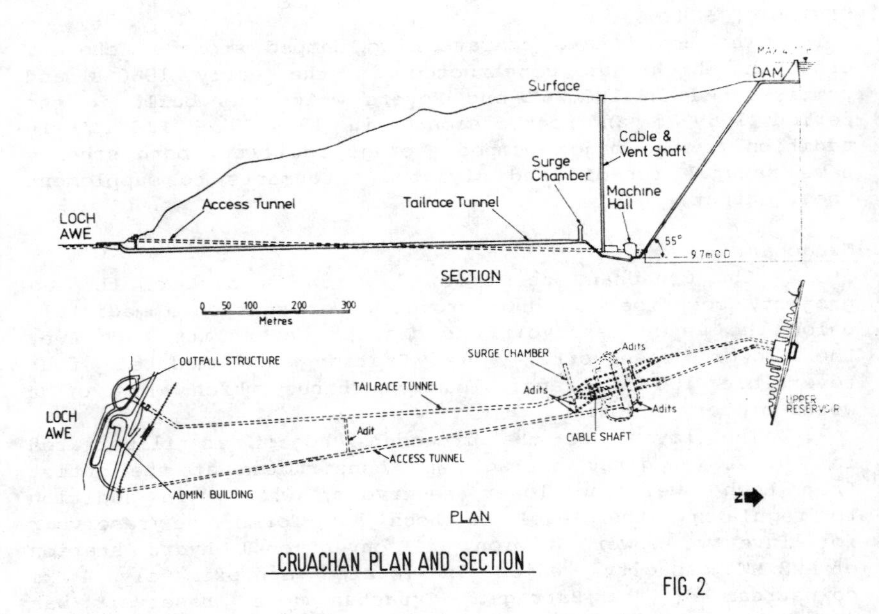

CRUACHAN PLAN AND SECTION

FIG. 2

4. The underground cavern was very carefully sited following a detailed assessment of the geology and optimisation of costs. Two 5 m diameter high pressure

shafts 402 m long were constructed at an angle of 55° to the horizontal from the upper reservoir to the cavern 335 m below, each of these shafts bifurcating into two 2.7 m diameter steel lined penstocks, 104 m long. Steel lined draft tubes from each of the machines were constructed to discharge into a single concrete lined tailrace tunnel of 7.0 m equivalent diameter, 951 m long, which incorporates a surge chamber and associated expansion galleries. A separate shaft constructed off the main access tunnel gives access to a high level gallery where the lifting gear associated with the 4 draft tube gates is located.

5. The underground cavern is 91.5 m long by 23.5 m wide and 38.0 m high and is serviced by a single unlined access tunnel of 7.3 m equivalent diameter, 1,097 m long, terminating at the loading bay at 0-00 MOD, some 42.0 m below the access tunnel portal at ground level. A 4.6 m diameter vertical services shaft carries cables and ventilation from the cavern to ground level at 331.0 m OD above. The transformers are housed in separate cells in a cavern adjacent to the machine hall with associated cable tunnels from the transformers to the main vertical service shaft. FIG 3 illustrates the layout of the machine hall and the adjacent tunnels, galleries etc.

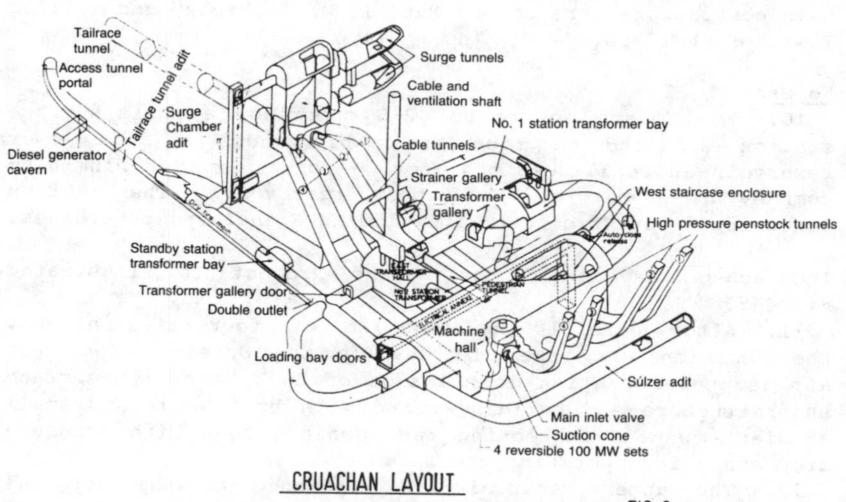

CRUACHAN LAYOUT

FIG. 3

6. During the conception of Cruachan and Foyers, which are both located in areas of high natural beauty, considerable attention was paid to the environmental aspects of layout and design. At Cruachan, the siting of the Power Station in an underground cavern was of major benefit and eliminated many significant environmental objections to the Scheme. The method of dealing with the appearance of the dam was important since it was a very prominent

feature in a wide mountain landscape. To provide a structure with clean unbroken lines when viewed from across the valley, the decision was taken to locate all the operational equipment in a chamber within the central block of the dam, thus leaving a clean elevation of the dam without the normal appendages of lifting equipment, gatehouses etc.

7. The formation of the access road up the hillside to the dam site left a scar which was slow to heal and it was necessary subsequently to soil and seed many of the exposed cuttings and embankments above and below the road. Trees were subsequently planted below the road line to screen, as far as was practicable, the rock cuts.

8. The Administration Building is located adjacent to the main tourist road to Oban and was afforded careful architectural design and high quality finishes. The Visitors' Centre, which was erected adjacent to the outfall structure and Loch Awe, was set low to avoid interfering with the views from the road across the Loch. This Scheme is a major tourist attraction in the area and the Centre has recently been rebuilt to cater for the continually increasing number of visitors.

9. The civil engineering works of the Scheme were designed by James Williamson & Partners, Glasgow and the main contractors were Edmond Nuttall of Camberley and William Tawse of Aberdeen.

<u>Foyers</u>

10. The Foyers scheme ref 2-3 comprises a shaft/surface station situated on Loch Ness which acts as the lower reservoir and utilises Loch Mhor, a former British Aluminium company reservoir as the upper reservoir. The station is equipped with two reversible 150 MW Francis pump turbines.

The scheme layout of the Foyers Project is illustrated FIG 4-5.

11. After careful investigation of four alternatives, the decision was taken to develop the Scheme based upon a shaft power station at the side of Loch Ness with surface superstructure and buildings, and with horizontal and small gradient tunnels connecting the machines to a high pressure drop shaft incorporating the surge chamber.

12. The upper reservoir is impounded by the original low masonry and embankment dams which were constructed in 1895. Apart from reducing the spillway crest of the 9m high masonry dam by 0.76 metres and repointing the masonry, a minimal amount of work was needed on this Dam. The 6 m high main embankment dam, 338 m long, and relied on a peat core and required strengthening. A crest road was formed, rip rap was tipped on the upstream faces over a filter blanket and a concrete wave wall constructed, with a 450 mm thick filter blanket, put down over the lower

areas of the downstream slope. A former blind dam with peat cut-off, between the masonry and embankment dams, was replaced by a moraine and rockfill embankment dam.

13. A 2.9 km long tunnel, D shaped, 6.9 m high by 6.2 m wide and concrete lined runs from the intake structure on the north shore of Loch Mhor to the drop shaft. A steel pipeline, 6 m diameter and 94 m long, crosses a deep valley, Glen Liath, between the two sections of the low pressure Tunnel.

14. The two reversible 150 MW machines are set 34.7 m below the level of Loch Ness in 19.05 m diameter shafts. Two 5.9 m diameter concrete and steel lined draft tube tunnels connect the machines to Loch Ness through the lower control work structures incorporating flap gates and screens.

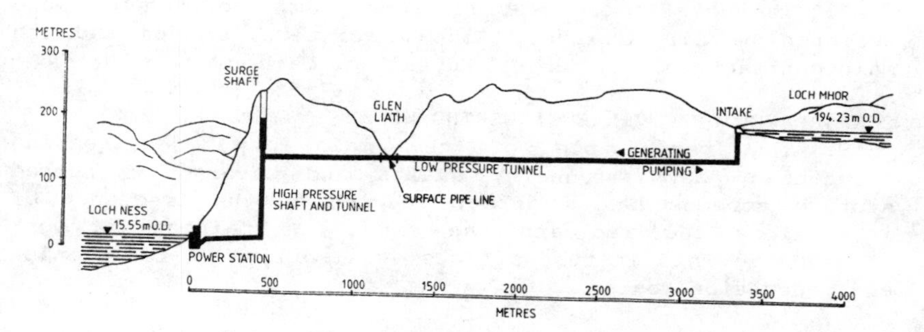

FOYERS SHAFTS AND TUNNELS

FIG. 4

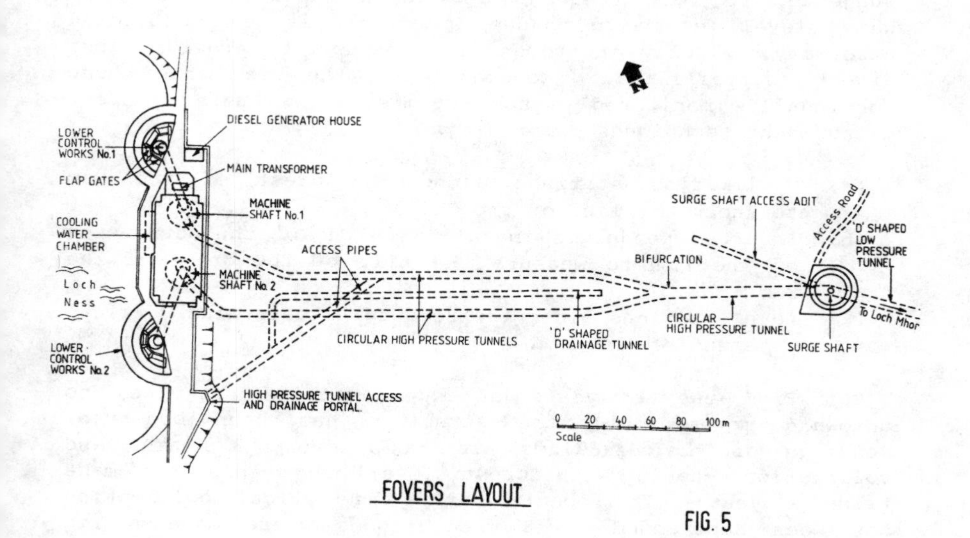

FOYERS LAYOUT

FIG. 5

15. To control the level of Loch Ness, the original masonry weir constructed at the outlet from Loch Ness where it discharges into the River Ness, was complemented by a control weir incorporating two undershot gates to ensure the level of Loch Ness and the discharge down the River Ness could be maintained within prescribed ranges irrespective of the operating regime of the Foyers Power Station.

16. The Foyers station is set down on the shore of Loch Ness. The spoil areas, around the power station, were landscaped and local species of trees planted to restore the local habitat. Following renovation of the old Foyers Aluminium Factory, additional landscape work, has been carried out. To minimise the impact of the overhead lines some 500 metres of the line were undergrounded from the main power station to the substation.

17. The civil engineering works of the Scheme were designed by Sir Alexander Gibb & Partners, London and the main contractors were Edmond Nuttall of Camberley.

PERFORMANCE OF CIVIL ENGINEERING WORKS

18. The performance of the main civil engineering structures, dams, tunnels, shafts and caverns of pumped storage schemes has a direct bearing on the frequency of inspections required and the extent of maintenance work necessary. This in turn affects availability of the Station and generation costs.

Dams - Cruachan Dam

19. The upper Dam at Cruachan is of the concrete massive buttress type, the Dam is 47 m high above foundation level with a spillway crest at 400.8 m OD. As this Dam is subjected to very frequent and rapid rises and falls in water level (eg daily changes of 5 m), it was considered necessary to fully instrument the Dam and to provide close direct surveillance. Movement of the Dam has been continually monitored since commissioning using direct measurement techniques comprising:

(a) Collimation stations along the crest of the Dam to check the line of the Dam
(b) Inverted pendulums installed in one of the buttresses of the Dam to measure the tilt of the crest of the Dam
(c) Level stations along the roadway to determine vertical movement of the Dam.

20. From the 23 years of instrumentation results, no untoward movement of the structure has been observed. Readings of the pendulums are taken every 3 months and collimation checks with crest levelling checks are made twice a year. In recent years, the optical collimation has been superseded by EDM techniques where movement in

the upstream and downstream direction is measured by electronic distance measurement instruments from stations established downstream of the Dam.

21. There is no significant leakage and despite its altitude, concrete damage due to freeze/thaw cycles has been minimal.

22. In 1985 during a routine inspection, the poured bitumen seals in the joints between the adjoining buttresses were found to be lower than recorded immediately following construction. No satisfactory conclusions could be drawn on the reasons for this migration of the bitumen in the plugs; the bitumen has now been refilled to original levels.

23. The main draw off and valving equipment on the Dam has operated well. The hydraulically operated intake control gates, which are located in a submerged chamber immediately above the soffit of the tunnel, have required very little attention other than routine repainting. All of the gates on the Dam are tested annually through their full travel and are partially moved every quarter.

24. Apart from a recent extended outage of the whole station for major electrical refurbishment, when the level of the reservoir was controlled by operation of the discharge regulator, the Dam has not spilled.

25. Overall, the Dam has performed very well and has required very little maintenance during its 25 years of service despite the fairly onerous climatic and operating regime to which it has been subjected.

Foyers Dams

26. The original masonry Dam on Loch Mhor has performed satisfactorily apart from repointing which is required periodically. The performance of the Embankment Dam has been monitored by piezometer, levelling, leakage measurement and by detailed inspection initially at monthly intervals and gradually increasing over 5 years to 6 monthly intervals. This Dam has performed well despite its questionnable design and the rapid fluctuations in level to which it has been subjected.

TUNNELS AND SHAFTS

Inspection

27. Based on the Board's policy and practice, main tunnel systems on hydro schemes are inspected at least once every five years by civil engineers well qualified in tunnel work. Experience on pumped storage indicates that, during the early years of operation, more frequent inspections should be made if opportunities, associated with plant outages arise to allow assessments to be made and defects to be timeously repaired. The configuration of shafts and the access provisions into tunnels has major bearing on the ease with which inspections can be made with related down-time penalties to the plant.

28. The configuration of the shafts at Cruachan, constructed at a 55° angle to the horizontal, has necessitated a purpose designed inspection/working platform trolley known as "The Queen Mary". This trolley is operated on a single haulage rope over a winch which is located in a special winch house situated between the horizontal sections of the tunnel immediately below the Dam.

29. The erection and subsequent removal of this equipment from the shaft is both time consuming and costly with attendant stringent safety requirements. The robust design of the Queen Mary has allowed major repairs to be carried out to the concrete lining and vindicates the decision to provide this equipment from the start of the project. The trolley allows access down each of the main shafts to a point above the bifurcation. Access below this level can be taken on satellite trolleys suspended from anchors in the invert of the shaft or by rope ladder. Access down each of the four 2.7 m diameter steel lined high pressure shafts, which continue at an angle of 55°, is on purpose made trolleys which are suspended through proprietory air powered winches from anchor points immediately below the bifurcation, installed as part of the original design. Using a experienced team of riggers it takes a minimum period of 8 days to assemble, rig and remove the Queen Mary including 2 days for the inspection of each shift. The cost of each inspection is around £40,000.

30. Extensive repairs had to be carried out to the tunnels and shafts at Cruachan during the first decade after commissioning.

31. As a result of this experience gained at Cruachan, a small Working Group comprising design and project civil engineers, operational staff, safety engineer was set up during the early stages of the Foyers Project to define in detail the inspection and maintenance facilities required so that these facilities could be built in during construction. This policy has paid handsome dividends in subsequent inspections and maintenance work over the past 15 years.

32. Inspections of the vertical surge chamber and high pressure drop shaft at Foyers are carried out using man riding cradles suspended from the ring beam at the top of the chamber and from the floor of the surge chamber respectively. Repair work on the vertical walls of the drop shaft is carried out using a purpose built cradle suspended from permanent anchor points provided during construction.

33. As a result of Cruachan tunnel experience, a conscious decision was made at Foyers to inspect the tunnels every year for the first five years after commissioning and to undertake as much remedial work as outage windows would permit. This proved to be a very advantageous policy for tunnel defects which, if left, propagate and develop rapidly and much expenditure can be avoided if they are tackled

as soon as practicable.

34. Over recent years, the Board have been employing alternative technologies for routine inspections of both dams, tunnels and shafts to minimise the cost of direct inspections and particularly to reduce the downtime on high merit plant. The advances in the techniques of underwater ROV systems (Remote Operated Vehicles) with onboard television and cameras has allowed this technology to be used successfully for particular inspections at both Cruachan, Foyers and other conventional schemes Ref 4. Experience to date indicates that the use of ROV's is a very valuable complementary facility for inspection work where the cost of de-watering and provision of temporary works or the safety aspects are significant provided that direct inspections have previously been carried out and have established first class datum bases for subsequent ROV inspections.

Repairs To Concrete Tunnel Linings

35. At Cruachan and Foyers repairs to the concrete lining of tunnels have been necessary after varying periods of service. Defects in the concrete linings normally arise from poor quality control of the concrete during construction.

36. For the large diameter tunnels, up to 7 m equivalent diameter at Cruachan, the use of pneumatic concrete placers, the logistics of the concreting train, difficulties in aligning shutters particularly in the inclined shafts and the unrealistic construction programme led to poor quality control and the formation of many cold joints and areas of weak or poorly compacted concrete. When exposed to the high velocity, bi-directional flow of water along with all the operating mode changes experienced, serious defects (up to 8.5 m long x 0.5 m x 0.5 m section) in the concrete linings quickly developed especially at construction joints. An interesting comparison with the Cruachan Scheme is the cost of maintenance which had to be carried out on the much longer tunnel of the adjacent Inverawe Hydro Scheme over the first 15 years' of operation. This latter scheme was built at the same time as Cruachan, carries the same water, used the same cement and aggregates and was designed and supervised by the same consulting engineers. On the Cruachan Scheme, during the first 15 years of operation £230,000 at current prices was spent on maintenance of the 950 m long tailrace tunnel whereas on the Inverawe Scheme only £44,000 was spent on the 5400 m power tunnel of approximately the same equivalent diameter. These defects have been repaired with various techniques: pumped concrete, hand-filled concrete, gunite and general trowelled repairs. The importance of providing good edge details and keys, undercut if possible, to repairs cannot be over emphasised. Suitable dowel bars secured into firm rock with local reinforcement is also considered necessary. Fibre reinforced

gunite has been used successfully on a number of these repairs and to date has exhibited excellent durability.

37. Poorly formed or cold joints in the lining invariably open up and deteriorate quickly as unravelling around the joint occurs. This type of defect can best be repaired by cutting out into sound concrete to form a good key, then filling with one of the proprietory fast curing mortars.

38. Invert damage has occurred in Foyers low pressure and Cruachan tailrace tunnels. Much of this damage may well have been due to constructional traffic and equipment being allowed over the surfaces when the concrete was still green, leading to abrasion of the cementitous material and aggregate and subsequent accelerated deterioration of the surface.

39. Flat inverts should be avoided, if at all possible, to minimise the risk of invert heave particularly during dewatering. The provision of a series of pressure relief holes through the flat invert of the Foyers LP tunnel overcame invert heave and extension of longitudinal cracking which was experienced shortly after the low pressure tunnel was commissioned.

Steel Tunnel Linings

40. The original treatment given to the steel tunnel linings at Cruachan and Foyers could not have been more different. Due to lack of adequate construction programme time, the steel linings at Cruachan had to be brought into service without being protected and subsequently were treated with conventional bituminous paint some 2 years later. After 12 years of service, paint had badly deteriorated the linings were corroding and a full shot-blast and re-coating of the linings with a multi-coat, coal tar epoxy pitch was planned. Working within the constraints of an operational underground power station in 55° degree shafts, 335 m deep, with all the related problems as well as the hazards of solvent release from the coating system, required a very precise and fully engineered contract. This work was held back until the main inlet valves were being removed for refurbishment which provided access and the outage window required to allow this work to proceed. The operations proved to be difficult and not without risk attracting stringent safety measures with high related costs.

41. At Foyers all of the linings were blasted and a multi-coat coal tar epoxy pitch applied during the original construction; this coating remains in excellent condition after 15 years of service giving total protection to all the steel linings.

42. All small diameter built-in pipes, such as drains and service pipework should be in stainless steel or be to an equivalent specification as the difficulties, problems and costs of carrying out subsequent maintenance is totally

dis-proportionate to the amount of work involved. At Cruachan, it has been necessary, as part of the recent major plant refurbishment programme, to replace a substantial proportion of the plant service pipework in stainless steel.

CAVERNS AND SHAFTS

Cruachan Machine Cavern

43. The rock cavern at Cruachan is 91.5 m long, has a span of 23.5 m and a maximum vertical height of 38.0 m. The cavern was very carefully located on the basis of a comprehensive rock investigation. The main generator transformers are located in a separate transformer gallery which is interconnected with tunnels to the main hall. The machine hall was excavated in the granite mass apart from a few blocks of weathered phyllite. The roof of the cavern was arched and concreted. The side walls and the end gables, the main access tunnel and a substantial length of the adjoining galleries were trimmed and left unlined. The rock faces are inspected every two years and the access tunnel with turning areas are inspected quarterly to give early detection of any changes in rock condition. These inspections are made by eye and, if appropriate, temporary works are arranged to permit a close "hands on" inspection of suspect areas of rock. Only one significant rock scaling and further rock bolting exercise has been necessary since the station was commissioned.

Foyers Machine Shafts

44. Rock conditions at Foyers were very poor since the station is situated in the shatter zone of the Great Glen Fault. As a consequence, the station concept was based on a shaft design.

45. The machine shafts at Foyers are lined with concrete and are provided with an inner lining of patent corrugated and coated steel sheet. Problems have been encountered at both Foyers and Cruachan keeping drainage ways clear of calcium salts which are picked up from the grouted rock and from adjoining concrete masses and precipitate on coming into contact with the carbon dioxide in the atmosphere. It has been found essential to maintain a regular rodding programme; high pressure water jetting has been used very successfully to clear persistent blockages in these drainage systems.

Industrial Hazards In Underground Power Stations

46. For all the Board's underground stations, detailed re-appraisals have been made of the design and related operational practices to meet the increasingly stringent safety requirements now being laid down by the Health and Safety Executive and Fire Authorities to protect all Board personnel. The Board's policy of allowing the Public to visit the Cruachan Station has increased the importance

of and requirement for safe working regimes.

47. An appraisal of Cruachan Power Station began in 1978, in conjunction with the Health & Safety Executive and the Fire Authority and culminated in the publication of an internal report which reviewed the improvements and modifications which were then judged to be necessary. A working party was set up to undertake a more detailed Study comprising representatives from the operational staff along with Civil, Mechanical, Electrical and Safety engineers from the design and project divisions and was recorded in detail in a Safety Status Report which defined the main hazards which could arise during normal operation of the Station and how these hazards were to be safely accommodated including the provision of the related protective measures. The types of hazard which were examined included:

(a) Fire (source, detection and protection);
(b) Smoke/products of combustion (including elimination measures);
(c) Flood (sources, discharge rates and times to fill cavern to critical levels and protection systems);
(d) Gas (sources, detection and warning systems);
(e) Chemical (uses and storage);
(f) Security, sabotage and vandalism;
(g) Explosion (possible sources);
(h) Other hazards (eg land slips, earthquakes, etc).

48. "Normal operation" of the station includes routine maintenance works required to keep the plant operable but specifically excludes major works which it was identified would normally be carried out under very carefully defined conditions and specially formulated procedures as dictated by their nature. The Safety Status Report was independently assessed by the Board's Safety Section and their findings and recommendations laid down in a "Safety Assessment Report". The recommendations and findings of the Assessment were well received by the operators and have now been incorporated into the normal operating regimes at Cruachan.

49. The selection and fire grading of materials used in underground installations is most important. Following a fire at Cruachan in 1985, caused by an explosion in a temporary compressor, the sheet lining of the cavern was sprayed with burning oil. Although the station filled with smoke requiring a full evacuation, the material did not catch fire or propagate due to the careful selection of the material by the original designers.

50. This type of study has subsequently been undertaken for the Board's other underground stations.

EVALUATION OF PERFORMANCE IN RELATION TO FUTURE DESIGNS

Concepts And Operational Roles

51. The operating regime of pumped storage schemes has been found to be very much more onerous than conventional

hydro operation especially in relation to the performance and maintenance of certain civil structures. The role of a pumped storage scheme has a dominant influence on inspection, maintenance and refurbishment. Those stations which are designed or are operated in the dynamic role are likely to have much greater maintenance than those schemes which are primarily used in an energy transfer role. These effects lead to the essential requirement of higher standards of construction than are acceptable on conventional hydro schemes.

52. The concept of the Cruachan and Foyers Schemes are fundamentally different in that the former is based on an underground power station and the latter is a shaft station. The performance of the two schemes provides a very interesting comparison. The shaft station at Foyers has been found to be much easier to operate and maintain than the underground station at Cruachan although it must be recognised that in most cases there is a significant economic advantage with an underground station. The advantages of an underground station are very important when schemes are to be built in areas of high environmental sensitivity.

Dams

53. The multiple buttress dam design for the upper reservoir on the Cruachan Scheme was developed from the earlier designs of a series - Sloy, Lawers, Giorra and Lubreoch. It has been subjected to much more rapid changes of water level than dams on traditional hydro schemes. This is an important requirement which must be taken carefully into account in selecting the most appropriate form of dam and in its design. The multiple buttress design has performed very well in this and other respects both on the Cruachan Scheme and on other conventional hydro schemes. Maintenance of the dams has been minimal.

54. With the increasing sensitivity and importance of environmental matters it seems probable that rockfill and earthfill dams are likely to be more acceptable than concrete dams particularly in sensitive areas such as the Highlands of Scotland. If a concrete dam is selected, it will form a major and dominant feature of the landscape and as such must be designed so that it is aesthetically and environmentally satisfactory. This was recognised in the case of Cruachan Dam which forms a dominant, interesting yet majestic feature of the landscape of the area.

55. Asphalt cores and upstream linings are likely to be increasingly adopted. An advantage of pumped storage is that it is normally possible to completely empty and inspect the upstream face of a dam, without too much difficulty. A disadvantage is the much more onerous

operating conditions which have to be resisted by the upstream membrane.

56. In the case of both Cruachan and Foyers, the lower reservoirs are in areas of natural beauty. Strenuous efforts were made to limit the range of water levels in Loch Awe and Loch Ness over which the Schemes would operate. There are big advantages in choosing very large lakes for the lower reservoirs in order to reduce the operating range of water levels. Overall the control of the two lower reservoirs for the Cruachan and Foyers Schemes has not led to any significant detrimental environmental effect, the control of Loch Ness being particularly successful.

57. There are now powerful and influential lobbies committed to resisting any plans for the development of conventional hydro or pump storage schemes.
The importance of ensuring that the environmental and conservation aspects are properly dealt with can not be over emphasised and the appointment of competent environmental advisers and landscape architect(s) at an early stage in the conceptual planning is essential.

Tunnels And Shafts

58. A major decision in the conceptual design of a scheme must be the number of machines served by any one tunnel. Single tunnels result in the whole station being taken out of operation if any inspection or maintenance work has to be carried out. The Cruachan Scheme is served by two high pressure tunnels each serving two machines and one tailrace tunnel serving the four machines. Overall down time of the station due to maintenance of tunnels and shafts has not been significant. However experience at Cruachan confirms to the desirability of providing dual tunnels, particularly on the high pressure side, for stations with four machines or more and especially for those schemes operating in a dynamic role.

59. Pumped storage has a much more onerous and damaging operating regime than conventional hydro. The extent and seriousness of defects is exacerbated on pumped storage particularly where any lapses in quality control of concrete occur (eg cold joints, honeycombing, shutter misalignment, etc).

60. It is considered that the reasons for the higher level of maintenance on the tunnel linings of pump storage schemes compared with conventional hydro could be due to a number of significant differences:

(a) The higher velocities adopted.

(b) The bi-directional flow of water.

(c) The number of mode changes in the station, each of which results in a pressure cycle and hence a stress cycle on tunnel linings and other works and equipment.

In the first ten years of operation, the Cruachan station experienced, on average, some 8,800 mode changes compared with Inverawe, where 500 cycles per year were recorded.

(d) Quality Control of Concrete

The Board's experience of pumped storage schemes emphasises the requirement for a much higher quality of concrete on pumped storage schemes for tunnel and shaft linings. On both the Cruachan and Foyers Schemes, pneumatic placers were used and this led to sub-standard concrete in tunnel soffits, lift joints and leading edges of the walls of tunnels. Tunnel lining experience over the last ten to fifteen years points to pumped concrete being a superior means of placing concrete provided that it is complemented with both internal poker and external shutter vibration. It is considered that impregnation of the surfaces of concrete linings should be carefully considered to improve the durability of concrete linings especially for pumped storage schemes operating in a dynamic role and at high heads.

61. Experience of the Board does not allow a clear demarcation to be defined between heads at which concrete linings are acceptable and those where steel linings are necessary. European experience, which is almost all with steel linings, has been good but the cost is high. Experience from Dinorwic of concrete linings operating under much higher heads than Cruachan and Foyers will be very revealing and helpful.

62. The inclined penstock tunnels at Cruachan (55° to the horizontal) were dangerous and expensive to construct and have been very difficult to inspect and maintain. In addition, the Cruachan tunnel linings have been inferior in performance to the Foyers linings. A conscious decision was made in designing the Foyers Scheme to eliminate inclined shafts and to depend on vertical shafts and generally horizontal tunnels. The construction, inspection, operation and maintenance of the Foyers tunnels has been good and has confirmed the wisdom of this configuration. The cost and time necessary to inspect the Cruachan shafts has been significant. More and more stringent and onerous health and safety requirements have been introduced over the years resulting in continually increasing costs of inspection and maintenance.

63. The tunnels at Foyers are very large (7 m equivalent diameter) and were subject to fairly extensive cracking during curing of the concrete. In future, consideration would be given to incorporating fly-ash into the design of mixes to slow down the rate of curing and to reduce shrinkage and cracking.

64. Much greater use of ROV's can be foreseen in the inspection of dams, tunnels and shafts. These vehicles have been found to be very valuable for quick inspection of tunnels and shafts without the need to de-water, provided

that a first class datum against which ROV observations can be directly compared has been previously established by direct inspection.

65. The tunnels of both the Cruachan and Foyers Schemes were driven by drill and blast techniques. Over the last 15 years great advances have been made in the use of tunnel boring machines and shaft sinking by augers. These techniques must be carefully taken into account in the design of the scheme from the conceptual stage since they are likely to lead to substantial savings in costs of the excavation and lining of tunnels and shafts.

Caverns

66. The great value of a first class and detailed site exploration including the driving of exploratory tunnels cannot, be overstressed especially, for underground works and in particular for the location of caverns. The cavern at Cruachan was well sited and did not present any major problems either during the construction or in its subsequent operation. Likewise the detailed site investigation undertaken on the Foyers Scheme generally gave very accurate prediction and guidance in the design and location of underground works. Money spent in these preliminary stages is very well spent and a good insurance policy for successful construction of the scheme.

67. Experience at Cruachan and during design studies for future schemes have led the Board to aim at providing separate caverns for transformers. The advantages of keeping the span of main caverns down to a minimum leads to very substantial economies and is well worth striving for. Adits, temporary tunnels and other construction caverns have all been found to be valuable for stores, simple workshops etc.

68. Over the past 10 to 15 years, health and safety requirements at Cruachan and the Board's other nine underground stations have continually increased and become more onerous particularly in relation to fire and smoke hazards. The very serious accident which occurred at Tonstadt in Norway when a transformer exploded adjacent to the main cavern and wrecked the control room points to the very clear requirement to locate control rooms outside caverns and on the surface. It is the authors' view that workshops and other personnel intensive areas should, as far as practicable, be located outwith the cavern in order to keep personnel in underground caverns to the minimum.

69. The need to study very carefully and specify escape facilities, not only for staff but contractors men and visitors must be incorporated into stations right from the very start of design. Cruachan has only one access tunnel; it is unlikely that a single access would be acceptable if a similar station were built today.

70. Stations such as Cruachan are of great interest to the public and there has been increasing pressure to

provide visitor facilities. Cruachan now entertains on average 50,000 visitors a year. Experience of handling this number of visitors presents formidable logistics problems for their management, escape, etc. In future stations, these facilities would be designed so that visitors can be completely separated from operational and working areas and ideally would be afforded separate, first class and direct means of escape. As a result of the Board's experience of handling visitors at Cruachan, all vehicles carrying visitors are electrically driven and for operational purposes only diesel engined vehicles are allowed in the cavern. One can foresee in the future only electric vehicles being permitted in future stations during normal operation.

71. Ventilation is a key aspect of underground stations. The need to prevent smoke filling of the cavern and preventing escape particularly via the access tunnels is vital in the case of a serious incident. Ventilation systems should be designed so that they clear the station of smoke by natural circulation even when power supplies are lost. The value of a first class ventilation system not only gives rise to good and safe working conditions but also has very substantial advantages during construction if built into and available from the early days of construction (eg as at Dinorwic).

72. The Board has experienced several serious incidents where underground stations have been flooded. The utmost care must be taken in designing drainage and emergency pumping facilities and systems, with very careful consideration being given to the maximum size of burst to be accommodated.

PROJECT STRATEGY

73. The Foyers Scheme was built within budget and within six months of the project programme despite one of the main machine shafts being completely flooded out for eight months during construction. The project strategy which was adopted at Foyers was very successful. It comprised a limited number of large contracts, giving a manageable number of interfaces for the Board to co-ordinate as project managers for the overall scheme. The Board's policy is to cover the indeterminate and difficult work in contracts (eg such as the cavern excavation contract) with special contract conditions. The majority of contracts, however, are relatively straightforward and as such are put out on a normal competitive tender basis (eg tunnels, shafts, dams).

74. Project programmes must be realistic with flexibility built into them to accommodate the problems which inevitably arise during the course of construction. At Cruachan, it appears that the project programme was too tight and this led to loss of quality control leading to the subsequent problems which have arisen on the civil engineering works. In contrast, the Foyers Project was a model of a good

programme for it was realistic and flexible and was able to accommodate problems, by adjustment and refinement as the work proceeded. It led to the contractors making strenous efforts to achieve the project objectives which were largely attained.

CONCLUSION

Overall, the Civil engineering works on both the Cruachan and Foyers Schemes have generally performed well, especially considering that Cruachan was very much a prototype scheme. However, the many lessons from the Cruachan Project were incorporated in the Foyers Project to great advantage and further lessons can be drawn from the Foyers Project for future schemes.

ACKNOWLEDGEMENTS

The Authors are indebted to the Company for permission to publish this Paper and wish to thank their colleagues for their help and encouragement in its preparation.

REFERENCES

1. William Young MBE, BSc, MICE and Richard Hove Falkiner BA, MAI, MICE, "Some design and construction features of the Cruachan Pumped Storage Scheme", Proceedings of the Institution Civil Engineers Nov 1966.

2. D J Millar BSc, CENG, FI MECHE, FIEE, A T L Murray, BSc, CENG, FIEE, C C Marshall, BSc, CENG, FICE, FCIT, FASCE and G G R Argent, BSc, CENG, FI MECHE and M Conse, "Foyers Pumped Storage Project", Proceedings of the Institution of Electrical Engineers, Nov 1975.

3. J H Lander OBE, MA FICE, FISTRUCTE, FNZIE, F G Johnson M ENG, FICE, MIWES, J R Crichton BSc FICE and M W Baldwin BSc FICE, "Foyers Pumped Storage Project: Planning and Design", Proceedings of the Institution Civil Engineers Part 1 1978.

4. F G Johnson M ENG FICE, MIWES and C K Johnston C ENG MICE, "Inspection of Underwater Structures by Remote Operated Submersible Equipment", The Institution of Civil Engineers Energy Engineering Group, Oct 1988.

Operation of Dinorwig pumped storage station on the UK National Grid system

J.R. LOWEN and A.J. STEVENSON, National Grid Company Division

SYNOPSIS. The present operating regime of Dinorwig was decided upon after extensive investigation into the system economic and security considerations. A description of these considerations is given in the paper together with a review of actual operating experience to date.

ABSTRACT

This paper comprises four parts.

In the first section the technical characteristics of Dinorwig are considered in the wider context of the British power system covering England, Scotland and Wales. The existing reserve provisions are also described.

The second part deals with the extensive studies undertaken by the CEGB to define the precise operating modes of Dinorwig, including a brief description of the computer program GOAL used for these studies. With limited water storage only supporting some 5 hours of full load generation, the competing claims of providing system reserve against demand underestimation and generation losses had to be compared with "in merit" generation for peak lopping and the potential advantages of giving automatic frequency control.

The third reviews experience of Dinorwig on the system so far, including a consideration of the impact of the 2000 MW HVDC link between France and Britain, which became fully available during January 1987.

The fourth part considers the various options for changing the role of Dinorwig in view of the forthcoming privatisation of the electricity supply industry.

TECHNICAL CHARACTERISTICS OF DINORWIG AND ITS ASSOCIATED POWER SYSTEM

Introduction

1. Dinorwig is the fourth pumped storage station to be built on the British power system and it is also the largest. It was preceded by an earlier station at Ffestiniog, (4 x 90 MW) which was commissioned by the Central Electricity Generating Board (CEGB) during the early 1960s. Both Dinorwig and Ffestiniog are geographically located in North Wales and are electrically connected, at 400 and 275 kV

Pumped storage. Thomas Telford, London, 1990.

respectively, to the National Grid 400/275 kV Supergrid network. The associated Scottish 275 kV network also connects with two more pumped storage stations at Cruachan (4 x 100 MW) and Foyers (2 x 150 MW).

2. Royal assent for the Dinorwig scheme was received by the CEGB in 1973 and work on the project began in early 1974. All six machines were commissioned by the summer of 1984.

Technical Characteristics of Dinorwig

3. The 1800 MW Dinorwig pumped storage power station is one of the largest in the world and is capable of providing additional output for its associated power system at a fastest rate of 0 to 1300 MW in 10 seconds. It is designed for a daily pump/generation cycle, at a target efficiency of 78%. To achieve a full upper reservoir overnight from empty takes about 6 hours of full load pumping. This, in turn, provides sufficient stored water for some 5 hours of full load generation, with the output reducing as the differential head decreases. Dinorwig is also capable of "black start", ie it is possible to start the station generating without external power supplies from the Grid system. The six 300 MW (nominal) generator motors operate at 18 kV, step up transformers change the voltage to 400 kV and energy is transferred at this voltage via two underground cables to the 400 kV outdoor substation at Pentir, about 10 km away.

Technical Characteristics of the Associated Power System

4. Thus, at Pentir 400 kV substation, Dinorwig pumped storage interfaces with the Supergrid system. This in turn covers the rest of Wales and all of England and is interconnected with the two Scottish Boards, South of Scotland Electricity Board (SSEB) and North of Scotland Hydro-Electric Board (NSHEB).

5. The combined maximum demand of the CEGB and NSHEB/SSEB was 53,978 MWh for the half hour ending 17.30 hours 12 January 1987.

6. Although Dinorwig makes a significant contribution to the whole British system, especially its dynamic performance, for commercial reasons its operation is optimised in the first instance with other plant connected to the National Grid system. Hence, the following paragraphs deal mainly with system demand and generating plant characteristics.

7. The composition of CEGB generating plant, as during early 1988, was as follows.

	MW	%
Nuclear Plant	5,069	9.4
Coal Fired	30,886	57.2
Oil Fired	8,417	15.6
Dual Fired	4,870	9.0
Hydro & Pumped Storage	2,195	4.1
Gas Turbine	2,517	4.7
	53,954	100.0

In addition there is access to supplies from France via the 2000 MW dc cross channel link, and Scotland via an ac interconnection with a capability of approximately 1000 MW.

8. Figure 1 illustrates the maximum/minimum demand relationships of the system during the financial year 1988/89. An annual peak on Tuesday 22 November 1988 of 46,875 MW compares with an annual minimum of 13,952 MW on early Sunday morning 31 July 1988, a ratio of 3.36:1. This compares with the daily maximum/minimum demand ratios of the order of 1.18:1 in winter and 1.32:1 in summer. Figure 1 also gives a general idea of typical seasonal daily demand profiles.

9. The electrical energy provision for England and Wales in 1988/89 was 248.1 TWh split between fuels as follows:

	%
Nuclear	14.4
Coal	81.5
Oil	4.1
	100.0

10. Figure 2 illustrates the contribution of the various generating plant categories to meeting the daily demand (including the role of pumped storage) on a typical winter weekday. It can be seen that all nuclear and most large coal-fired units operate continuously whilst the remaining plant has to perform cyclic duties.

The Provision of System Reserve Capability Prior to Dinorwig

11. The policy of providing system reserve capability on the British Power System (NSHEB, SSEB and National Grid) has evolved from the electrical isolation of this system from the continent of Europe. In the 1960s and 1970s there was a 160 MW direct current link to France, which has now been superseded by a new 2 GW Link, but there has never been a synchronous link between the British Isles and mainland Europe.

12. The legal requirement in the United Kingdom, laid down in the Electricity Supply Regulations 1937, is for a 1% tolerance around the declared system frequency of 50 Hz. Hence the statutory limits are between 50.5 and 49.5 Hz, but

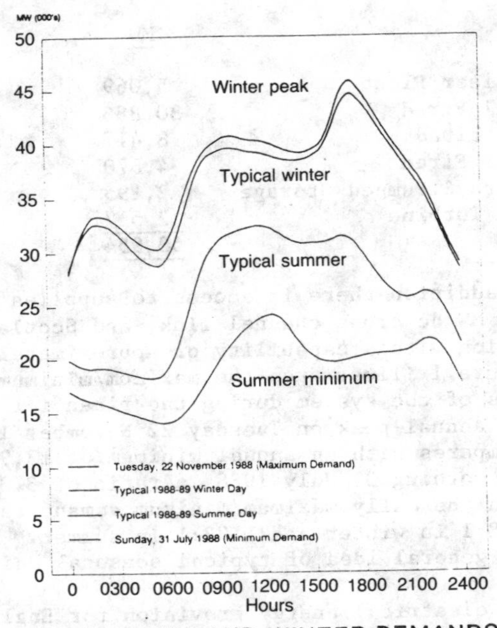

FIG. 1 SUMMER AND WINTER DEMANDS 1989 (including days of Maximum and Minimum demand)

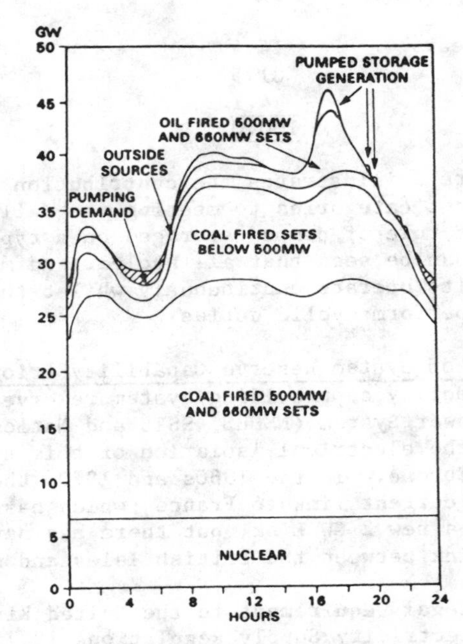

FIG. 2 PLANT CONTRIBUTIONS FOR TYPICAL WINTER DAY DEMAND

the National Grid works to a tighter operational target of keeping the frequency normally between 50.2 and 49.8 Hz (and the Scottish Boards follow suit).

13. The largest single generating units operating on the British System at present are 660 MW sets and supergrid substations and main transmission circuits are usually switched in such a way so as to avoid the risk of losing more than one 660 MW unit at a time. However, there have been instances when 2 (or more) large units have been lost from the system simultaneously.

14. National Grid short-term demand forecasting performance is generally of a high standard and is closely associated with accurate weather forecasts obtained under special contract from the Meteorological Office. Typically, most errors are in the range of 1 to 2% of demand.

15. System reserve capability also has to cope with plant output shortfalls (other than plant losses referred to under 13) and these are usually no larger than 500 to 750 MW.

16. Finally, all generating plant and demand estimates apply to metered half hours. This in turn requires a higher spot demand value to be covered within the half hour, which can be as high as 400 MW during sharp peaks.

17. The policy of providing system reserve on the British System is therefore based on the need to cover any likely combination of contingencies outlined (under 13 to 16 above) in the most economic manner at an acceptable risk level of not meeting the demand in full.

18. The policy also relies on the established capability of the British System to withstand sudden instantaneous large demand increases or generation output losses by calling on stored, mainly thermal, energy of the generating plant already contributing to the system. This will normally suffice for a relatively short period (2 to 3 minutes) until the scheduled reserve comes into action to augment and ultimately replace the initial but non-sustained response.

19. The necessary scheduled system reserve to meet the foregoing requirements was provided, in the pre-Dinorwig era, from a spinning reserve allocation of 1000 MW for the whole British system, backed by a standing reserve of some 500 MW.

20. The spinning reserve of 1000 MW consisted of the whole of the earlier Ffestiniog pumped storage, 360 MW, at synchronous speed (on no load), 80 MW contribution from the two Scottish Boards and 560 MW on steam plant, partially loaded to some 80% of capacity (ie some 2,800 MW of steam plant being involved in this duty). The then existing seven CEGB Grid Control Areas shared the 560 MW of steam reserves but not always on an exactly equal basis.

21. The standing - (gas turbine) - reserve came to another 500 MW, thus ensuring a total of 1500 MW of reserve capability on the British System. This was further supplemented by some 800 MW of standby reserve, made up mainly of some of the older gas turbines, not capable of quick starting.

FIG. 3 PATTERN OF PUMPED STORAGE OPERATION DURING TYPICAL WINTER DAY DEMAND

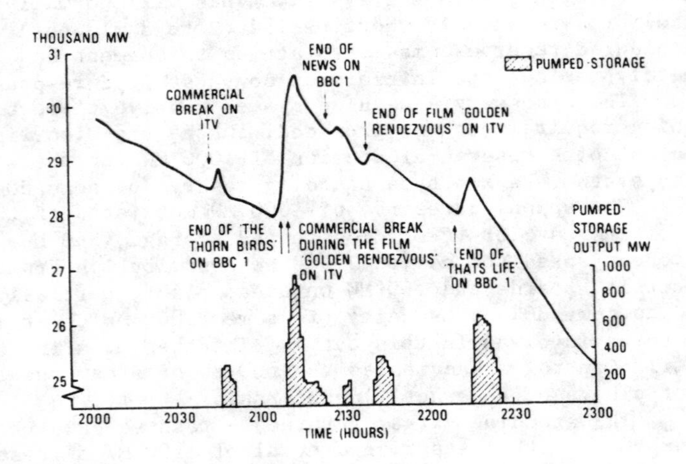

FIG. 4 DEMAND CURVE FOR THE EVENING OF SUNDAY 22ND JANUARY 1984 WHEN 'THE THORN BIRDS' ON BBC1 CAUSED DEMAND INCREASE OF 2600 MW AT 2107 HOURS

22. Overnight the spinning reserve was reduced to 400 MW plus Ffestiniog in an interruptable pumping mode. Similarly at the time of the weekday winter peak demand the spinning reserve was held by Ffestiniog in spinning mode.

23. Thus, when the very much larger pumped storage plant at Dinorwig was authorised the CEGB Operations Department gave serious consideration to eliminating steam plant from spinning reserve holding altogether, except for the overnight pumping periods, and this development is included in the second part of this paper.

STUDIES UNDERTAKEN TO DEFINE OPERATING MODES OF DINORWIG

Cost of System Reserve

24. The studies undertaken assumed System Reserve to be provided in two components:

(a) Scheduled Reserve - plant connected to the system able to achieve full output in 5 minutes. This reserve can be provided either by pumped storage plant synchronised to the system in the 'spin gen' mode or on partial loaded steam plant which itself requires redistribution of output on the marginal cost units as described in 20.

(b) Back up reserve - held on Gas Turbines at standstill able to supply full output in 5 minutes.

25. In general terms, the cost of the system reserve can be considered to comprise two elements (see Fig 5a).

(a) The capability aspect - is the annual cost of providing extra capacity on the system to give the necessary safety margin.

(b) The actual cost of the energy when the plant scheduled for reserve duty is called upon in the event.

26. Previous work in the field of determination of Reserve policy had concentrated only on the capability aspects. The cost of providing extra capacity on the system comprises the additional no load costs, the costs of starting up and shutting down the extra capacity on a daily basis and the cost of redistribution of output from the marginal plant to the reserve plant. The contributions of these elements will be significantly different depending upon the type of plant contributing to reserve eg steam plant has relatively high start up and energy redistribution costs when compared with pumped storage plant whilst quick start gas turbines have very low cost for capability but high energy cost penalty.

27. In contrast to the low cost of capability, gas turbine energy cost is very high (some 3 to 4 times the cost from steam or pumped storage plant) and there was, therefore, a requirement to consider the actual costs incurred when the

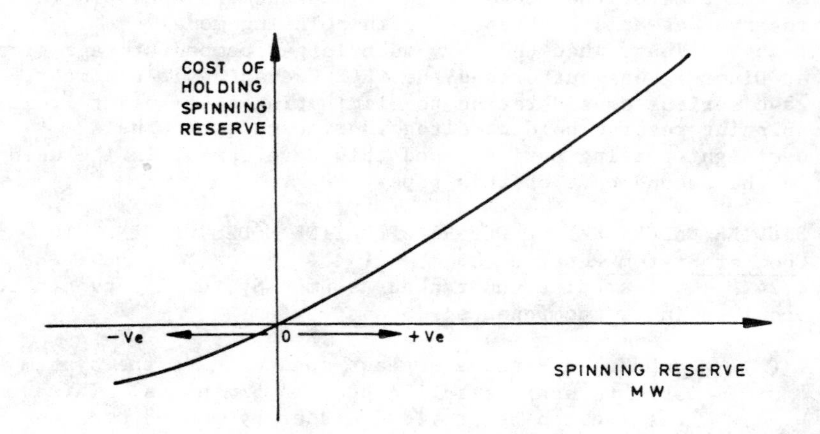

FIG.5A GENERAL COST CHARACTERISTIC OF HOLDING SPINNING RESERVE

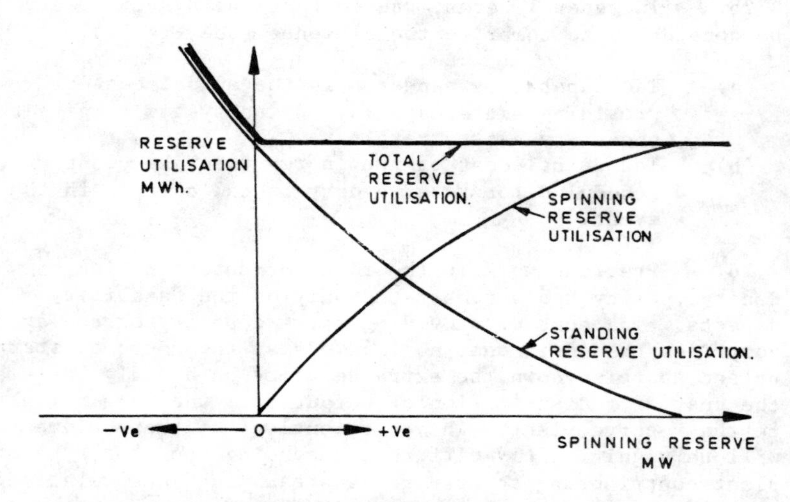

FIG. 5B GENERAL CHARACTERISTICS OF RESERVE PLANT UTILISATION

plant scheduled for reserve was called upon in the event. This, in turn, required a survey of the magnitude and frequency of these calls on the system reserve to actually produce reserve energy. In general reserve capacity utilisation is dependent upon shortfalls in generating plant output and the underestimation of consumer demand, in particular the combination of the two elements. Taking into account that output of generation can exceed prediction, demand can be overestimated and the significant day to day variations in magnitude, it is clear that the level of reserve utilisation will vary significantly and that a probability technique was necessary.

28. The data analysed for this purpose were the differences between estimated plant margins at the final scheduling stage (some 2/3 hours before the event) and the actual plant margins experienced in the event. Plant margin is defined as the excess (or deficit) of plant output actually or scheduled to be available over the half hourly integrated system demand (actual or predicted). By analysing the difference in plant margin, the data revealed the difference in the combined plant and demand position between the final scheduling stage and the event, and for all negative samples gave a direct measure of the energy that would need to be provided by the reserve. The working day was divided into some 9 periods, each was analysed and the resulting probability of various levels of margin change determined. By using the combination of magnitude of change and its probability it was possible to determine the mean expectation of energy required from the reserve and its proportions from the various elements of reserve depending upon their magnitude. Data was collected for some two years of recent operating experience and were statistically sorted into three main groupings:

(a) Daytime periods during British Summer Time (BST).
(b) Daytime periods during Greenwich Mean Time (GMT) from late October to late March, but excluding the daily peak period between 1600 and 1800 hours on weekdays.
(c) Weekday daily peak periods during Greenwich Mean Time (GMT).

It was found that there was little statistical justification for considering further sub-groups.

Cost of Generating Electrical Energy

29. The most important factor in all these studies was, of course, the cost of producing the necessary electrical energy to meet requirements throughout all the sample days examined. This was derived from the heat rate of each individual generating unit multiplied by the cost of the marginal heat which in turn is closely related to the price of the particular fuel supplying this heat.

30. Allowances were also made in these studies for heat requirements for starting up plant on a once daily cycle, where applicable, and in some cases for repeated start-ups and shut-downs within the daily operating cycle. Similarly, heat expended on keeping boiler plant in a state of readiness during the shut-down periods was also taken into account.

31. The cost of generating output obtained from pumped storage was debited at the appropriate overnight marginal cost of production, with due allowance for the overall efficiency of the pump/generate cycle of that plant.

32. It can thus be seen that the overall cost of energy produced can be modelled for many regimes of system reserve allocation varying in both magnitude and composition between holding it on pumped storage and/or on partially deloaded steam plant. On those occasions when the pumped storage or steam reserve could not meet all the requirements, the shortfall would be allocated to gas turbine plant up to the limit of the known availability of that plant for the sample day.

33. Hence by a careful choice of the variable study parameters optimum cost solutions were found for the major BST and GMT groupings of periods listed under 28. The weekday daily winter peak period is a very short duration, large magnitude peak which gives a low reserve energy utilisation in relation to the magnitude of the margin difference. Hence, the energy cost penalty is small in relation to the cost of providing reserve capability and the duration of the period of risk is small, both aspects resulting in the reduction in the level of reserve over this period. Figure 7 illustrates, graphically, how energy costs study results were interpreted and also shows the risk of failing to meet demand in full associated with various levels of gas turbine plant back up totals.

34. Figure 5b shows the general characteristic of utilisation of reserve for a given level of utilisation. Figure 6 is the practical results and demonstrates how the required reserve energy contribution by gas turbines is inversely related to the amount of system reserve provided on pumped storage (or possibly on partially deloaded steam) plant, with the maximum use of gas turbines at zero or even negative system reserve holding on the scheduled plant.

Recommended System Reserve Holding

35. After careful interpretation of the results of the studies for various conditions and composition of reserve holding (figure 7) the following policy for System Reserve was adopted.

(a) Hold all scheduled system reserve capability on pumped storage plant (except during overnight pumping periods).

(b) Provide 1300 MW of scheduled reserve capability during GMT throughout the day, except between 1600

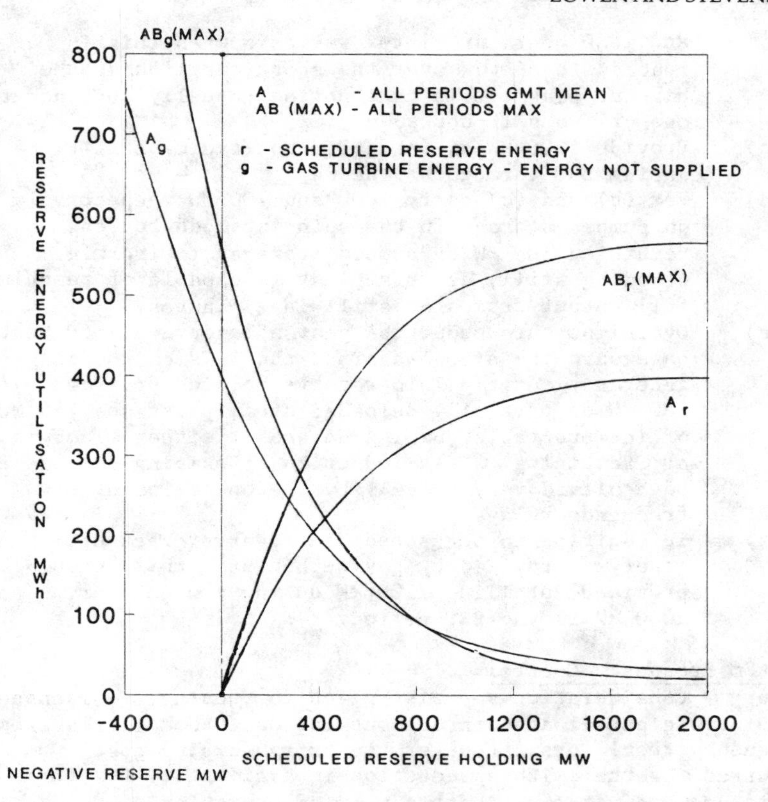

FIG.6 RESERVE ENERGY UTILISATION CURVES

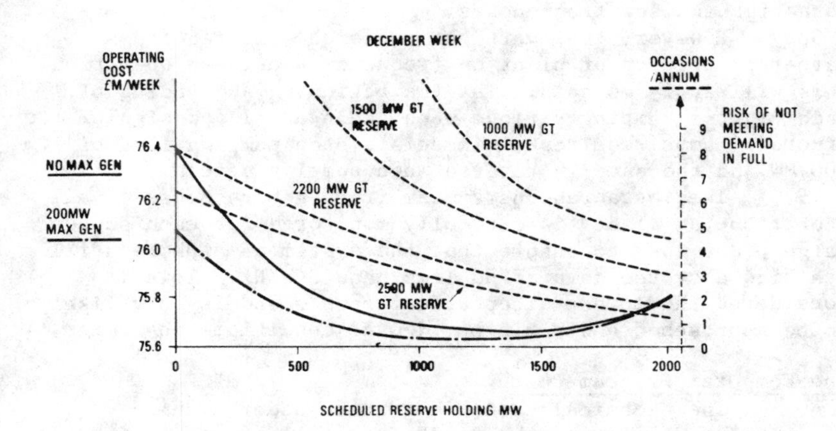

FIG.7 COMPARISON OF RISKS AND COSTS OF VARIATION IN RESERVE POLICY

and 1800 hours on winter weekdays when this is reduced to 360 MW over the short duration of the highest peak within that period (usually confined to one or two half-hours).

(c) Provide 1000 MW of scheduled reserve capability during BST throughout the day.

(d) For (b) and (c) carry 1000 and 700 MW respectively on pumped storage in the spinning mode but the remaining 300 MW on pumped storage at Dinorwig standing still, as this plant is capable of reaching full output from standstill in 2 minutes.

(e) Overnight throughout the year a reserve of 660 MW to safeguard the system against the loss of the largest generating unit. This reserve holding provided by 380 MW on partially deloaded steam plant and 280 MW of interruptable pumping demand at either Dinorwig or Ffestiniog. This rejection of pumping demand can be achieved very quickly by the operation of low frequency relays.

(f) In addition to the scheduled reserve, back up reserve (provided by gas turbines at rest) should be provided totalling 2200 MW during the GMT period and 2000 MW in the BST period.

System Frequency Control

36. Consideration was also given to running two machines at Dinorwig part-loaded throughout the day, to act as system frequency regulators. However the extra pumping cost incurred together with a reduction in availablity for in merit use did not justify this mode of operation.

37. In order to satisfy the dynamic requirements of the power system it is therefore essential that at least 50% of the thermal plant on the system is operated with the governors in the mode that gives frequency correction for both high and low frequencies.

38. However, the overnight position will require a larger proportion of plant on frequency sensitive operation. This will apply especially at the beginning and ending of each night's pumping periods when the hydraulic design of the Dinorwig pumps requires an immediate step pumping load of 300 MW and the same in reverse when pumping ceases.

39. The instantaneous system frequency response of all "contributing plant" is carefully monitored for each sudden large plant loss to ensure that the system as a whole still exhibits a system gain of no less than 3GW/Hz. This is considered the lowest acceptable response and is only likely to be approached under minimum demand conditions in summer.

The Computer Program GOAL

40. The technical/economic studies undertaken as described under paragraphs 24-39, were all carried out using the computer programme GOAL (Generator Ordering and Loading).

41. This program was designed to simulate actual operation of individual generating units, from optimum-timed scheduling of such units on to the system with subsequent load despatch, to eventual optimum shut-down. This is used in operational forward planning as well as for very short term application for joint use in the National and Area Control Centres.

42. One design objective has been to include all constraints in system and plant performance necessary to achieve viable minimum cost operation.

43. The program includes a model of the system demand profile and takes into account generator dynamic parameters, generator heat rates and heat costs (including start-up and no-load costs), load dependent maintenance costs, external transfers and transmission constraints.

44. A deterministic approach is taken to find a minimum cost solution which satisfies all the constraints. In preparing the generation schedule for peak periods, the program takes into account time varying start-up heat costs. Similarly over the trough periods consideration is given to the relative merits shutting down or running through generators at the margin.

45. Incremental and decremental costs are next calculated for each time interval and the daily demand profile (or part thereof) is considered to assess the economic use of available pumped storage plant not allocated to scheduled reserve duty. For this assessment minimum and maximum storage levels and overall cycle efficiencies are required, together with the initial and final storage levels.

46. The main outputs are generator loadings and scheduled reserve requirements over each time interval, system marginal cost merit orders for scheduling and loading of plant, and incremental and decremental costs for each time interval.

Economic Benefits from Pumped Storage Plant

47. The holding of all daytime scheduled system reserve on pumped storage plant is assessed to save some £20M per annum, when compared with holding equivalent amounts of reserve on partially-loaded steam plant.

48. The economic benefits of using pumped storage plant in the "in merit" mode as peak-lopping generation and as trough-filling additional overnight demand are less amenable to accurate prediction; such benefits depend very much on the system marginal cost differential between night and daytime operation.

49. In addition to the economic benefits of removing peaks from the daily load shape there is also a saving in repair and maintenance costs on steam plant relieved from onerous starting, stopping and load following duty.

DINORWIG OPERATING EXPERIENCE

Plant Performance

50. Generally the plant performance so far has come up well to design requirements.

51. Throughout the year there is a definite weekly cycle of pumped storage utilisation, the upper reservoir being full by Monday morning and empty by Friday night.

52. Up to three Dinorwig machines are allocated to scheduled system reserve duty and the other machines used commercially for "in merit" generation.

53. A study has been undertaken to determine the actual use of pumped storage energy on a day to day basis. A typical winters day pumping/generating regime is illustrated in figure 4 and the results of the study are given below.

Reserve	44%
Programmed in Merit	25%
Television Demand Pick-Up	21% (see paragraph 64)
Other	10%

Automatic Start Facilities

54. The machines at Dinorwig allocated to scheduled system reserve duty, spinning in air, can be brought to generation either manually (on receipt of a lamp telegraph signal from the National Control loading engineer), or automatically by low frequency initiated relays. Each machine can be selected to low frequency (lf) relay in the range 49.4-49.9 Hz to change the mode of operation from spin gen to generation or from pump to spin pump.

55. The following facilities are being utilised at present:

(i) At 49.8 Hz lf relay operation will bring the lead machine at Dinorwig from spinning to 160 MW generation in 10 secs.

(ii) If this machine is already generating the lf relay at 49.8 Hz will cause it to pick up another increment of 80 MW if possible.

(iii) At 49.75 Hz further lf relay operation will bring the second machine from spinning, to 160 MW generation in 10 seconds.

Additionally, there will be governor action (see 63) to vary the above outputs with frequency.

56. These lf relay settings are supplemented by a similar installation at Ffestiniog pumped storage station with 2 x 90 MW machines, spinning in air, as a pair on a lf relay setting of 49.80 Hz and the other two 90 MW machines are set to come in at 49.75 Hz.

57. After an automatic start (as per 55 or 56) the machines are restored manually to the spinning mode when system conditions permit.

58. These automatic pumped storage actions are backed up by two tranches of gas turbines which are fitted with automatic lf relay initiated start up sequences effective to full load within 5 minutes. These consist of some 900 MW set at 49.7 Hz and 1400 MW at 49.6 Hz spread geographically throughout the system.

Cross Channel Link

59. The 2000 MW Direct Current link between France and Britain was fully commissioned in January 1987. The link comprises two bipoles each rated at 1000 MW, thus the largest single infeed risk on the system has been raised from a 660 MW generator to a 1000 MW bipole, this risk being well within the "1320 MW in 10 seconds" capability of Dinorwig.

60. However, the Anglo-French DC link adds some diversity to the geographical location of system reserves. Instead of the connection of all scheduled system reserve in North Wales, an area likely to suffer during extreme weather conditions, the cross channel link when not fully committed to commercial transfers of 2000 MW may also serve as back-up system reserve. There are technical provisions for a very rapid rate of change of link transfer if an emergency push button is operated in either the relevant National Grid or EdF control centres to obtain assistance up to a pre-arranged emergency limit. Such assistance is of course available for use in either direction of flow to support a partner in need.

61. The increase in largest single infeed risk from a 660 MW generating unit to a 1000 MW bipole led to a review of system reserve requirements in 1987. It was decided that the 360 MW level of reserve held during the GMT darkness peak period should be increased to 1000 MW. An associated factor in this decision was the falling price of oil, which led to a more compact merit order and a reduction in the economic benefit of in-merit generation at Dinorwig.

System Frequency Control Performance

62. As stated in paragraph 36 it is not intended to use Dinorwig for dedicated system frequency control. However, the facility to achieve an effective output in some 10 seconds from the spinning mode has improved the general frequency performance on the British system. This is further aided by a governor droop setting selected to 1% on the machines operating as system reserve. Hence they will respond, when actually generating, by 60 MW up or down for a fall or rise in frequency of 0.1 Hz.

63. The quick response obtainable from Dinorwig has made a significant contribution towards the containment of frequency following disturbances on the system. This has contributed to a reduction in the need for expensive gas turbine operation.

64. Dinorwig has proved to be particularly valuable in dealing with rapid demand pick-ups following the ending of popular television programmes. This was demonstrated on Sunday 22 January 1984 when a sudden demand rise of 2600 MW occurred and some 1000 MW of generation was contributed by pumped storage (see figure 4).

65. Overall, there has been a marked improvement in System Frequency Control Performance. The weekly standard deviation of five minute spot frequencies has improved from about 0.085 to 0.075 Hz when comparing 1982 with 1983, and has been sustained since.

IMPACT OF PRIVATISATION

66. Under the Government proposals for privatisation of the British Electricity Supply Industry, the CEGB is to be split into three successor companies. The generation assets of the CEGB are to be shared between the National Power and PowerGen Companies, whilst the National Grid Company will inherit the transmission assets. The special role of pumped storage plant in meeting system requirements has been recognised, and both Dinorwig and Ffestiniog are to be owned and operated by the National Grid Company.

67. At the present time, the National Grid Company are considering the establishment of a separate business for pumped storage plant within the company. Pumped storage plant will have to compete in an environment in which sales of energy are undertaken on a contract basis, with a market for both electricity supply and reserve provision.

68. The process of identifying the role of Pumped Storage plant post privatisation is ongoing, and it is not possible to provide further details in this paper. The situation is expected to be developed throughout 1989 and further information should be available for submission to the Conference in April 1990.

Lay-out criteria, in situ tests and operational experience of the Kühtai pump-turbine groups

H. SCHMID and R. ERLACHER, Tiroler Wasserkraftwerke AG, Innsbruck, Austria

INTRODUCTION

The 780 MW Sellrain-Silz power scheme, as shown in figures 1 and 2 is located approximately 40 km west of the city of Innsbruck, between Stubai and Ötztal valley and has a catchment area of 140 km². The lay-out and the

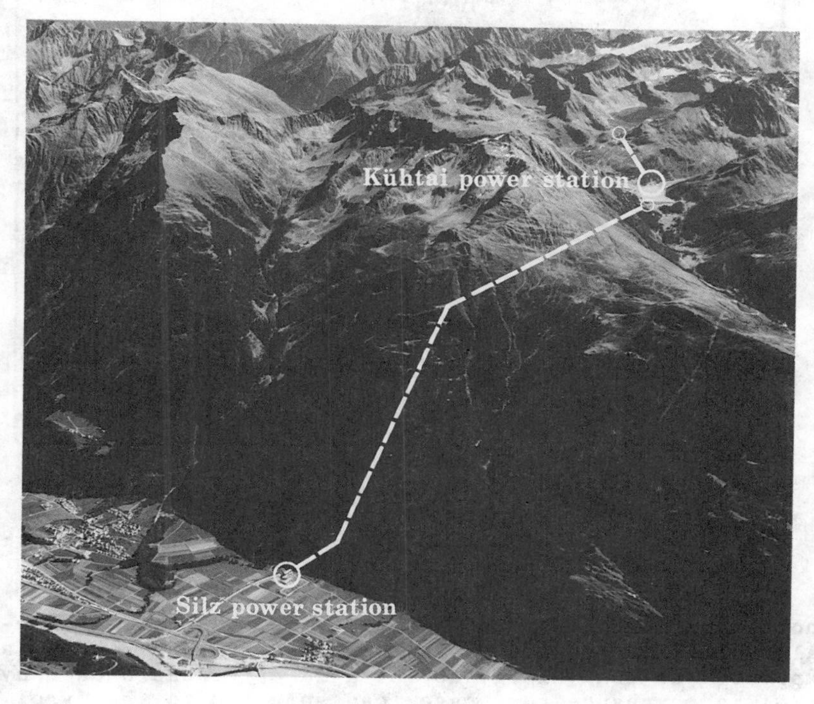

Fig. 1 Aerial photograph of the SELLRAIN-SILZ power scheme

overall design of the whole scheme was done by the builder TIROLER WASSERKRAFTWERKE AG. The construction work started in 1977 and since 1981 the scheme is under full operation.

The main purpose of the upper station Kühtai, which is equipped with two reversible units (pump-turbines), is the annual filling of 88 % of the 60 Mio m³ Finstertal reservoir in addition to the inflow from its small natural catchment area. Water is provided by an adduction system to the intermediate Längental reservoir and is pumped to the Finstertal reservoir with a seasonal variable head between 319 m and 440 m.

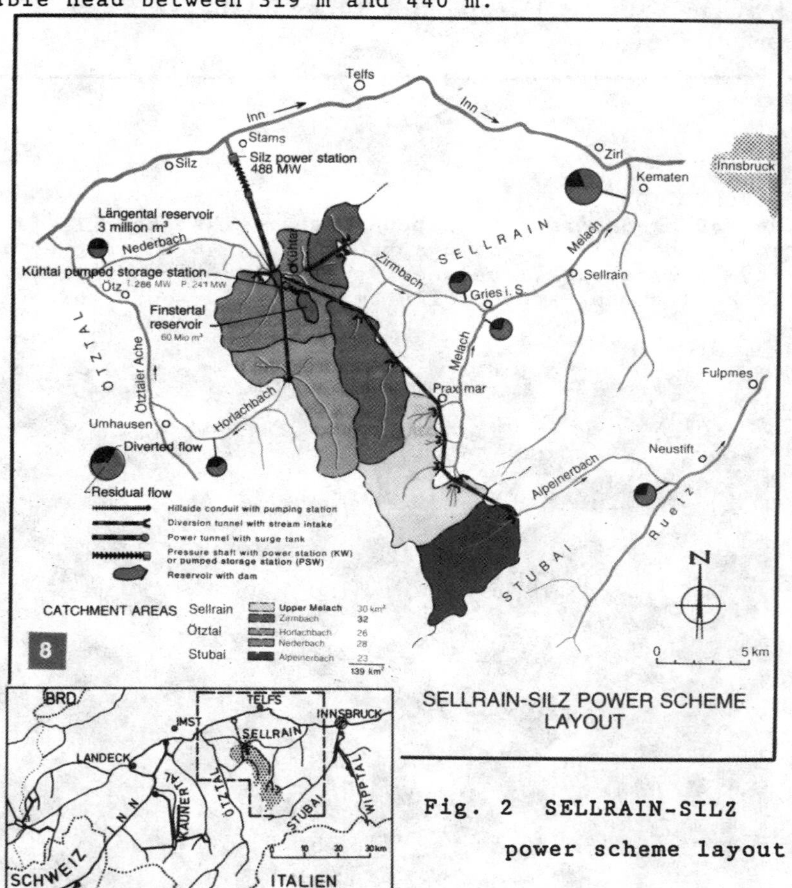

Fig. 2 SELLRAIN-SILZ power scheme layout

Short-term pumped storage is superimposed to the seasonal cycle. The entire generated power of this scheme is exported as peak energy to the grid of southern Germany (FRG) in exchange for base load supplied to the Tyrol. Special topographic conditions and a minimum backpressure of 48 m, required in the pumping mode, led to the design of a 82 m deep shaft power station as shown in figure 3, with a diameter of 30 m, excavated and lined after a new construction method.

Each of the two reversible pump-turbine groups consists of a one-stage Francis pump-turbine, an air-cooled motor-generator and a starter motor.

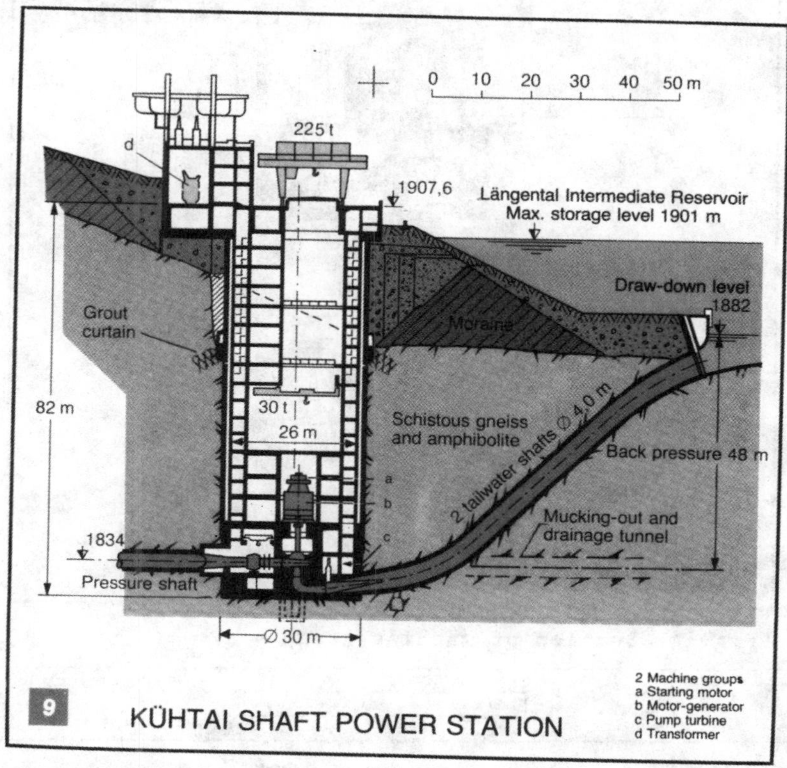

Fig. 3 KÜHTAI shaft power station

The arrangement for the vertical axis shafting system shows in figure 4 four guide bearings, a bearing for the pump-turbine, close to the runner, a lower and an upper generator-motor bearing and above the starter motor one additional guide bearing. The upper generator-motor bearing is a combined guide-thrust bearing construction. All four guide bearings and the thrust bearing are tilting pad bearings. The machine groups are designed for the specifications, given in figure 5; in generating mode both machines together have an output of 289 MW and in pumping mode they operate with 250 MW. The machines can change from pumping mode to max. turbine output in less then 90 seconds. In figure 5 are further informations given concerning discharge, dam constructions, capacities of the dams and size of the dams.

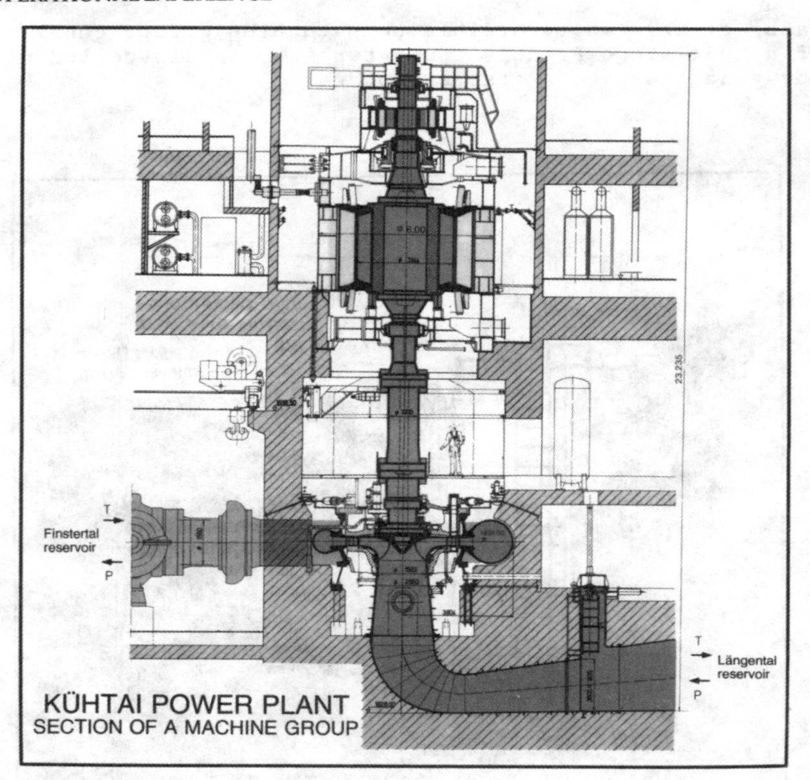

Fig. 4 Section of machine group

KÜHTAI POWER STATION

Shaft-type building:

excavation diameter	30 m
depth	82 m
diameter of inner lining	26 m

Equipment

two vertical axis pump-turbine motor/generator groups with starter motor

output of a generator/transformer group	176 MVA
speed of turbine/generator group	600 rpm

Run-up times:

from still-stand to max. turbine output	75 s
from still-stand to max. pump output	175 s
from max. pump output to max. turbine output	90 s

Specifications:

Conditions	Discharge	Power
1 group:		
- generating mode	max. 40.4 m^3/s	max. 151 MW
- pumping mode	max. 34.5 m^3/s	max. 125 MW
2 groups:		
- generating mode	max. 80.0 m^3/s	max. 289 MW
- pumping mode	max. 67.6 m^3/s	max. 250 MW

FINSTERTAL RESERVOIR

Capacity	60 million m^3
Rockfill dam with asphaltic concrete core Hmax. 150 m, total volume	4.46 million m^3

LÄNGENTAL INTERMEDIATE RESERVOIR

Capacity	3 million m^3
Earthfill dam with asphaltic concrete core Hmax. 45 m, total volume	0.4 million m^3

Fig. 5 Plant layout, specifications

LAY-OUT CRITERIA, TECHNICAL SPECIFICATIONS

Pump storage machines, especially reversible units have shown in the past, that they produce significant hydraulic induced random forces on the runner due to pressure pulsations between runner and spiral casing. This particular item was studied in detail, together with the producer of the hydraulic part of the units (VOITH, FRG). During scaled model tests beside the evaluation of all other characteristic data, the measurements of the hydraulic induced forces were important key points. Based on informations from this scaled model tests, the design and the dimensioning of all main construction parts of the foundation could be started. It was decided to manufacture the whole hydraulic machine except the shaft, in stainless-steel of type GX5CrNi 13.4, to make an elastically embedded spiral casing with high stiffness perpendicular to the shaft axis. The spiral casing is fixed on the foundation with high tensile steel anchors through an upper and lower base ring construction (see axonometric drawing of the spiral casing in figure 6).

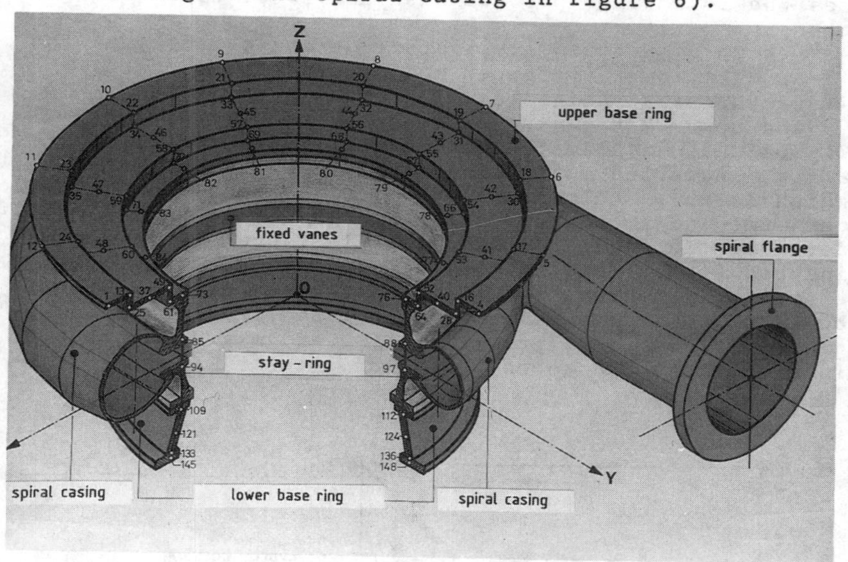

Fig. 6 Axonometric drawing of the spiral casing

The shaft diameter and the guide bearing was designed to fulfill the requirements of maximum allowable deflections of the runner under worst loading conditions (i.e. emergency shut-down from full load) this means that the clearance in the lower and upper sealing of the runner is wide enough, so that no grazing between runner and casing can occure. To limit the forces due to the product of cross section times net head pressure, acting on the spiral casing, between globe valve and spiral casing a pressure balanced stuffing box was designed.

The lay-out and the dimensioning of the sliding leaf gate, located at the end of the draft tube, was made such that it can resist the high pressure pulsations in the draft tube. In case of an emergency shut-down of the machine the closing time of the wicket gates and the globe valve cause a rise in the speed from 600 rpm up to 935 rpm; our requirement for the design of the rotating shafting system was to get a critical speed for the bending mode which is above 935 rpm with a safety margin not less than 10 %. The first critical speed for this particular design of the shafting system is mainly influenced by the different components of the motor-generator, its connections to the building and the building itself. All the members in the chain of stiffnesses had to be designed, so that they fulfill the requirements on the critical speed. Thermal effects, i.e. temperature distributions on the supporting structure during full operation conditions result in high radial forces, which act on the surrounding building. To limit this high forces the following precautions had to be taken into account:

- set up of a fixed temperature during assembly of the spider (lower and upper motor-generator support)
- set up of a radial tensile force in the lower and upper motor-generator support during assembly

To balance and limit such thermal induced radial forces a fix installed monitoring system was set up by TIWAG, which is able to measure the actual forces in the spider-arms and to control the heating elements on the spider-arms. In figure 7 the allowable range of the thermal induced forces and the switch points for the heating elements are shown. For verification purposes and to build up some knowledge about installed new technologies, different analytical studies and a great number of in situ tests had to be performed.

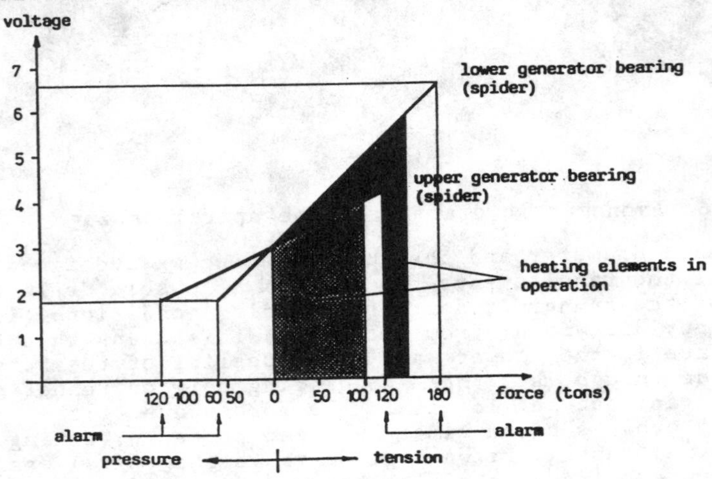

Fig. 7 Amplifier output vs. thermal induced force

ANALYTICAL STUDIES

A typical example of such analytical studies will be shown for the FEM-calculations of the shafting system, which were carried out to evaluate the critical speeds for the bending mode and the deflections, forces and stresses under certain loading conditions. The mathematical model, consisting of massless beams and lumped masses in the nodes, supported by massless springs and dampers (springs are assumed to be stepwise linear, dampers are velocity dependent and of type Rayleigh damping model) is shown in figure 8. The main steps of such an analysis for a multi degree of freedom system are given.

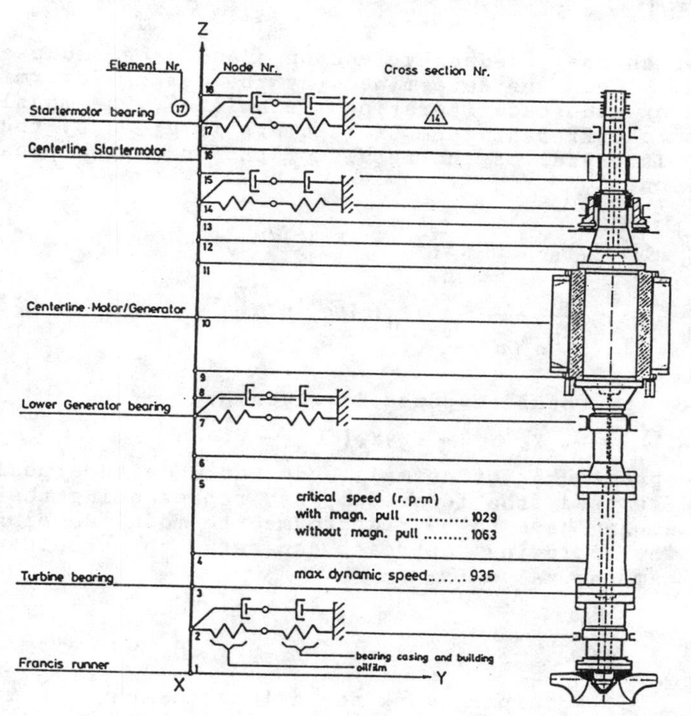

Fig. 8 Mathematical model of the shaft (FEM)

Multi degree of freedom system (MDOF)

Equations of motion

$$M\ddot{v} + D\dot{v} + Cv = f(t) \tag{1}$$

where:
- M ... diagonal mass matrix
- v ... displacement vector
- D ... damping matrix
- C ... stiffness matrix
- f ... force vector

Transformation to normal coordinates leads to N uncoupled equations

$$v(x,t) = \phi(x) \cdot Y(t) \tag{2}$$

ϕ describes the eigenvector in spatial coordinates
Y describes the generalized coordinates (time dependent)

Substition of (2) into equ. (1), and solved for the homogeneous part gives the well known eigenvalue equation.

$$\left[C - \omega^2 M\right] v = 0 \tag{3}$$

From which the frequency vector ω and the mode-shape matrix ϕ can be determined (by the use of determinant search or subspace iteration technique). The modal response to different type of loadings is given by the Duhamel Intergral as the result of the uncoupled equation of motion

$$\ddot{Y}_n + 2\zeta_n\omega_n\dot{Y}_n + \omega_n^2 Y = F_n(t)/M_n \tag{4}$$

and

$$Y_n(t) = \frac{1}{M_n\omega_{Dn}} \int_0^t F_n(t)\, e^{-\zeta_n\omega_n(t-\tau)} \sin \omega_{Dn}(t-\tau)\, d\tau \tag{5}$$

so that the total response is given by

$$v(x,t) = \phi_1 Y_1(t) + \phi_2 Y_2(t) + \ldots.. \tag{6}$$

If the procedure of normal mode or mode superposition method is used, the local dampers, representing the oil-film damping have to be transformed to modal damping values. The following approach can serve to calculate the modal damping value

$$\zeta_n = \frac{W_n}{4\pi L_n} \tag{7}$$

with W_n ... damping work for z local dampers in mode n

$$W_n = \sum_1^z \pi\omega_n d_i \phi_i^2 \tag{8}$$

where d_i ... damping resistance at node i
ϕ_i ... eigenvector at node i in mode n

The max. kinetic energy for the n th-mode of the whole structure (with N nodal points) can be written in the form

$$L_n = \frac{1}{2}\sum_1^N m_i\phi_i^2\omega_n^2 \tag{9}$$

where m ... mass at node i
N ... number of nodal points with mass

The final result for ζ_n is

$$\zeta_n = \frac{\sum_1^Z \omega_n d_i \phi_i^2}{2 \sum_1^N m_i \ (\phi_i \omega_n)^2} \qquad (10)$$

The computer calculations showed, that the lowest critical speed for this linear elastic mathematical model, depending on the assumptions for the oilfilm stiffness are:
- without magnetic pull (on the generator) .... 1069 rpm
- with magnetic pull .... 1029 rpm

The following flexibilities of the supports and the rotating mass led to this results:

- flexibilities
  - turbine bearing : 0,7 μm/kN
  - lower generator bearing : 0,7 μm/kN
  - upper generator bearing : 0,7 μm/kN
  - starter motor bearing : 1,3 μm/kN
- masses
  - total rotating mass : 280.000 kg

For the worst loading condition (combination of unbalance on the rotor and on the runner, stochastic force on the runner and short circuit of adjacent poles) the calculated radial forces on the supports are:
- turbine bearing .... 2500 kN
- lower generator bearing .... 4000 kN
- upper generator bearing .... 3000 kN
- starter motor bearing ...... 330 kN

The bearings, the surrounding structure and the building is designed to withstand also such faulting conditions, so that the integrity of the system is assured.

IN SITU TESTS

Some of the assumptions for analytical studies could be verified through dynamic testing on the full scale structure (machine and building). The main concern was viewed on the stiffness of the building and the supporting structure in such regions where in case of a fault the highest forces occure. The whole structure (building and machine) was excited with an electro magnetic or electro mechanical shaker (see figure 9) within a frequency range of zero and 35 Hz to find the most significant dynamic characteristics (stiffnesses vs. frequency, eigenvalues, mode shapes and damping).

Fig. 9 Electro mechanical Shaker

Figure 10 shows the measured radial displacement vs. frequency for two different locations (fifth and sixth floor). For the inner lining of the shaft-type building the resonance peak at 5,9 Hz is representing the first bending mode ("beam mode") of the whole shaft-building, the frequency 14,95 Hz (dominant at the sixth floor) is representing the first radial in plane mode of a circular thin walled ring or cylinder. With a similar measurement equipment (Shaker, FFT etc.) the bending modes of the nonrotating shafting system (with fixed guide bearing shoes) were evaluated. This allowed us to find the critical speed, the bending mode shapes and the damping characteristics for the first three critical speeds. The results of such dynamic tests showed us that the overall design, the constructions and the assumptions for different calculations agreed within acceptable margins. After some time of full operation a few minor faults on different parts of the machines led to some sophisticated tests on the machines. One of them will be shown as an example.

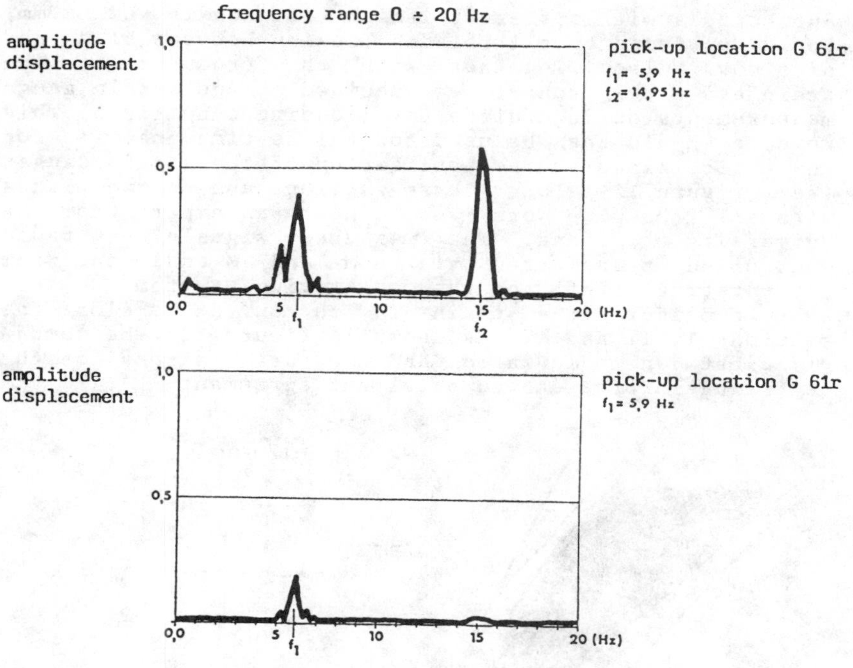

Fig. 10 Transfer functions of the shaft-type building

During a periodically performed inspection on a runner in 1983, cracks as shown in figure 11, were found.

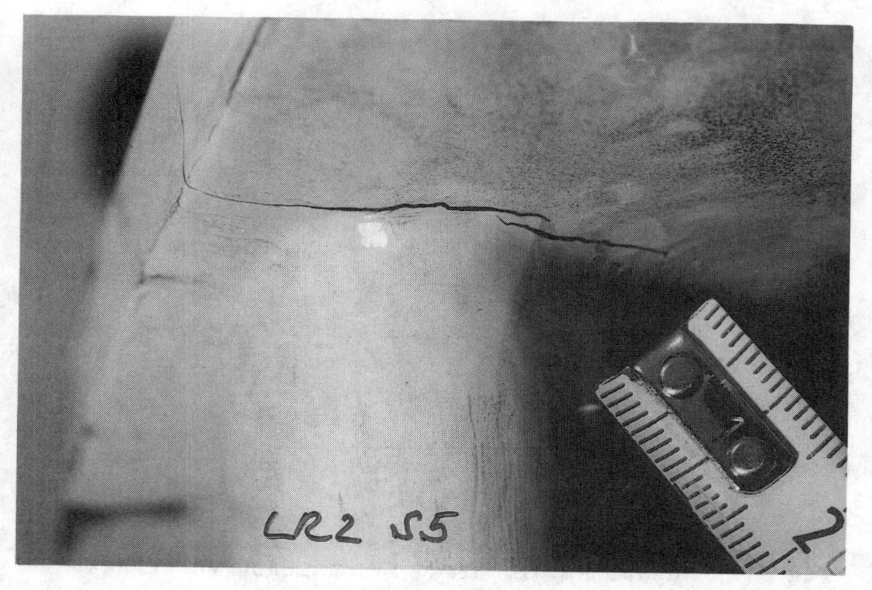

Fig. 11 Runner Nr. 2, blade Nr. 5 crack lengh 58 mm

Such cracks with different length (between 15 and 58 mm) could be found on all blades (region between blade and side cover-disc). Together with the producer of the stainless-steel runner we decided to do strain gauge measurements during different loading conditions. This results should then be used for a life-time analysis for the entire runner. The application of the strain gauges (see figure 12) along three sections and on two blades with all the wire work was done by an expert from the University of Vienna. The measured signals were radio controlled transferred via PCM- and FM-equipment from the rotating shaft to the monitoring system. We were able to measure the strain in the curvature along the sections I, II and III as shown in figure 12. The comparison between calculated and measured stress for the centrifugal force showed excellent agreement.

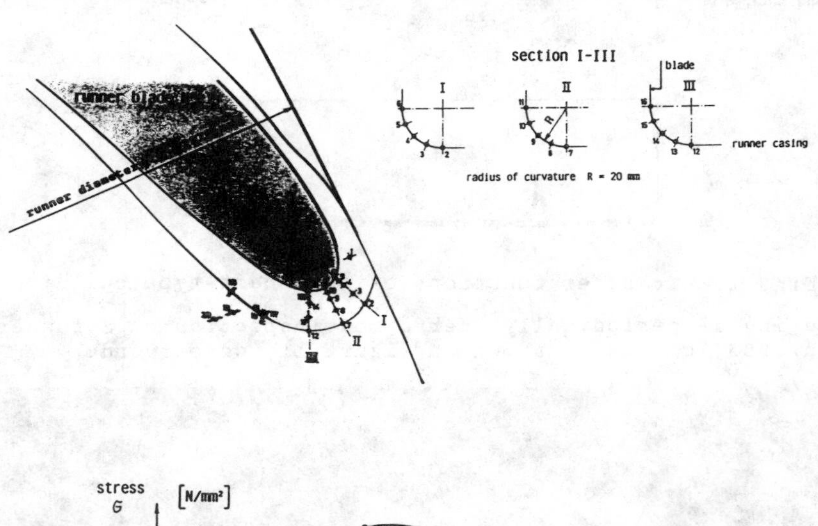

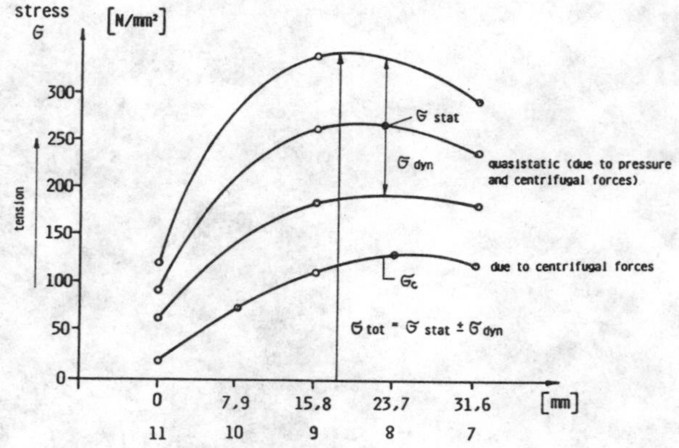

Fig. 12 Runner Nr. 1, blade Nr. 1 measured stress distribution along contour II

The measurements and the additional performed life-time calculations turned out the dominant influences of the radius of curvature between blade and side cover and also the flexibility of the side cover. It was found, that the radius of the fillet was originally between 6 and 14 mm in this particular regions. The calculations showed, that a radius of 20 mm is necessary; a repair of the cracks and a correction of the radius is meanwhile carried out on all three runners. The correctness of this decision could be verified during a six year period of full operation and several inspections.

VIBRATION MONITORING SYSTEM

Both machine groups are installed with a vibration monitoring system. This equipments are able to detect any unusual vibration behaviour of the rotating shaft. The vibration pick-ups are mounted on all four bearings. There are two different types of pick-ups used:

- measurement of the relative movement of the shaft to the casing with two proximity pick-ups (S_{max} - displacement in µm)
- measurement of the absolute case vibration (V_{eff} velocity in mm/s)

The monitoring system for both machines, the racks and the positions of the pick-ups are shown in figure 13.

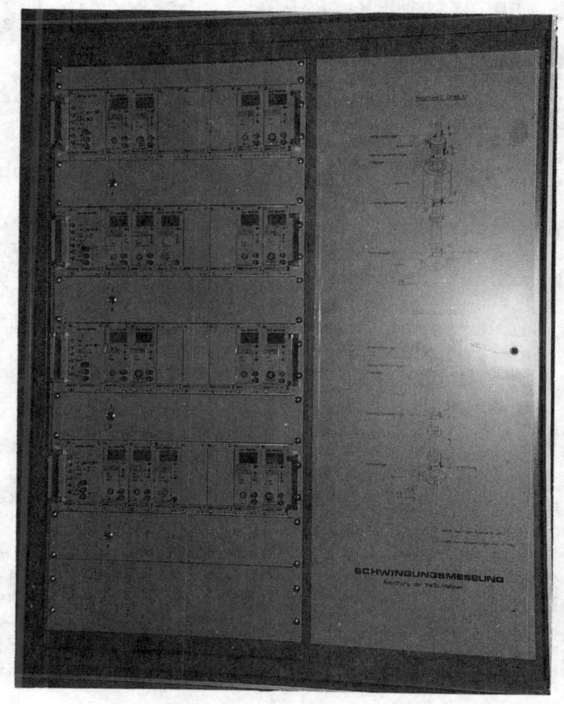

Fig. 13 Monitoring system

Continuous registration of the measured signals in a micro-computer, alarm levels and relais of the monitoring system are completely incorporated into the machine control system.

OPERATIONAL EXPERIENCES

It must be said, that both units have an excellent availability; for instance 99,4 % within the last year. Since the start of operation in 1981 up to August 1989 the power plant had totally 26.870 hours of generating and pumping mode and 15.982 starts (turbine or pumping mode).

Two characteristic calculated average values for each unit per day are:

5,5 hours of turbine or pumping mode and
3,2 starts (turbine or pumping mode).

Between 1981 and 1988 we had a few major faults and a couple of minor defects. Some of the disturbances are typically for systems with high vibration levels and will be characterised with a few words:

- cracks in a bypass piping system (elimination by better design, better weldings and stress free assembly);
- cracks in electronic prints of the angel transmitter of the wicket gates (elimination by elastically embedding the prints within the casings);
- cracks on two runners (elimination by increase of the fillet radius from 6 mm up to 20 mm);
- high vibration levels on the stair cases, which are typical pre-fabricated concrete production parts (elimination by passive isolation of the different segments);
- guillotine cracks on the spindles of the popped valves in the bypass system of the globe valve (elimination by better design and high-tensile material);
- damages on piping systems and faults on supports due to high vibrations (elimination by new designed supports);
- cracks on generator pole connections (elimination by better design and proper assembly);
- damage of a graphit sealing on one pump-turbine through a metal part, which was probably left during assembly;
- damage of two seismic pick-ups of the vibration monitoring system on the turbine bearing due to high perpendicular accelerations (elimination by installation of piezo-electric acceleration pick-ups)

It can be summarized, that all the extensive scaled model and full scale tests, the additional investigations and calculations during the design stage, the assembly controls, the very precise and complete start-up program and a highly scilled technical operating staff give satisfactory plant conditions and an excellent availibility.

Sixteen years operating and maintenance experience of the 1080 MW Northfield Mountain pumped storage plant

A. FERREIRA, Northeast Utilities, USA

INTRODUCTION

The 1080 MW Northfield Mountain Pumped Storage Plant, owned and operated by Northeast Utilities (NU) began commercial operation with all four units in late 1973. The first unit, Unit No. 4, began operation on November 30, 1972, and the other three units were completed sequentially the following year; Unit No. 2 in February, 1973; Unit No. 3 in July, 1973; and Unit No. 1 in October, 1973. The plant has continued in successful commercial operation since that time furnishing peaking capacity, economic energy generation, fast reserve and load following service to the New England Power Pool (NEPOOL) and the other interconnected power pools. Each of the four units produced 250 MW for a station total of 1000 MW until 1988, at which time the units were uprated by 20 MW each for a new plant total of 1080 MW.

This paper presents the significant milestones of the Northfield Mountain Plant's operation and maintenance experience over the period 1973 to 1989. The plant's economic and reliability contributions to the interconnected electrical system are described. In providing the dynamic capacity, energy, and reliability benefits to the system, the plant has experienced unique wear-and-tear on specific equipment and structures.

Routine and major maintenance on these plant components are described with comments and suggestions regarding the original design concepts and philosophies and in consideration of the System Dispatcher's increasing reliance on the fast-reacting flexible operating characteristics of these hydro-electric machines.

PLANT DESCRIPTION

The plant is owned and operated by Northeast Utilities (NU), and investor-owned utility. NU is the largest utility participant (annual 1989 load of 4825 MW) in the New England Power Pool (NEPOOL) which experienced an annual peak of 19,722 MW in 1989. NEPOOL is a voluntary association of 92 individual public and investor-owned utilities (almost all the electric utilities) in New England. Its principal function is to plan, coordinate, monitor, and direct the operations of the region's major generation and transmission facilities. The electric facilities of all the member companies are operated as though they comprise a single electric system, in this way obtaining the maximum benefits of power pooling. The NEPOOL organization consists basically of three groups: the planning arm, New England Power Planning (NEPLAN); the operating and dispatching arm; New England Power Exchange (NEPEX); and the billing arm, New England Power Billing (NEBILL).

NEPEX has the responsibility for the central dispatch of power throughout New England. The NEPEX System Dispatchers, using computer facilities, monitor and direct the operations of all major generators and transmission lines 24 hours a day, every day. The Dispatchers make minute-by-minute decisions in providing electricity to customers at lowest possible costs and meeting prescribed reliability standards.

The Northeast Utility Service Company (NUSCO) performed the overall economics, evaluations, and planning, and, during construction, provided the engineering and construction management liaison. The detailed engineering was commissioned to Stone & Webster Engineering Corporation of Boston, Massachusetts, who were also given the on-site construction management responsibility. The contractor firm of Morrison-Knudsen Northfield Associates constructed the civil works, including the tunnels, underground powerhouse, the upper reservoir dam and dikes. The four Francis-type reversible pump-turbines and spherical valves were furnished by Baldwin-Lima-Hamilton. The generator motors, starting motors, transformers, unit breakers, and control equipment were furnished by General Electric Company of Schenectady, New York.

The units, initially rated at 250 MW generating are now rated at 270 MW.

PLANT DATA

Turbine type: Reversible Francis, single stage, vertical shaft (4 Units)
Turbine output: Initial 250 MW, Present upgrade 270 MW; Total Plant Output 1080 MW.
Unit Speed: 257 rpm (pumping and turbining)
Pump input: 250 MW

Generator-motor: Vertical shaft, 28 pole, 13.8 KV, wound rotor type

Starting motors: Wound rotor, 26 pole, 13,000 HP, liquid rheostat
Transformers: 500,000 KVA, 65oC, forced oil, water cooled, 345 KV/13.8 KV (2 transformers - one for each of two plant units, located underground in powerhouse vault)
Connection to Switchyard: Two pipe-type cables, 345 KV, oil filled.
Maximum gross head = 828 ft. (252.4 m)
Minimum gross head = 775 ft. (236.3 m)
Upper reservoir capacity = 10,500 MWH.

LOCATION

The plant is located in the state of Massachusetts, Franklin County, United States of America, within the towns of Northfield and Erving (refer to plate 1). It is approximately 100 miles (161 km) west of Boston and 200 miles (232 km) northeasterly of New York City. The 300 surface acre upper reservoir is on top of Northfield Mountain about 1 1/4 miles (2 km) east of the Connecticut River. A dam across the river forms a twenty mile (32 km) long pond which serves as the lower reservoir. The plant's underground powerhouse is connected to the river (lower reservoir) by a single mile-long (1.6 km) concrete-lined tailrace tunnel. A single high pressure concrete-lined inclined shaft, separating into four steel-lined penstocks at its lower end, connects the upper reservoir to the powerhouse. Water levels in the upper reservoir normally fluctuate between twenty to thirty feet (6 to 9 m) per day with about a 44 foot (12.1 m) maximum on a weekly basis. The lower reservoir normally fluctuates about three feet (1 m.) on a daily basis and up to 5 feet (1.5 m.) on a weekly basis.

DESIGN PARAMETERS

Total plant output - 1080 MW (270 MW per unit)
Total plant pumping input - 1000 MW (250 MW per unit)
Normal load pickup rate - 50 MW/minute.
Emergency load pickup rate - 1000 MW/minute
"Cold start" to maximum capacity, each unit - 3 minutes maximum (units actually tested out @ 1 1/2 to 2 minutes)
Full Automatic Operation - Yes
Start/Stop (normal) - Local by Plant Operator upon call from System Dispatcher.
Load/Unload - Remote, by NEPEX System Dispatcher, via Automatic Generation Control (AGC) equipment.
Pumping Shutdown - normal by Plant Operator; automatic sequentially by upper reservoir full elevation controls.

OPERATIONS AND DISPATCHING

The operation of the plant units is completely automated. Following instructions on load demand via telephone by the power exchange System Dispatchers, the Plant Operators, by push button control, bring the required numbers of units on line. The units are synchronized to the system automatically and each unit is brought up to the minimum load of 125 MW. The plant load controller jointly loads the on-line units to the desired MW output at which point they are turned over to the System Dispatchers and placed on AGC operating status. Loading and unloading of the units is performed by the AGC as needed for system frequency control and as desired and called for by the System Dispatchers to meet system economic energy and power transfer within the NEPOOL system and in the interconnected systems.

Unit controls are also available in-situ at each machine as well as in the plant control room. These at-unit controls are used only during special maintenance work on the units. Normally the operating functions are performed by the Operators in the plant control room. Unit outputs and some monitored unit operating characteristiscs are inputted into the station computer. Hourly MW readings of unit generation and pumping, reservoir levels, river flows, and special operation are manually logged.

In addition to the 1080 MW of the Northfield Mountain Plant, there are two other pumped storage plants on the NEPOOL system. The first pumped storage plant built in the United States, the 31 MW Rocky River Plant (another NU plant) in 1932, and the 600 MW Bear Swamp Plant of New England Power Company, which came on-line in 1974, are also dispatched by the power pool. The units of these two other pumped storage plants are dispatched on a block loaded basis, i.e., the units are not used to follow load. The units of the Rocky River Plant are relatively small (24 MW and 7 MW) - as such, they do not load follow. The two 300 MW units of the Bear Swamp Plant are not used for load following because of special hydraulic surge concerns in the plant's penstock system.

The Northfield Plant's pumping and generating schedules are developed by the NEPEX Forecaster. In forecasting the operation, NEPEX predispatches the plant for the maximum benefit to NEPEX from both an economic and a reliability standpoint. The prepared forecast is for the period midnight to midnight of the forthcoming day. In addition to the pumping and generating MWH, the desired pond elevations for the appropriate hours are designated. The operating schedules and elevations are issued to System Dispatchers during the 1600 to 2400 shift. The process actually begins at the end of each week when the Forecaster develops a weekly Northfield Plant pumping and generating schedule. The schedule runs from 7:00 a.m. Monday to the following Monday. This rough schedule is refined on a day-to-day basis by the Forecaster as actual system operations bear out or conflict with the original assumptions. The fundamental assumptions are the anticipated pumping and generating incremental and decremental dollar rates that will exist. In arriving at these, the current rates of the thermal plants on the system are considered in the light of expected scheduled overhauls, forced outages, expected weekend maintenance and estimated hourly loads. The Forecaster has a continuing record of current pumping and generating costs which are provided by the Pool Coordinator. These are obtained using an on-line interchange negotiation program to determine the dollar value of the blocks of proposed Northfield generation during the daytime heavy load hours and the expected pumping costs in the early morning off-peak hours.

In refining the day-to-day dispatching schedule for the plant, the Forecaster and the Dispatchers must abide by the physical plant constraints of minimum and maximum elevations of the upper and lower reservoirs and in tracking the balance of combined water availability (in terms of energy production) in these reservoirs. The lower reservoir, as noted, is part of the Connecticut River wherein a stringent operating concern is to prevent the river flow downstream of the lower reservoir dam (the Turners Falls Dam) from increasing beyond limits set by the U.S. Army Corps of Engineers, the organization responsible for flood control operation of the river and basin. Additional lower reservoir constraints include operation of a 59 MW conventional hydro plant system and critical elevations relative to recreational use of the pond. With the plant joint load controller accepting the control pulses from the digital control system, the plant on-line units are treated as a separate unit. In this way, the different categories of system reserve carried by the plant can be tracked. This refers to the 5-minute spinning reserve category for which the unloaded portion of the generating units qualify and for which the units, also qualify because they can drop load instantly while in the pumping mode.

In addition to the system plant normal response rate of load pickup of 50 MW/minute, a faster response rate is also available for meeting minor hour-to-hour system problems. This is a 200 MW/minute response to control signals from the System Dispatcher.

Finally, the dispatching procedure also provides the NEPOOL system with a 500 MW quick pickup, including the start-up of off-line units. These latter operating functions are utilized for large system disturbances such as loss of large thermal or nuclear units or loss of imported power.

PLANT EFFICIENCY

Real time pump-turbine unit efficiency based on measurement of flows has only recently been initiated at the plant. An acoustic flow measurement system (AFMS) was installed on Unit No. 3 in the fall of 1989. A corroboration of the accuracy of the AFMS was performed using a dye dilution flow measurement coincident with readings taken on the AFMS. Having established the necessary confidence level, acoustic flow meters are planned to be installed on the other three plant units starting this year. Real-time pump-turbine unit efficiencies will be calculated for each of the machines over the entire operating output range from the minimum load output of 125 MW to the maximum output of 270 MW.

The best efficiency point (BEP) of the operation will be established for each unit. More importantly, the overall efficiency curve for each unit will be developed. The affirmed efficiency characteristics will allow the plant owners, as well as the NEPEX dispatcher, to determine the reduction in the overall plant efficiency resulting from operating the units at outputs other than BEP.

The industry has not as yet developed an established and confident methodology for quantifying the benefits to the system by having the fast reacting hydro units perform load following functions. However, it is a clear and accepted concept that substantial benefits do, in fact, accrue from this procedure. Accordingly, the System Dispatchers do continuously call for unit operation at outputs that depart from the BEP. The presumption is that the resultant reduced plant efficiencies are more than offset by the value to the system derived by the load-following and frequency service of the plant units. The system value for having hydro units perform load following has its basis in the improved operating efficiencies of those system thermal units which have been replaced in this duty. By not having to depart from their own more efficient heat rates to follow load (either by increasing or decreasing unit output), the system thermal units can experience lower operating costs with attendant system production cost savings.

Unit efficiencies are a first-line method of indication of the need for pump-turbine unit maintenance. Unit efficiency is influenced by a host of factors directly impacted by mechanical and hydraulic wear and tear. The major items in this aspect, with respect to the pump-turbine, are the increased clearances resulting from seal wear, misalignment and wear of the wicket gates, and increased roughness and distorted shape of turbine blades due to cavitation erosion and roughened blade conditions resulting from continuous cavitation welding repairs. From a civil structures standpoint, reduced efficiencies can result from increased hydraulic roughness of the water passage conduits which may indicate concrete and/or rock discontinuities from serious spalling and rock falls, trash rack and supporting steel distortions and failures.

A common way of keeping check on the overall plant efficiency at pumped storage plants has been the calculation of the pump/generate (P/G) ratio. This has been continuously calculated essentially since the plant has gone into commercial operation in 1973. The calculation has been performed on a monthly basis, using the total pumping energy input logged on an hourly basis for the month divided by the total energy generated for the month. The calculation equates the water used for generation to that used for pumping by restoration of the upper reservoir water level to the same starting and ending elevation in the period.

Tables 1A, 1B, and 1C depict the history of the P/G ratios over the plant years of operation. The degradation of the P/G ratios from the early years when it was running about 1.35 to values in the 1.39 to 1.40 range was a strong factor influencing the decision to initiate the major unit overhaul program in 1984.

It has been the author's experience that major unit overhauls have not been the industry practice for hydro units until, essentially, the advent of pumped storage maintenance. Unless there was a specific occurrence of machanical or electrical machine or structure failure, it has not been the industry practice to schedule a major unit overhaul wherein the entire unit has been dismantled. Interestingly enough, the author made a special effort in the late '70's and early '80's to canvas several major turbine manufacturers for their recommendation regarding the need and benefits of a major unit overhaul, specifically for pumped storage plants. There was not one manufacturer that indicated the need for such a major maintenance overhaul, i.e., stripping the machinery train down to check wear-and-tear of seals, bearings, loose generator wedges, and the like, as well as to check alignment of critical components. The interest arises from the fact that turbine manufacturers today, clearly in reacting to the experience of the pump-turbine major overhauls performed by several U.S.A. utilities, are now indicating the desirability of this type of major maintenance procedure. The length of recommended time between major overhauls, of course, is still quite flexible and vague, having its basis in size of units, number of units in the plant, effect on system reliability, and, importantly, upon the methodology and credibility of unit efficiency computations.

PLANT MAINTENANCE

It has been the standard and normal practice by Northeast Utilities to schedule an annual maintenance overhaul of each of its major hydro station machines. Maintenance is considered a planned program of inspections of the machinery, structures, and instrumentation, and the necessary correction of deficiencies. The inspections and corrective work are scheduled and performed in accordance with the experience, operating history, and in consideration of the economic payback of the program. The basic purpose of the scheduled annual maintenance is to:

- Reduce number and duration of unplanned outages
- Improve reliability of machinery and support systems
- Increase availability of service of the units
- Reduce system production costs

Until three years ago, the scheduled annual maintenance for the Northfield units was two weeks per unit. The work normally performed during the two week outage interval involves wicket-gate adjustments (reduction of top and side clearances), recalibration and testing of equipment and controls, electrical tests on stators and rotors (Doble, slot discharge, etc.), replacement of worn bearings, cleaning and replacement of air, water, and oil filters, and repair of cavitation pitting damage. The latter repair procedure has been one of the factors at Northfield influencing the length of annual outage. Following the first two years of operating experience, the runner blades in the vicinity of the low pressure turbine discharge corner and low pressure pump entrance edges showed pronounced evidence of cavitation erosion pitting. The initial stainless steel protective segment overlayed on the cast steel was not sufficient in area and depth. Continual stainless welding has been necessary to repair the runners in this area.

This work, in combination with other factors, determined the feasibility of extending the routine maintenance outage to three weeks. In order to maintain the same annual unit availability factor and still provide for the necessary extended outage, the routine maintenance schedule is now based on a three-week outage on an 18-month interval. As an example of other items that can be accomplished in such a three-week outage, a complete upper reservoir drawdown is planned for the three-week unit outage scheduled for this coming October 1990.

MAJOR UNIT OVERHAULS

The practice of performing major overhauls on pumped storage units is not universal, certainly not yet in the U.S.A.

A number of plant managers have, however, scheduled and performed such overhauls and the general feeling has been that the work has proved beneficial. Accordingly most such plants are now planning repeat future overhauls on schedules that are in the 10 to 15 year range. In concert with other major pumped storage plants in the U.S.A., the Northfield Mountain Plant undertook major unit overhauls on each of its four units.

The principal indication of the need to perform some measure of restoration work was the reduction in overall plant efficiency as reflected by the increase in the plant pumping to generate (P/G) ratio. An inspection of the P/G ratios in Tables 1,A, 1,B, and 1,C shows that the P/G ratio began to increase on a steady basis shortly after full plant operation began in 1973. Figure 1, depicts the deterioration in P/G ratio on an annual basis from the low 1.34 to 1.35 levels in 1974 to the 1.40 level by 1982.

A research study was performed in the mid-80's to determine whether the operating usage of the plant units had been undergoing any changes that might be a major factor in the reduced plant efficiency. It was thought that perhaps an increased usage of the plant's units by System Dispatchers for load-following purposes might be resulting in more off-bestefficiency-point operation. The study results have not been overall conclusive; the data did not reflect any dramatic change in usage of the units since the early plant operation. The conclusion generally was that the increase in plant P/G ratios was due to physical wear and tear on the machinery components, most likely in turbine seals.

The decision to perform the major unit overhauls was based on accomplishing the following objectives:

1. Re-establish the early plant lower P/G ratios, i.e., to reduce mechanical and hydraulic losses.

2. Inspect turbomachinery components to check condition of machines.

3. Inspect components to prevent major unscheduled outages.

4. Assure long-term reliable operation.

5. Assess mechanical damage to machinery components resulting from use of the units for flexible operation (dynamic-duty experience).

As Northfield Mountain was one of the first plants to undertake a major unit overhaul, there was little industry experience to guide procedures and schedule. Early estimates of removal of the generator rotor, headcover, and pump-turbine from the pit, repair or replace wicket gates, and perform seal repairs were expected to be in the 35 to 40 week range. Continued plant manager planning, including solicited inputs from manufacturers, consultants, and specialized turbomachinery maintenance organizations, resulted in revised scheduled outages of twenty to twentyfour weeks. The first unit to be worked on, Unit No. 4, was actually overhauled in slightly over 18 weeks, attesting to a well-planned and executed work program. The units were overhauled as follows:

Unit No.	Dates	Length of Outage
4	1984	18 1/2 weeks
3	1985	17 weeks
1	1986	17 weeks
2	1987	17 weeks

The current long-range maintenance scheduling for the plant calls for an 18-week scheduled outage for the No. 4 unit in 1994. Plant managers are, however, considering the additional work of a major generator rewind which is expected to add some 6 to 12 weeks to the schedule. Generator rewind work will generally impact the pump-turbine work as it tends to interfere with turbine pit and crane availability.

The work accomplished in the major overhaul work included the complete disassembly of the unit, repair or replacement of components to original specifications, and reassembly to proper balance and alignment.

A. Pump-Turbines

1. Cavitation repair on runner blades
Replacement of rotating wear ring; machining to size.
Water passages coated with polyester to reduce corrosion and erosion and to reduce surface turbulence; Equalizer lines repaired; Sand blasted stay rings.

2. Head Cover and Discharge Ring
Stationary wear rings replaced; Gate bushing seats line-bored for improved alignment; Coated wetted portion of head cover with polyester.

3. Turbine Shaft-
Replaced packing sleeve

4. Servomotors-
Bored cylinders to remove scratches; Installed new piston rings.

5. Wicket Gates-
New stainless steel wicket gates were installed and doweled to existing gate arms.

B. Thrust Bearing-

Replaced shoes; Installed additional RTD's to monitor temperatures of the bearing shoes.

C. Generator Stator-

Cleaned, removed bars to inspect sidewalls, replaced bars as needed; Applied conducting varnish; Conductive Room Temperature V injected to assure contact between the ground wall of the bars and the stator iron; New wedges installed; Performed slot discharge tests.

D. Field-

Bottom interpole jumpers replaced with flexible types.

E. Ancillary Equipment-

Equipment cleaned, inspected, control and protective series tested and calibrated.
Speed Sensing Generator and starting motor reinstalled following cleaning and inspection.

Calculation of the pump/generate ratio subsequent to the major overhaul work (refer to Tables 1 A, 1 B, and 1 C) substantiate the beneficial results of the work. The overall plant efficiencies are now back to the early plant levels of 1.35. This represents an overall plant improvement of 1.40/1.35 = 1.037. Using the conservative figure of 3%/year improvement, when evaluated at a net generation value of $30 per MWH, this represents an annual value in the vicinity of $1 M/year.

$30/MWH x (.03) x 1,116,000 MWH/year*

* Average of '88 and '89 output.

Among the items to be further analyzed regarding the variance in the P/G ratios as computed monthly and quarterly is the fact that the lowest (more plant efficiency) P/G ratios appear to take place during the summer months. (Refer to Figure 1. A more intensive and detailed analysis of the plant's operation and dispatching history will be undertaken to determine the extent to which degradation in P/G ratios results from plant operation versus that which takes place due to mechanical wear and tear.

PLANT OPERATION

The plant staff has remained essentially the same over the past 16 years since the plant went commercial in 1973. There have been some additions in the form of increased technical and administrative staff persons.

The activities of the plant's recreation and environmental programs and facilities have also been added to the plant manager's responsibilities. The organization is depicted in Figure 3.

The dispatching of the plant units as regards generation and pumping load levels is still governed and scheduled by the System Dispatchers. The plant operators retain the option of which units to place in operation.

The practice of spreading the operating service to each of the units uniformly is still maintained, pending the installation of acoustic flow measurement systems to the units later this year (1990). Calculated individual unit efficiencies may influence operating priorities of the machines in either the generate or pumping modes. In addition to establishing the actual real time operating efficiencies of the units for improved overall plant economies, it is expected that the flow measurement systems will provide the efficiency characteristics over the entire operating range. This information should assist the plant owners in recovering the operating costs of off BEP dispatching for NEPOOL system benefits.

DAM, DIKES, AND INTAKES

Settlement of a small section approximately 50 ft. (15.6 m.) in length, of the main dam was recorded starting shortly after dam filling in 1972. By 1978, the vertical deformation was approximately 1 ft. (0.3 m.) and projected to be approximately 1.6 ft. (0.45 m.) after 50 years. To maintain the 7.5 ft. (2.28 m.) freeboard, the plant had lowered the maximum pool elevation by 1 ft. (0.3 m.). This section of the dam was repaired in 1979 to its original grade. Settlement data taken periodically show that the dam's primary consolidation is virtually complete. Because of their lower heights, the dikes along the northerly sections of the reservoir have consolidated by much lower rates and present no problem. Generally, the main dam and dikes have remained in excellent condition with no other observable longitudinal cracks or differential crest settlements.

The rock intake area, i.e., the channel connecting the main body of the upper reservoir with the intake, shows no signs of rock deterioration or spalling of concrete. A separate water intake tower, constructed for possibility of future water supply transferred to a 10-mile distant reservoir has shown no signs of distress and the concrete plug in the foundation-laid pipeline is watertight.

POWERHOUSE

The rock walls of the powerhouse have retained their regularity and have remained exceptionally dry. Minor seepage in the underground chamber parking area next to the powerhouse has been controlled and presents no problem. Surface and some groundwater seepage into the 15 ft. (4.5 m.) diameter ventilation shaft on the northerly side of the powerhouse caused buildup of ice on the stairwells and on the hoist steel structure supports. The hoist has been removed and the ventilation shaft is entered only for inspection purposes, periodic inspection, and minor maintenance in the surge chamber.

RESERVOIR DRAWDOWNS

There have been three complete upper reservoir dewaterings since initial filling in 1972. In April, 1974, the reservoir was dewatered for a first "baseline" inspection. No reservoir shoreline or dam and dike problems were observed. Buildup of silt in the intake channel was noted. It presented no problems at that time and was estimated to be at the 5 to 10 ft. depth range (1.5 to 3.0 m.). The high pressure concrete lined intake shaft showed no problems.

In 1979, it was decided to inspect the reservoir again. This time, the water was again drawn down by generating with all four units to check that plant output at lower elevations. The final drawdown was performed by Unit 2 (whose penstock is a straight in-line layout with the high pressure shaft). Following shutdown, it was found that the silt buildup in the upper reservoir intake channel had velocity-eroded at the final lower reservoir levels and was carried into the penstocks and scroll cases of each of the units. Extensive hand excavation and hose sluicing through the scroll case drains was required, entailing considerable time and effort. Dams, dikes, and intake structures were in relatively good condition as observed in the dewatered reservoir.

During the last upper reservoir drawdown in April/May of 1985, special low reservoir level unit operations were devised to preclude carrying the silt into the penstocks and units again. Underwater surveys had been indicating silt depths of 30 to 40 ft. (9.1 to 12.2 m.) in the channel. Lower velocities in the intake channel by controlled releases with the last unit on-line, however, resulted in successful dewatering. The silt depths in the reservoir are still well below the dead volume level and are not expected to present any volumetric water problems: The silt comes in principally during the spring freshet runs in the lower reservoir (Connecticut River). Plant operators should take special note of this silt moving during planned dewatering inspections since increased plant outages can result from the unforeseen time it takes to hand clean silt buildup in the machinery chambers - which are not normally designed for this circumstance.

ACCESS TUNNEL

The bedrock from which the underground structures were excavated remains extremely competent. The access tunnel was not lined with concrete nor did it require and steelsets. Some rock bolting was done to insure safety of construction and operating personnel. From time to time minor rock spalls have been noticed. In 1987, an extensive access tunnel scaling job was done. The work was straightforward with rock scaling work performed on the walls and roof of the tunnel by conventional equipment. Special shields did have to be constructed to prevent damage to the two exposed 345 KV pipe cable runs and the station electric service lines. Pavement protection and traffic protection procedures were also required. These latter items resulted in slower rates of scaling progress than were initially estimated. Included in the work at the time was some shotcreting of the underground parking area (the four-way junction) immediately adjacent to the powerhouse cavern as well as extensive rock bolting and associated steel strapping in this area. Safety of station personnel as well as the public was the principal reason for this work. A total of about 50 cubic yards (65.4 m.) was scaled off the access tunnel walls. It is expected that extensive scaling will not be required for another 8 to 10 years.

ENVIRONMENTAL IMPACTS AND CONCERNS

The Northfield Mountain Plant is an outstanding representative of a hydroelectric facility that can be built with a miniscule impact on the visual environmental intrusion of a rural farmland countryside. Soon after the on-site construction scars healed, viz., construction sheds taken down, roads paved, rubble and rock deposition areas loamed and seeded, and laydown and field storage areas regrassed, it is not possible for anyone driving by the site area to recognize that a 1,080,000 kilowatt electric facility exists in the area, and is generating power and energy into the electric system. The only visible evidence of an electric facility, as seen from the roadside, are the switchyard transmission strain towers connecting the plant output to the 345 KV transmission lines running north and south from the project area. The transmission design engineers now conclude, in hindsight, that the switchyard might also have been located near the top of the mountain rather than at the access tunnel adit area. The mountain top site would have further reduced the minor visual impact of the transmission circuits.

The upper reservoir on top of the mountain is accessible to the public via company recreation program bus tours and by persons walking the constructed mountain-side recreation trails. With drawdowns in the forty foot range, the upper reservoir is not available to the public. The entire perimeter is chain-link fenced off to prevent access for fishing or boating. A hilltop viewing platform area accessible by the bus tour and by walking provides a public northerly viewing of the reservoir basin.

During construction, the major noise of rock excavation for the powerhouse and connecting tunnels presented no problem. Upper reservoir clearing, dam and dike construction, and intake channel rock excavation were far enough away from public and residential areas to present any dust or noise problems. The constructed river edge watertight embankment confined all tunnel rock excavation traffic and associated construction activities of the tailrace structures so that river recreation boating continued with no impact.

The road connecting the lower project construction area and the upper reservoir area, mainly for construction communication and supervision, was laid out in a normal countryside curved alignment to prevent a hard straight line highway approach. Currently, wooden guard rails and a standard asphalt road surface make a normal country appearance. The Recreation headquarters building and the small tailrace pump house are the only project structures visible to the public. The eight-foot high embankment surrounding the switchyard, the hillside replanted with native pines, and the pleasant well-kept grassed areas make the plant area blend in well with the normal New England countryside. The velocities of flow into and out of the tailrace tunnel via the tailrace structure are low enough (aided by the diverging entrance-exit canal) so that there is no detrimental effect visually or on river boat traffic. With all four units coming on line simultaneously, or the plant experiencing a four-unit load rejection on pumping, the surge in the tailrace canal was designed (and it checked out) as a gentle one and one-half foot (0.5 meter) rise and fall. To prevent boaters from nearing the tailrace structure, mainly for fishing, a floating boat barrier is laid out every summer across the canal entrance. The rise and fall in the Connecticut River, the ponded twenty mile stretch of which serves as the lower reservoir, is normally about three feet on a daily basis. Because the daily recharge pumping is usually less than the daily generator water use, the river elevation increases from Monday to Friday about 5 feet (1.5 m.) and is returned to the normal lower starting elevation by Monday morning following the heavier weekend pumping.

RECREATION PROGRAM

Northeast Utilities decided, in response to the licensing requirements and in continuance of the company's normal policy of making certain hydro project lands available to the public, to provide recreation and environmental facilities as a formal part of the pumped storage plant. The field construction office was designed so that it could be converted to a Recreation and Environmental Center, now the center of year-round activities. Exhibits depict past and present day activities of the local Connecticut River valley including logging, ice-harvesting, and water power.

A riverboat, the Quinnetukit II, provides public cruises in summer along a twelve-mile section of the Connecticut River from the Riverview Picnic Area at the Northfield tailrace to Barton Cove downstream. The Barton Cove Campground, with available tent camping sites, is located downstream on a mile-long peninsula offering picnicking, group camping, canoe and boat rentals, fishing, and nature trails. Further upstream is the Munn's Ferry campground, accessible by boat only, with tent sites and a camping shelter. Twenty-five miles of carriage-width trails are available year round at the mountainside near the Recreation Center. In summer, the nature trails are lined with wildflowers and follow hillside and bubbling brooks. In winter, the trails, covered with snow, provide well groomed cross country skiing, backed up by instruction and equipment rentals. (Refer to Figure 4).

The Center staff personnel offer bus tours into the underground powerhouse cavern during the spring, summer, and fall months, and to the upper reservoir summit area. A myriad of school programs for children of all ages is provided, including farm tours and orienteering. An agricultural and wildlife management area is maintained to offer and encourage wildlife studies of nature and migrating species. At the downstream dam, the Turners Falls Fishway is active every spring for anadromous shad and salmon annual migration. Facilities for viewing these and other river species are made available to the public.

The recreation facilities enhance the pumped storage plant's quiet acceptance as a part of the area's contribution to electric reliability and public recreation and environmental activities.

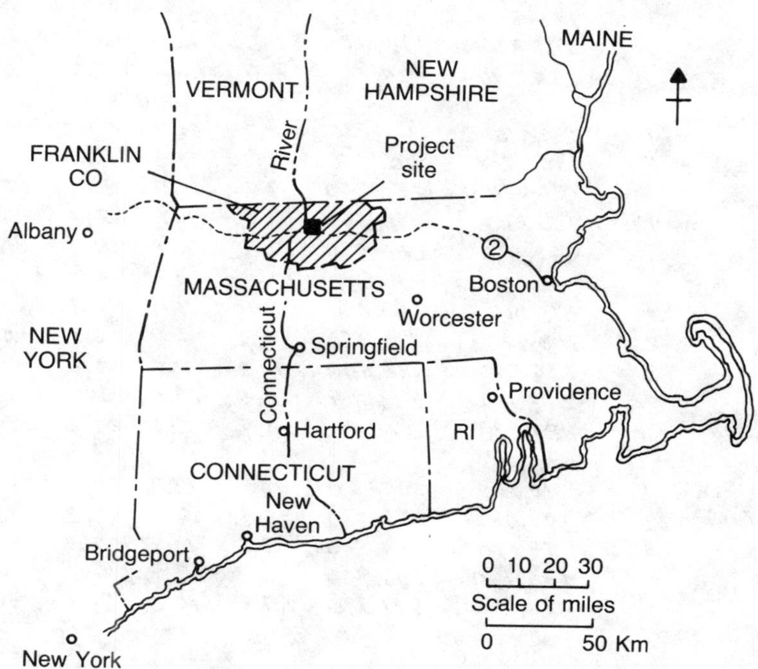

Location map

TABLE 1A
Northfield Mountain Plant - Generation & Pumping Data
1974 TO 1979

YR.	SEASON/ QUARTER	MWH Generation (G)	MWH Pumping (P)	Plant Factor Gen. %	Plant Factor Pump %	Plant Factor Total %	P/G Ratio
	D,J,F-1	274,966	369,920	12.6	16.9	29.5	1.34
'74	M,A,M-2	229,930	302,110	10.5	13.8	24.3	1.32
	J,J,A-3	307,800	402,200	14.1	18.4	32.5	1.31
	S,O,N-4	310,990	418,100	14.2	19.1	33.3	1.34
		1,123,686	1,492,330	12.8	17.0	29.8	1.33
	1	238,210	319,450	10.9	14.6	25.5	1.34
	2	209,260	284,840	9.6	13.0	22.6	1.36
'75	3	240,190	319,670	11.0	14.6	25.6	1.33
	4	276,710	375,140	12.6	17.1	29.7	1.36
		964,370	1,299,100	11.0	14.8	25.8	1.35
	1	294,670	400,550	13.5	18.3	31.8	1.36
'76	2	169,160	226,540	7.7	10.3	18.0	1.34
	3	145,330	195,060	6.6	8.9	15.5	1.34
	4	143,930	192,110	6.6	8.8	15.4	1.33
		753,090	1,014,260	8.6	11.6	20.2	1.35
	1	189,550	263,830	8.7	12.0	20.7	1.39
'77	2	152,020	205,670	6.9	9.4	16.3	1.35
	3	168,730	220,240	7.7	10.0	17.7	1.31
	4	150,810	198,350	6.9	8.9	15.8	1.32
		661,110	888,090	7.8	10.1	17.9	1.34
	1	175,010	240,200	8.0	11.0	19.0	1.37
'78	2	150,980	210,950	6.9	9.6	16.5	1.40
	3	236,290	318,540	10.8	14.5	25.3	1.34
	4	219,110	306,730	10.0	14.0	24.0	1.40
		781,390	1,076,420	8.9	12.3	21.2	1.38
	1	219,110	305,260	10.0	13.9	23.9	1.39
'79	2	107,030	138,130	4.9	6.3	11.2	1.29
	3	159,350	221,580	7.3	10.1	17.4	1.39
	4	145,760	206,920	6.7	9.4	16.1	1.42
		631,250	871,890	7.2	10.0	17.2	1.38

TABLE 1B
Northfield Mountain Plant - Generation & Pumping Data
1980 - 1985

YR.	SEASON/ QUARTER	MWH Generation (G)	MWH Pumping (P)	Plant Factor Gen. %	Plant Factor Pump %	Plant Factor Total	P/G Ratio
	D,J,F - 1	191,600	263,060	8.7	12.0	20.7	1.37
80	M,A,M - 2	234,200	330,500	10.7	15.1	25.8	1.41
	J,J,A - 3	222,760	305,670	10.2	14.0	24.2	1.37
	S,O,N - 4	199,200	284,140	9.1	13.0	22.1	1.42
		848,260	1,183,370	9.7	13.5	23.2	1.40
	1	146,010	197,170	6.6	9.0	15.6	1.35
'81	2	144,910	205,550	6.6	9.4	16.0	1.42
	3	157,360	217,910	7.2	9.9	17.1	1.38
	4	146,690	208,010	6.7	9.5	16.2	1.42
		594,970	838,640	6.8	9.6	16.4	1.41
	1	135,520	192,390	6.2	8.8	15.0	1.42
'82	2	179,360	249,620	8.2	11.4	19.6	1.39
	3	246,390	337,650	11.3	15.4	26.7	1.37
	4	184,560	265,830	8.4	12.1	20.5	1.44
		745,830	1,045,490	8.5	11.9	20.4	1.40
	1	142,690	201,570	6.5	9.2	15.7	1.41
'83	2	247,380	342,020	11.3	15.6	26.9	1.38
	3	288,080	397,780	13.2	18.2	31.4	1.38
	4	214,340	304,900	9.8	13.9	22.7	1.42
		892,490	1,246,270	10.2	14.2	24.4	1.40
'84	1	189,880	264,610	8.7	12.1	20.8	1.39
	2	171,409	243,910	7.8	11.1	18.9	1.42
	3	252,670	345,550	11.5	15.8	27.3	1.37
	4	247,930	345,740	11.3	15.8	27.1	1.39
		861,889	1,199,810	9.8	13.7	23.5	1.39
	1	186,610	254,810	8.5	11.6	20.1	1.37
'85	2	195,130	272,300	8.9	12.4	21.3	1.40
	3	253,460	351,710	11.6	16.1	27.7	1.39
	4	271,550	371,810	12.4	17.0	29.4	1.37
		906,750	1,250,630	10.4	14.3	24.7	1.38

TABLE 1 C
Northfield Mountain Plant - Generation & Pumping Data
1986 - 1989

YR.	SEASON/ QUARTER	MWH Generation (G)	MWH Pumping (P)	Plant Factor Gen. %	Plant Factor Pump %	Plant Factor Total %	P/G Ratio
	D,J,F - 1	256,040	353,000	11.7	16.1	27.8	1.38
86	M,A,M - 2	239,900	335,090	11.0	15.3	26.3	1.40
	J,J,A - 3	238,700	322,470	10.9	14.7	25.6	1.35
	S,O,N - 4	212,810	294,200	9.7	13.4	23.1	1.38
		947,450	1,304,760	10.8	14.9	25.7	1.38
	1	224,630	303,930	10.3	13.9	24.2	1.35
'87	2	240,580	329,670	11.0	15.1	26.1	1.37
	3	268,290	363,940	12.3	16.6	28.9	1.36
	4	228,060	310,120	10.4	14.2	24.6	1.36
		961,560	1,307,660	11.0	14.9	25.9	1.36
	1	277,000	376,810	12.6	17.2	29.8	1.36
'88	2	246,350	332,930	11.2	15.2	26.4	1.35
	3	290,300	393,090	13.3	17.9	31.2	1.35
	4	271,360	368,740	12.4	16.8	29.2	1.36
		1,085,010	1,471,570	12.4	16.8	29.2	1.36
	1	284,124	390,322	13.0	17.8	30.8	1.37
'89	2	288,107	392,745	13.2	17.9	31.1	1.36
	3	294,865	395,029	13.5	18.0	31.5	1.34
	4	280,413	380,427	12.8	17.4	30.2	1.36
		1,147,509	1,558,523	13.1	17.8	30.9	1.36

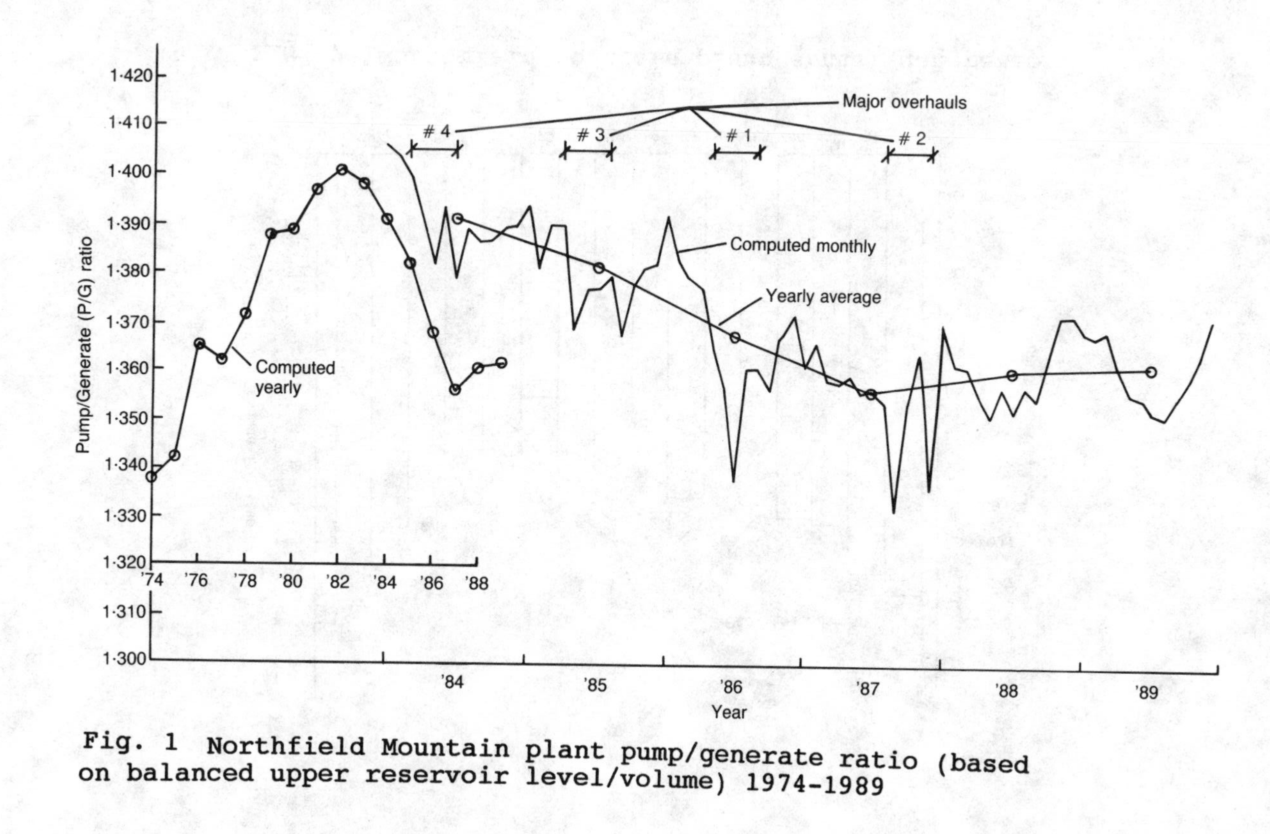

Fig. 1 Northfield Mountain plant pump/generate ratio (based on balanced upper reservoir level/volume) 1974-1989

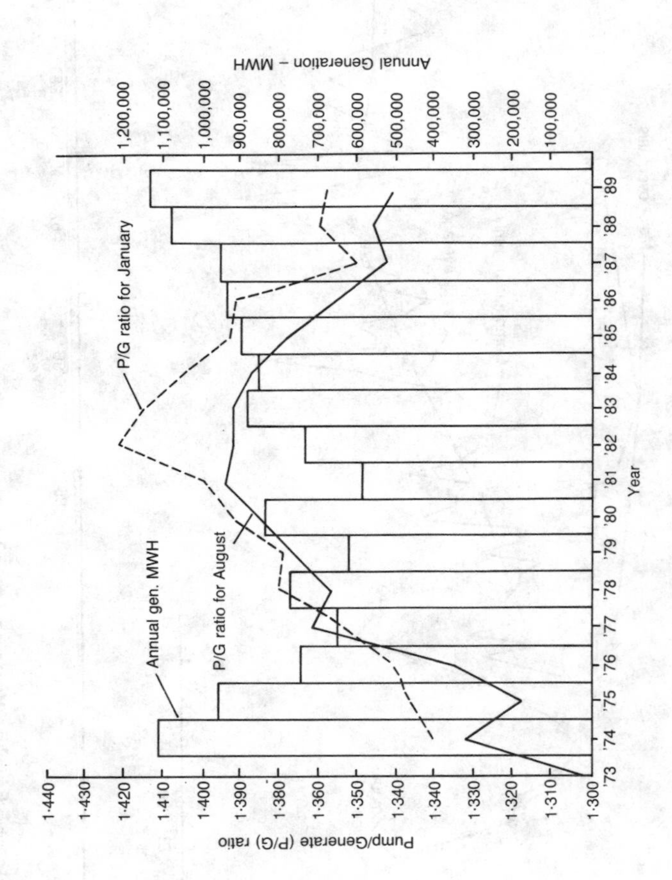

Fig. 2 Northfield Mountain plant annual generation

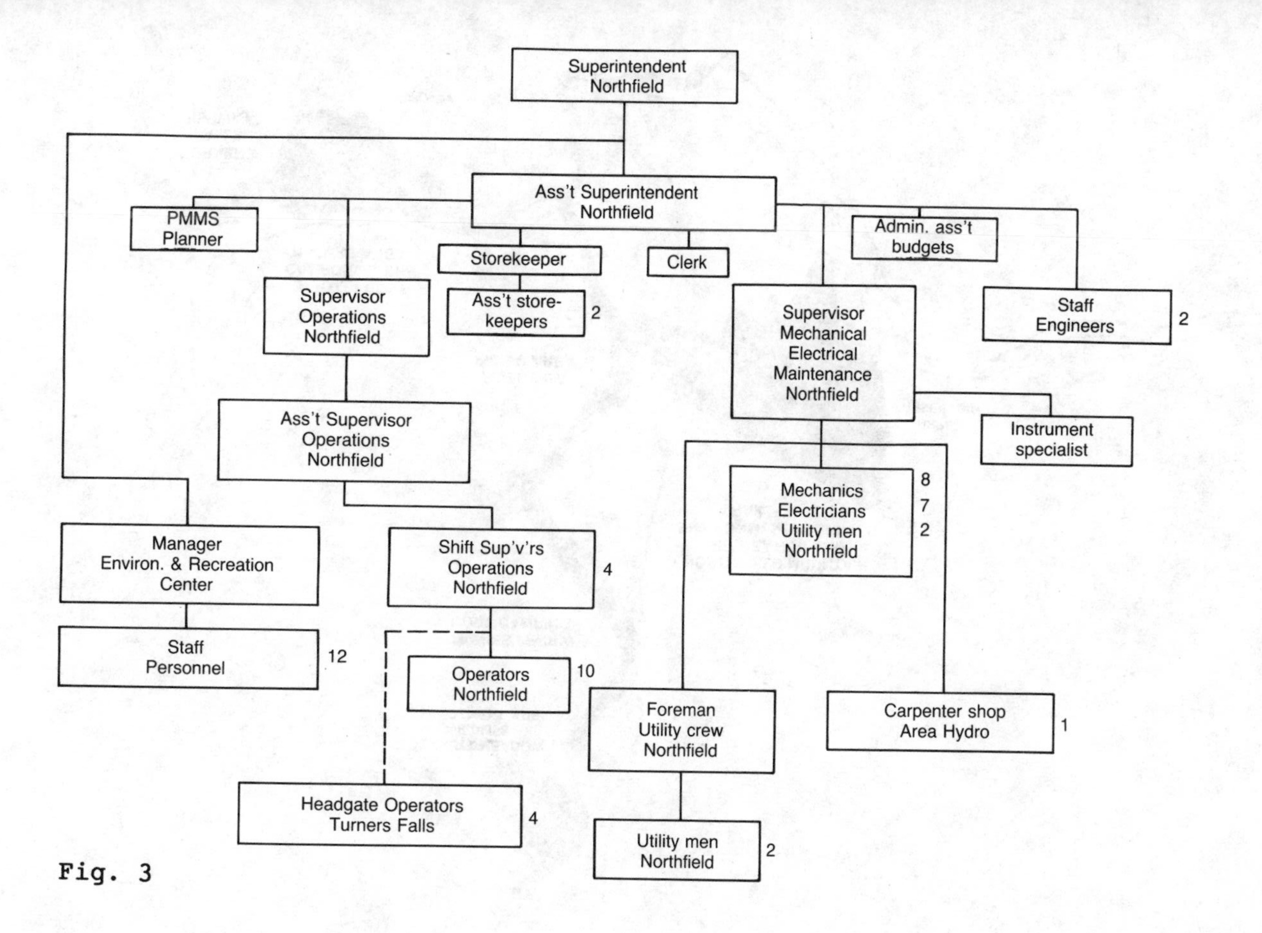
Superintendent
Northfield
Ass't Superintendent
Northfield
PMMS
Planner
Storekeeper
Clerk
Admin. ass't
budgets
Supervisor
Operations
Northfield
Ass't store-
keepers
2
Supervisor
Mechanical
Electrical
Maintenance
Northfield
Staff
Engineers
2
Ass't Supervisor
Operations
Northfield
Instrument
specialist
Mechanics
Electricians
Utility men
Northfield
8
7
2
Manager
Environ. & Recreation
Center
Shift Sup'v'rs
Operations
Northfield
4
Staff
Personnel
12
Operators
Northfield
10
Foreman
Utility crew
Northfield
Carpenter shop
Area Hydro
1
Headgate Operators
Turners Falls
4
Utility men
Northfield
2

Fig. 3

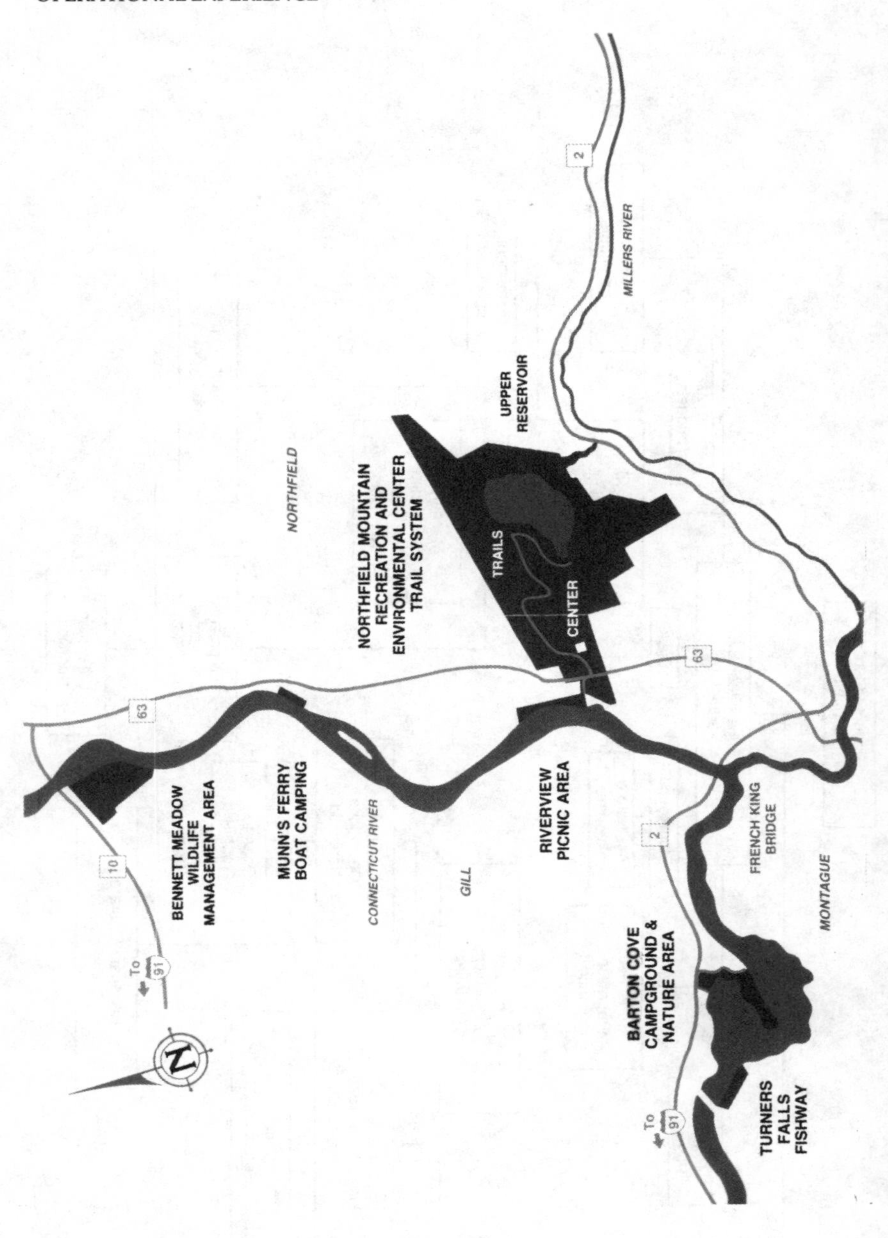
NORTHFIELD MOUNTAIN RECREATION AND ENVIRONMENTAL CENTER TRAIL SYSTEM
UPPER RESERVOIR
NORTHFIELD
MILLERS RIVER
2
TRAILS
CENTER
63
63
BENNETT MEADOW WILDLIFE MANAGEMENT AREA
MUNN'S FERRY BOAT CAMPING
CONNECTICUT RIVER
GILL
RIVERVIEW PICNIC AREA
2
FRENCH KING BRIDGE
MONTAGUE
10
To 91
BARTON COVE CAMPGROUND & NATURE AREA
To 91
TURNERS FALLS FISHWAY
N

Fig. 4

Discussion

M.J. HIGGINS, ***Electricity Supply Board, Ireland***

The contribution of the only pumped storage plant in Ireland to the provision of Primary Spinning Reserve on the ESB system has been critical. It has enabled the use of 300 MW units on a system, which, to date, has a peak demand in the region of 2500 MW and valley of 800 to 900 MW, without pumping load. Example of the response achieved - overheads with underfrequency initiation and fast wind down signal initiation from large thermal units - were given and the necessity to achieve very high level of reliability emphasized.

Although Turlough Hill Pumped Storage Station is only approximately 50 miles from Dinorwig, it is across the Irish Sea and operating in a completely different environment.

Turlough Hill pumped storage station was being operated on an isolated system with peak demand in the region of 2500 MW and minimum value demand of 800/900 MW therefore without pumping load. Thermal units of 270 and 305 MW size were being operated on this isolated system. Turlough Hill was the only pumped storage station on ESB's system and since commissioning in 1973/74 many modifications have been made to enable it to provide the service that the operation of the isolated system demands. Many of the ideas for modifications have come from discussions with the staff of the UK pumped storage stations. Initial training for the operating staff of Turlough Hill was carried out at Cruachan. The main inlet valve opening times had been reduced from 55 s to 7 s without interfering with the closing times. Major control modifications have also been made and protection/supervision systems have been upgraded.

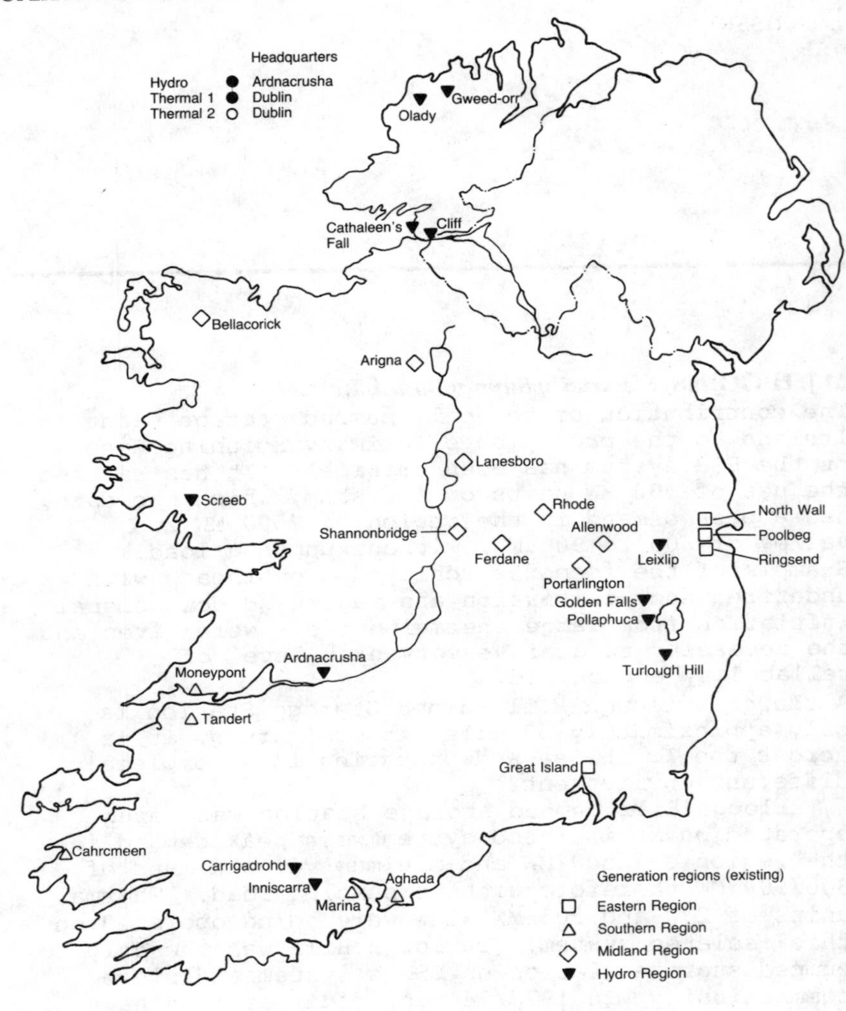

Fig. 1 Electricity supply board, Ireland

Turlough Hill units respond automatically through under frequency relays and, more recently, through receipt by carrier system, of fast wind-down signals from large thermal units, which have load wind-down facilities for certain faults.

A. BJORNSGAARD, ***Kvaerner Eureka, Norway***
The Author mentioned the repair of cracks after welding, is PWHT or other treatment carried out on the weld?

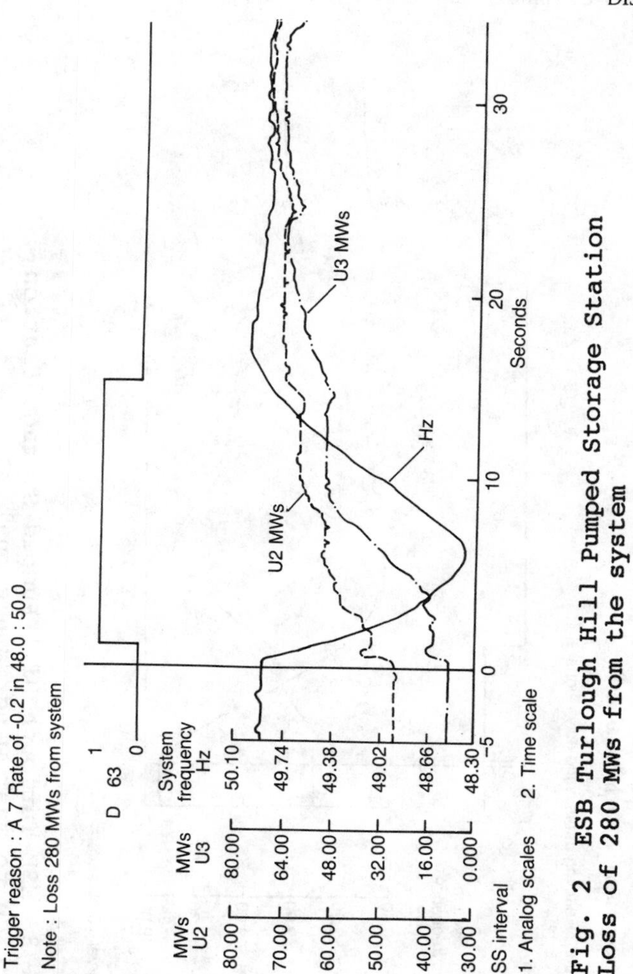

Fig. 2 ESB Turlough Hill Pumped Storage Station
Loss of 280 MWs from the system

F.G. JOHNSON, *formerly of North of Scotland H.E. Board*

The Author considers that pump turbines have a much more onerous and damaging operating regime than conventional hydro. This involves large expenditure to reduce the maintenance costs. Does the Author think the tunnel lining damage is a function of the frequent changes in water direction or of the water velocity. What is the cost of 1 MW out of operation for maintenance, in peak time and what is the mean value?

C.P. STRONGMAN, *Merz and McLellan, UK*

Mr Schmid explains how the critical speeds of the Kuhtai reversible sets depend on the assumptions

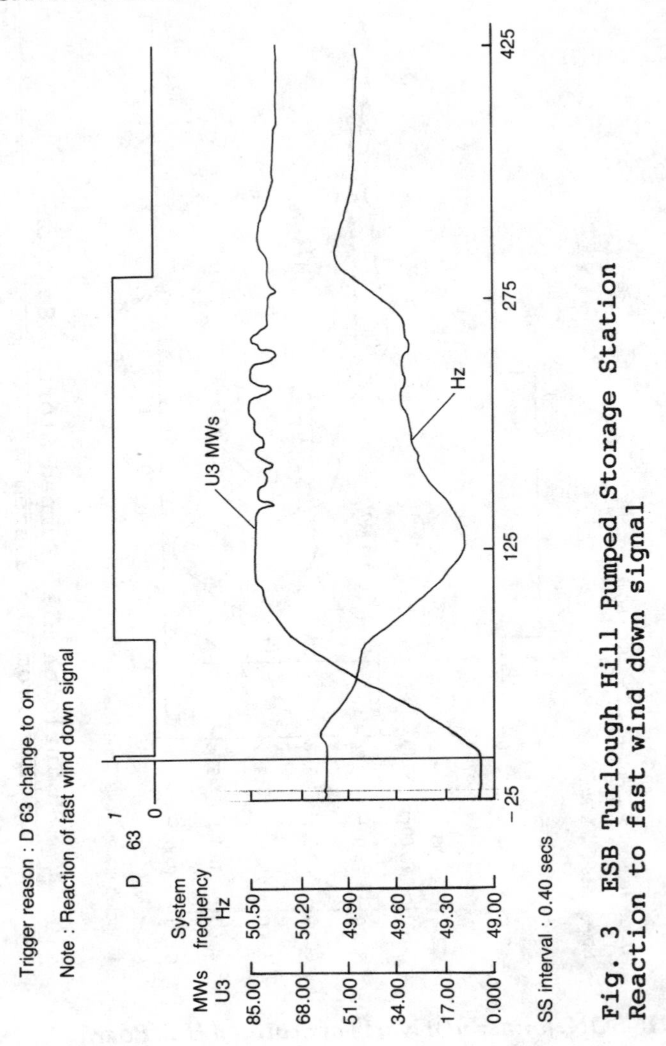

Fig. 3 ESB Turlough Hill Pumped Storage Station Reaction to fast wind down signal

made regarding the unbalanced magnetic pull of the motor-generator. Could he explain the origin of the assumed unbalanced magnetic pull. Was it considered to be due to a slight eccentricity of the motor-generator rotor or to some of the poles being short-circuited?

Is it not the case that, since the protection will remove the excitation as the speed rises on full-load rejection, only the critical speed obtained without unbalanced magnetic pull will be significant? (Admittedly in the case of Kuhtai there is little difference between the two figures stated).

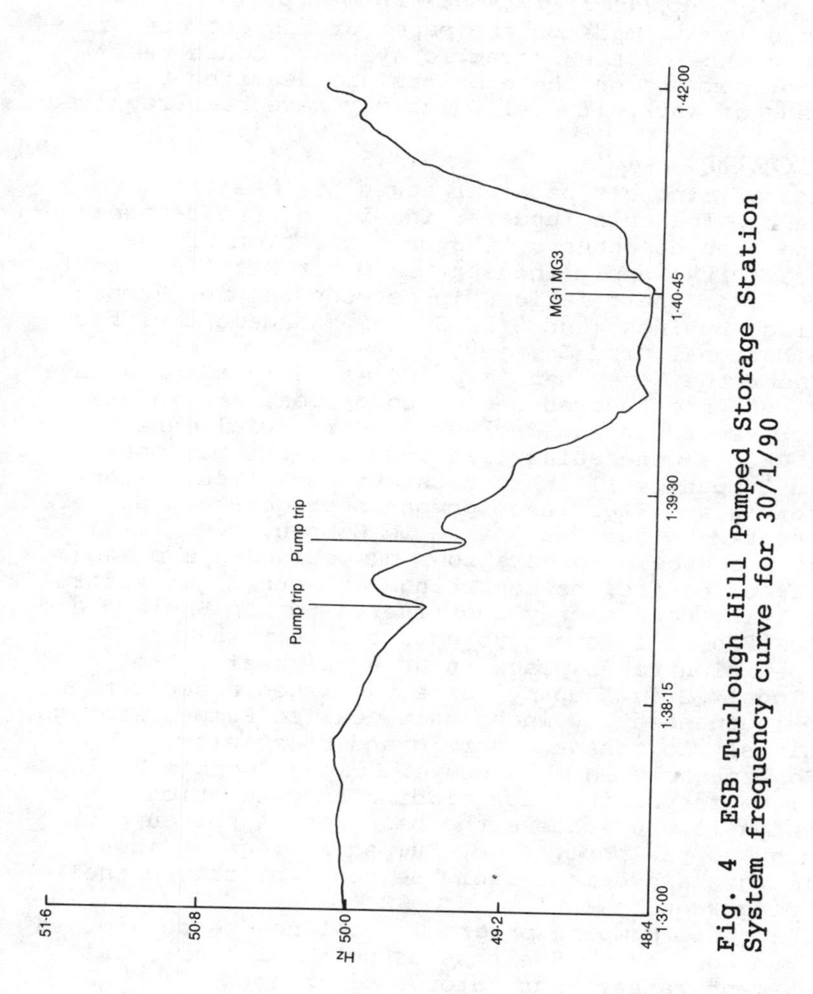

Fig. 4 ESB Turlough Hill Pumped Storage Station
System frequency curve for 30/1/90

R.W. BUCHANAN, ***James Williamson & Partners, Glasgow***

In Fig. 3, which shows a typical winter's day operation, night pumping is much less than the six hours storage provided in the reservoirs. Is the full six hours seldom required?

In a future pumped storage scheme with an underground reservoir the reduction of storage provision from six hours to, say, four hours pumping could represent a saving of up to 10% in capital cost. Could the Authors, from their Dinorwig experience, hazard a guess whether the reduction in scheme value would be greater or less than this figure?

T.H. DOUGLAS, ***James Williamson & Partners, Glasgow***

No mention is made in the paper of inspections or maintenance of the hydraulic system. Could the Author comment on these points and describe the extent of work, if any, which may have been required.

J.R. LOWEN, ***Author***

Since writing his paper Mr Lowen has been able to make the following updates in view of the changes being brought about by the privatisation of the Electricity Supply Industry in Great Britain, and in particular the relationship between the new Pumped Storage Business and Grid System Management within the National Grid Company.

Under the Government's proposals the CEGB has been divided into four successor companies. National Power, PowerGen and Nuclear Electric will inherit the CEGB's generating plant, whilst the National Grid Company will inherit the transmission system. Dinorwig and Ffestiniog pumped storage stations were allocated to the National Grid Company.

As a source of generation, pumped storage plant will, of course, be competing with other generators supplying both energy and reserve to the England and Wales Pool. In order to ensure that this competition takes place on an equal basis, the National Grid Company has established a separate and independent business unit, called Pumped Storage Business, to manage Dinorwig and Ffestiniog.

Consequently in the future, Pumped Storage Business will be responsible for bidding in generation, on a unit basis, to achieve the best return for supplying energy to the pool. Also, Pumped Storage Business will purchase demand from the pool to achieve their pumping requirements.

The basic pumping programme will now be determined by Pumped Storage Business using their commercial judgement rather than through operational optimisation against the demand shape.

It was, of course, recognised that the management of a complex power system requires more than the basic supply of energy. Additional services, known as ancillary services, are also required. These include system reserve, reactive power, frequency response and black start capacity. It is intended that these services may be offered by all generators and that the National Grid Company will purchase such services as required.

The pumped storage units are able to offer all these services and they are particularly effective in meeting the requirements for system reserve and frequency response. Pumped Storage have offered to provide these services at specific prices, and these together with the prices submitted by other generators are now used by National Control to determine the allocation of reserve and frequency response duty to the generators on the system. The new commercial structure has brought the wider economic benefits of pumped storage plant into sharper focus. Consequently pumped storage now provide, for example, prices for the following ancillary services: reactive power, variable frequency control, automatic low frequency fast start, automatic low frequency pump deload and trip, the load smoothing capability of a despatchable pumping load and rapid mode change facilities, together with black start capability.

All these ancillary service capabilities are now offered at specific prices and National Control will use all these costs to determine, together with other generators' offered prices, the operational duty to be allocated to pumped storage and in particular its minute to minute mode of operation.

The story of Dinorwig and Ffestiniog to date is one of success and I am sure it will continue into the future in the new electricity market-place in the UK. The plant has performed well against its design criteria and its flexibility has proved an enormous benefit in managing the power system to give our customers a quality service.

F.G. JOHNSON and C.K. JOHNSTON, *Authors*

In reply to J. Remondeulaz, we consider that the main cause of tunnel lining damage is the number of mode changes experienced, not the water velocity or bi-directional flow of water.

The cost of loss of plant availability for planned maintenance on pump storage stations is very dependent on the time of year and requirements of the system. A typical loss of around £300/MW/day could be anticipated.

In reply to C.P. Strongman, experience with the hydraulic performance of the surge chambers at both Cruachan and Foyers has been good and no problems have been encountered. Both chambers have coped very well dynamically with the enhanced role which the stations have had to play, although this was to be expected as the chambers were designed to accommodate the dynamic conditions and not the much increased number of mode changes to which they have been subjected.

In reply to R.B. Kydd, in the light of the experience from defects in the tunnels of the Cruachan scheme, where it was some six or seven years before the problems were properly tackled, at Foyers, tunnel inspections were and are undertaken whenever plant outages allow and defects rectified - on the basis of a stitch in time saves nine. This latter policy has much reduced the extent of the maintenance work that has had to be carried out on these tunnels and is commended. Tunnel maintenance requirements have been very much heavier on the company's pumped storage schemes than their conventional hydro schemes - by one or two orders of magnitude. Maintenance on dams, intakes, outfalls and stations has not been significantly different to that on hydro schemes.

Tunnels on conventional hydro schemes are given detailed inspections, normally on a five year basis and any significant defects which are revealed are rectified some time during the following 12 months, depending upon the seriousness of the defect. Over the past 15 years, we have inspected tunnels of pumped storage schemes, at least one every five years and frequently more often as plant outages have allowed. Over the last ten years, we have employed ROVs increasingly for subject after first establishing a comprehensive and detailed base by direct inspections against which we can accurately and confidently compare ROV results.

A. FERREIRA, *Author*

In reply to F.G. Johnson, the quantification of added costs (maintenance) to perform dynamic operating benefits is not complete. We expect to define these better by 1991-92. Tunnel lining (concrete) damage does have its basis in high water velocity and is affected considerably by flow reversals plus transient pressure loadings. The current value of lost capacity for outages is about $95/kW year.

In reply to R.B. Kydd, when possible and economical the visual personal inspection of tunnel systems is preferred to the use of ROVs. At Northfield, reservoir dewatering will be done this October during three week, 18 month, planned maintenance outage.

In reply to T. H. Douglas, inspections of the hydraulic system are performed about every five years during a scheduled dewatering of the upper reservoir and the tunnel conduit system. No major work has yet been required for repairs. The inspections, results and required periodic maintenance will be covered in a paper to be complete early in 1991.

H. SCHMID, *Author*

In reply to C.P. Strongman, the origin of the assumed imbalanced magnetic pull on the rotor may be explained as follows

(a) due to a balance quality Q4 or Q6.3 (which are common values for hydro units) a certain amount of eccentricity of the centre of gravity of the rotor is related

(b) if the axis of the rotor and the axis of the stator are not in coincidence, or if there are alignment problems, the amount of eccentricity will produce magnetic pull on the rotor due to imbalanced magnetic forces between runner and stator

(c) if both runner and/or stator are not of exact circular shape, it will result in imbalanced magnetic forces between runner and stator

(d) in case of a short-circuit between adjacent poles, a significant imbalance of magnetic forces will produce a force vector, which rotates with the speed of the machine - in such a case significant gravity-dependent imbalanced forces will occur.

The assumptions in the mathematical model for the analysis of the critical speed are only of type (a) to (c); the magnetic pull is introduced into the mathematical model as a negative spring, which acts along the length of the poles.

As long as the critical speed is above the operating speed and the influence of the magnetic pull to the critical speed is not significant (which is true in general if the shafting system and the bearings are stiff enough) there would be no need to calculate both critical speeds (without and with magnetic pull).

In all cases, where the critical speed is close to or below the operating speed it is necessary to calculate the critical speed without and with magnetic pull. It cannot be assumed, that the protection system will be quick enough, if the rotating system passes the region of the critical speed. Only a few revolutions in this region are necessary to produce unacceptable deflections and bearing loads.

Pumped storage accumulation and generation in the Netherlands

H. VAN TONGEREN, L. GILDE, J.A. DE RIDDER, A.L. VAN SCHAIK, C.J. SPAARGAREN, D.P. DE WILDE and E.R. TEGELBERG, Ballast Nedam Group

SYNOPSIS
A feasibility study was carried out by a consortium of public and private engineering companies for a man-made, above-ground pumped storage reservoir, as part of a government study into peak-shaving. Two possible locations were selected. Both the storage capacity between 5 and 40 GWh and the generating capacity between 500 and 2500 MW were simultaneously optimised. For each of the two locations a preliminary design was made for the optimum solution.

INTRODUCTION

1. After the first energy crisis in 1973, which was accompanied by an enormous increase in oil prices, many important developments took place. The world's industrial countries started a feverish quest for alternative sources of energy and possibilities to save energy, so that they would no longer be dependent on the traditionally dominant oil producing and exporting countries.

2. The most remarkable developments which took place in the industrial countries during the last decade are the following:
- the exploitation of relatively expensive offshore oil and gas fields;
- the development of alternative power sources, such as the sun, wind and water;
- the development of energy management, in order to use energy more efficiently;
- the development of a strategy, including a diversification of the power sources for the generation of electricity.

3. The Dutch Government, too, strongly stimulates these developments, despite the recent slump in oil prices.
With regard to energy management and strategy, the Pumped Storage Accumulation and Generation Plant (PAC) ('Pomp Accumulatie Centrale') is an interesting option; most important industrial countries already have one or more of such units.

4. The demand for energy in the form of electricity has increased again during the last few years and this development is expected to continue well until the end of this century. The increase in the demand for electricity is accompanied by an increase in its fluctuations. This means that a significant part of the power stations will only be utilized at full

capacity during a limited period of time (see figure 1). The tailoring of the generating capacity which is to be supplied by the existing power stations to the demand can, however, be less desirable for technical and economic reasons. The differences between the continuous basic supply and the demand at a certain moment can be covered, completely or partly, by the supply from pumped storage power stations.

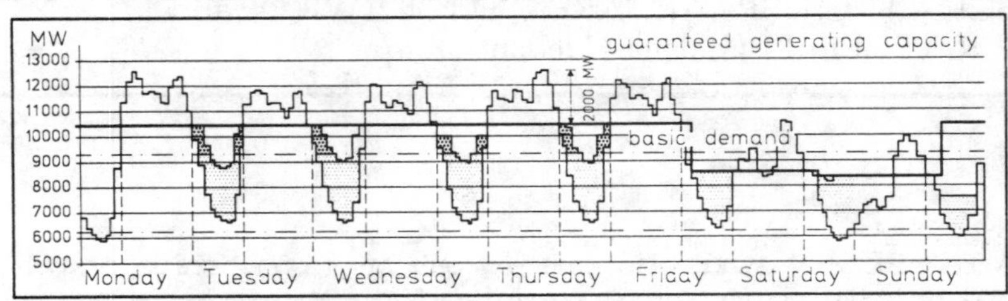

- Figure 1: Example of a demand curve of the electricity grid

5. Up till now, the pumped storage accumulation and generation method has proved to be by far the most economic of the various possibilities for storing energy. Obvious advantages of such a system are:

- the stored energy can be supplied almost immediately;
- the adaptation to fluctuations in the demand is almost limitless;
- the installations can be easily operated and have a high operational safety.

6. Since the Netherlands have no natural possibility of getting water to supply an elevated reservoir without assistance, a solution must be sought in which the water is pumped up. Experience abroad indicates that even then a profitable process is possible.

7. In addition to the possibility of buffering the variable supply of alternative forms of energy in the future, it can now also be economically attractive to store electrical energy which is cheaply available during certain periods (night and week end) and to sell this energy at, of course, a higher price during periods of increased demand.

HISTORIC DEVELOPMENTS

8. Between the years 1910 and 1927, the first energy-storage plants were built in Italy, Germany and Switzerland. Their generating capacity was limited to a couple of megawatts. Plants with a generating capacity of 20 to 40 MW were built in Germany between 1928 and 1938. After 1949, numerous projects with pump turbines were carried out in France, Italy, Germany, Great Britain and Switzerland.

9. In 1979, the total installed generating capacity in pump turbine power stations within the European Community was estimated at approximately 10,000 MW; approximately 7,250 MW was still under construction.

10. In 1980, in the United States, approximately 20% of the hydraulic power stations were equipped with pump turbines. The

Table 1: **Examples of existing pump turbine power stations.**

Location		Head (m)	Capacity (MW)
* Plate Taille	- Belgium	47	135
* Gabriel y Galan	- Spain	59	110
* Torrao	- Portugal	52	74
* Hiwassee	- Tennessee, USA	63	76
* Ludington	- Michigan, USA	100	1842
* Smith Mountain	- Virginia, USA	58	108
* Jocassee	- S.Carolina, USA	105	200
* Midono	- Japan	72	245
* Mazegawa	- Japan	100	288

same can be said for Japan where a turbulent development took place which started as early as 1975.

11. The modern pump turbines have a considerable generating capacity, 250 - 350 MW per machine. For a head of 40 - 500 m., the radial type, the Francis machine, is by far the most common. The interest in heads with a range of 30 - 100 metres has strongly increased during the last few years; as a consequence, the machines appropriate for this range have experienced a strong development resulting in a continuously increasing diameter of the runner and an improvement of the output.

12. In table 1, some examples are given of projects which have been realized in the meantime and which are equipped with vertical Francis machines, which fit very well into the Dutch situation, especially with regard to the available head.

DEVELOPMENTS IN THE NETHERLANDS

13. In the Netherlands, there have been few plans to extract energy from hydropower. Until recently, the natural head of our rivers proved to be insufficient to exploit a power station with enough output. Even tidal power-stations cannot be taken into consideration in view of the limited tidal range.

14. At the end of the seventies, an idea was launched in the Netherlands for large-scale utilization of wind-turbine plants in the national electricity production system. As a result of the strongly fluctuating wind supply, this kind of application caused technically and financially insurmountable problems.

15. At the end of 1979, Ir. L.W. Lievense presented a memorandum, entitled "Voorstel brandstofbesparing door toepassing van geaccumuleerde windenergie tezamen met piekegalisatie van het landelijk electriciteitsnet" ('Proposal for saving energy through application of accumulated wind energy in combination with peakshaving of the national electricity grid'). In the Lievense Plan, a solution is proposed to introduce a link between the electricity gained from wind energy and the electricity supply system in the form of an energy buffer. Acting as a buffer system, the water storage method was chosen, in which electrical energy can be temporarily transformed into potential energy by pumping up water to an elevated reservoir, which is later followed by the inverse process.

16. The uniqueness of the reservoir to be built in the

Netherlands is represented by the absence of a natural difference in altitude and by its construction in a lake or in a shallow part of the sea, since there is no room on land.

17. In the middle of 1980, the Government ordered a counselling committee to study the above-mentioned memorandum. In May 1981, this committee issued its report called "Windenergie en Waterkracht" ('Wind Energy and Hydropower').

18. In view of the positive results of the study, it was recommended to undertake with the greatest possible speed the preparatory studies and procedures for making the necessary policy decisions. As a supplement to this report, the 'Appendix Windenergie en Waterkracht' ('Appendix Wind Energy and Hydropower') was issued in January 1983.

19. On the basis of these reports, the Government disclosed their views. These were laid down in "Windenergie en Opslag" ('Wind Energy and Storage'), July 1982. In this report, the Government announced that a demonstration wind energy power station was to be built as soon as possible (at this moment, this power station is already in operation in Friesland province) and that wind energy must be developed further.

20. It was also decided that feasibility studies were to be carried out with respect to aboveground and underground Pumped Storage Power Plants (PAC and OPAC, respectively), as well as of a Compressed Air Accumulation Gas Turbine Power Station. These studies would serve as a basis for any Government decisions to be taken. The "Nederlandse Energie Ontwikkelings Maatschappij B.V." (NEOM) ('Netherlands Energy Development Company') was the commissioner on behalf of the Ministry of Economic Affairs. The feasibility study of the Pumped Storage Power Plant (PAC) was carried out by a project group which has been set up especially for this purpose and which was chaired by NEOM. Participants in the PAC Project Bureau are "Rijkswaterstaat/Rijksdienst voor de IJsselmeerpolders" ('Department of Public Works/Public Services for the IJsselmeer Polders'), "Hollandsche Beton Groep N.V.", "Ballast Nedam Groep N.V." and "Ingenieursbureau Lievense".

21. The aboveground Pumped Storage Power Plant could be located either in the IJsselmeer or on the Dutch coast.

22. The first phase of the study was directed at making a choice of one or more preferential variants. These variants were checked for reliability, feasibility and investment expenses. Safety, environment and planning aspects formed an integral part of the study.

23. In the second phase of the study a more detailed design took place, including a provisional choice for the location and optimization of the dimensions.

THE PRINCIPLE OF THE PAC

24. It is economically attractive to store electrical energy which is cheaply available in certain periods (night and week end) and to sell this energy at prices that are, of course, higher during periods of increased demand. In Pumped Storage Power Plants, this process takes place by temporarily trans-

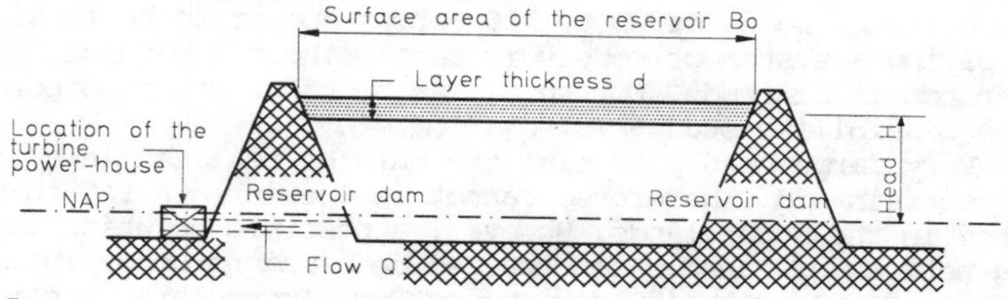

- Figure 2: Basic principle of a PAC system

forming electrical energy into the potential energy of water, followed by the inverse process.

25. Energy storage takes place by pumping up a water volume V to a head H within the reservoir to be formed by the dams and with water surface area Bo. The layer thickness is d = V/Bo (see figure 2).

26. In a later stage, the same water flows down and in doing so it puts turbines into operation that are connected to motor generators. The stored energy is thereby released in which process (limited) losses occur during the various energy transformations.

27. It is a prerequisite that the balance of the selling price and the purchase price of the electrical energy, taking into account the occurring losses, is positive.

28. Based on a potential energy volume of 20 Gigawatthour (GWh) and a guaranteed generating capacity of 2,000 Megawatt (MW), these electrical units are, in order to gain a better understanding, "translated" into units in which only water volumes (V,Q) and the head (H) occur as variables.

29. By using these simplified relations, one can draw up the table 2 shown below from which an impression of the dimensions for a PAC is obtained.

30. As has already been stated, the use of an invariable value for the height of head H is a simplification. It is not used in the actual study. Extensive calculations indicate that a relation layer thickness/head which is 0.15 to 0.30 leads to economic solutions.

31. Heads of 1 - 10 metres are not realistic for a PAC. These heads are only suitable for a river power-station or tidal

Table 2: Relationship between surface area and layer thickness.

Reservoir surface area Bo (km^2) as function of the layer thickness d (m)

H (m)	V (m^3)	Q (m^3/s)	d=1	d=5	d=10	d=20
1	7200 x 10^6	200,000	7200	-	-	-
10	720 x 10^6	20,000	720	144	72	-
100	72 x 10^6	2,000	72	14.4	7.2	3.6
1000	7.2x 10^6	200	7.2	1.44	0.72	0.36

power-station. Heads of up to 100 metres appear to be feasible by applying a system of reservoir dams without a natural difference in altitude. For this reason, they were incorporated in the feasibility study.

32. Very large head potential can only be realized in a very hilly terrain and, therefore, cannot be chosen as a PAC to be located in the Netherlands. However, a possible solution is to use a network of tunnels constructed deep underground. This system is known as the "Ondergrondse Pomp Accumulatie Centrale" ('Underground Pumped Storage Power Plant') (OPAC).

PUMP TURBINES

33. Pump turbines are a development of the more traditional turbines, in which paddle wheels or runners driven by water (or gas, but that is not important in this context), transmit a torque to a motor generator. Turbines can be classified as follows:

- If the driving water discharge is parallel to the axis of rotation of the runner, this is referred to as the axial type. Example: the Kaplan machine.
- If the driving water discharge is at an angle of approximately 45° to the runner and the axis of rotation, this is referred to as the semi-axial type. Example: the Dériaz machine.
- If the driving water discharge is in plane with the runner and directed towards its centre, this is referred to as the radial type. Example: Francis machine.
- If the driving water discharge is in plane with the runner and directed according to its tangent, this is referred to as the tangential type. Example: Pelton machine.

34. The first three types are overpressure machines and can be equipped in such a manner that they can also operate as a pump. Such a machine, a pump turbine, is highly suited for use in a PAC.

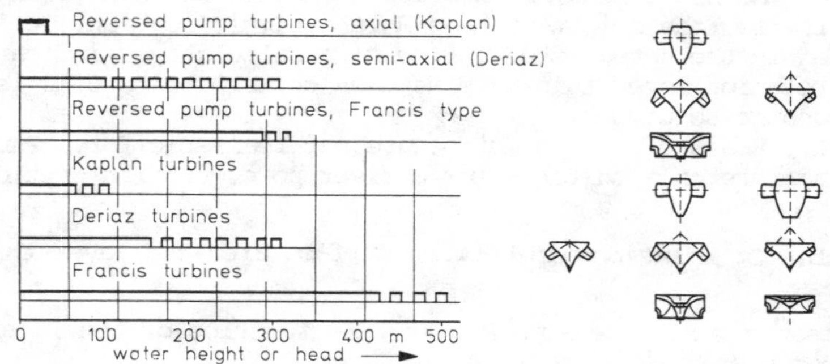

- Figure 3: Comparison of the maximum working heights of the Kaplan, Francis and Deriaz turbines and pump turbines

35. The areas of application are roughly as follows (figure 3)

Axial	H < 15 m
Semi-axial	H = 20 - 50 m
Radial	H = 40 - 500 m

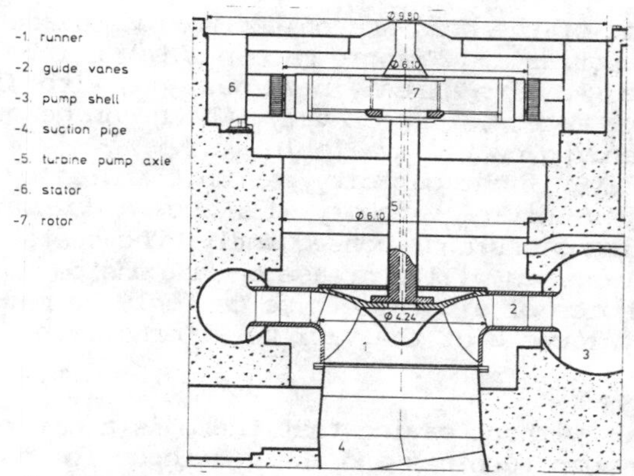

Figure 4 : The interior of the pump turbine power station at Plate Taille

36. In principle, all types can be equipped both with a vertical or a horizontal axis of rotation, whereas a vertical axis is preferred for the Francis machine.

37. While the pump turbine is the real "heart" of the PAC system, the civil engineering aspect is to a large extent also determined by the choice of turbine type (See figure 4). The most important basic conditions for the dimensions of the casings are determined on this basis.

38. The following are worth mentioning:

- the minimum internal and external dimensions;
- the shape of the inlet and outlet ducts;
- the depth of the runner in connection with the risk of cavitation, and, consequently, the foundation depth;
- the requirements for the mechanical installations that are specific to each type of turbine, such as the facilities for draining the turbines, valves, sluice valves, maintenance facilities, etc.

39. In view of its importance for the proper operation of a pump turbine, the topic of cavitation during the pumping phase will be discussed in a little more detail. If the underpressure on the suction side of the pump rises to a level at which, at the prevailing water temperature, the vapour pressure of the pumped water is not reached, the water will vaporize. The vapour bubbles that are carried off with it end up in the runner, where they will implode under the prevailing pump pressure. This phenomenon is called cavitation, and is accompanied by an irregular flow pattern (vibration).

40. When the vapour bubble implosions take place near the materials' surfaces, there will be more wear at that spot; the runner blades in particular can become seriously affected.

41. Sufficient depth of the runner is therfore necessary to prevent this phenomenon from occurring. Guidelines based on theoretical and practical knowledge, which were laid down by the I.E.C. (International Electronic Commission), are already available. Nevertheless, model tests are needed in order to make the final decision. The capacity of a pump turbine depends

on the height of the head and on the maximum diameter of the runner. The Dériaz and Francis machines taken into consideration for the PAC have already been designed with diameters of six and eight meters, respectively, which can generate 50 to 250 MW, depending on the available head.

42. Normally, if the capacity per unit increases, the investment costs per unit of capacity will show a downward tendency, while the dimensions and, consequently, the costs of the appropriate power-houses will increase quite considerably. Therefore, the choice of a turbine type can only be made by comparing the total costs of the various alternatives.

POWER-HOUSES

43. It goes without saying that there is a clear relationship between the pump turbines and the corresponding power-houses:

- the minimal dimensions are imposed by the type of pump turbine selected.
- the head, in combination with the type of pump turbine, determines the depth of the runner, and, with it, the level of the foundation of the casing.
- the mechanical and electrical installations that are specific for each type of pump turbine require facilities in the casing.

44. Other factors, listed below, can be a reason to opt for dimensions larger than the minimum values:

- the height of the (horizontal) earth and water pressures;
- bad foundation properties of the subsoil;
- the requirement that the top of the casing must remain above the water surface at all times;
- the requirement that the casing must be capable of being completely drained for maintenance.

45. In principle, the casing can be situated at two different locations with respect to the dam:

Within the dam (figure 5):

46. In that case, the casing is part of the reservoir dam and, therefore, part of the dam itself; the stability requirement necessitates large dimensions that increase strongly progressively with the head.

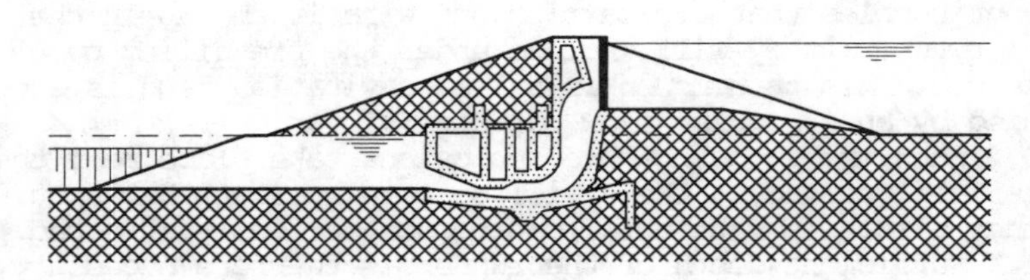

- Figure 5: Power house built entirely into the dam

Outside the dam (figure 6)

47. Initially, the dimensions are not determined by the external loads, but by the type of pump turbine selected.

Transport of water between reservoir and pump turbine takes place through waterducts.

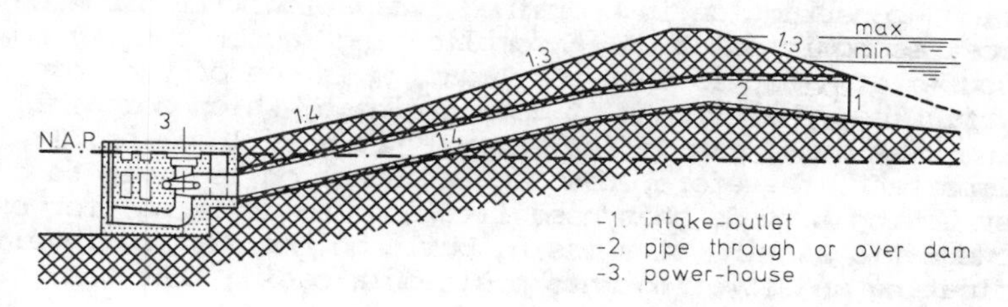

- Figure 6: Power house situated outside the dam

48. If the head is small, a location within the dam is an economical solution; when the head increases, a location outside the dam in combination with waterducts appears to be a more economical solution.

49. The waterduct is subject to high soil and water pressures, the latter occuring both internally during water transport through the pipe, and externally when the pipe is drained. Furthermore, it is possible that a water hammer occurs as a result of large changes in discharges within a short period of time. This can be caused by operating the closing devices and the switching on and off of the pump turbine. Finally, one should take into account the possibility of enforced deformations because of (differential) settlement within the dam. Their diameters being 8 to 10 metres, the dimensions of the waterducts are relatively large. This choice was made because it is important for economic reasons that the rate of flow in the waterducts and the related energy losses should be limited.

50. Concrete, steel or combinations of these materials are appropriate construction materials. For casings, concrete is the only appropriate material. To give some idea, it should be noted that the external casing dimensions will be in the order of 50x30x40 = 60,000 m^3 per casing; 50 to 60% of this volume consists of concrete.

51. Although the scale is impressive, and despite the compli-

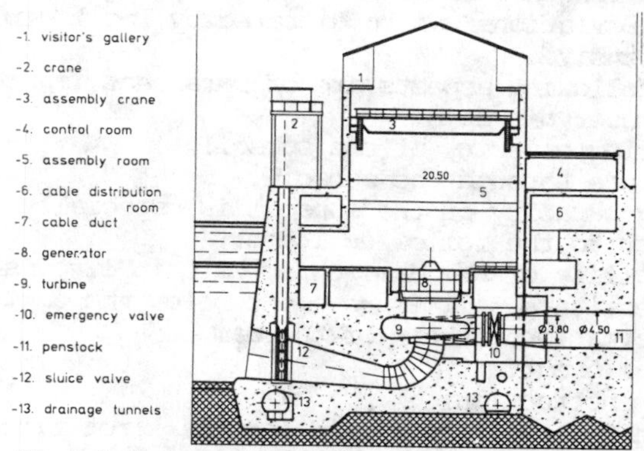

Figure 7: Cross section of the pump power-house in Plate Taille

cated design of the flow ducts, it can hardly be called a complicated concrete structure (figure 7). In the Dutch situation, construction in a drained excavation will often be the most economic solution. A complicating factor for the powerhouses of a PAC is that the location is one of the (many) variables. The effects of deep drainage, which certainly is the case here, can differ strongly at each location. It is important, therefore, that ample attention is given to construction methods that need little or no draining, for example pneumatic sinking of a casing built on the surface level, or floating and lowering into position a casing that was built elsewhere.

52. When a PAC is in operation, large amounts of water are periodically abstracted from and then drained back into the surrounding waters. The consequences of this process greatly depend on the local circumstances:

- if a PAC is situated in a closed-off water area of limited size, such as the IJsselmeer lake, there are small fluctuations in the level of the lake;
- if the flow can be directed towards an existing channel of sufficient size in a larger regime, such as the North Sea, the effects are negligible;
- in the case that there are only shallow waters in the vicinity of the PAC, measures need to be taken to ensure rapid spreading of the water's flow and the accompanying attenuation of the flow-rates.

53. As indicated previously, the flow is roughly inversely proportional to the head if the capacity remains the same. In hydraulical terms, the systems with a larger head have less impact on their environment.

RESERVOIR DAMS

54. The reservoir of a PAC in the Netherlands can be realized by means of earthfill dams. There are several ways of doing this. With the materials available in the Netherlands a lined sandfill dam will be the first option.

55. From the point of view of costs, one should aim at minimum dam lengths. This is realized by means of a circular reservoir. Deviations can be dictated by local circumstances, but are expensive.

56. The following aspects are of paramount importance for the design of reservoir dams:

- the bearing capacity of the subsoil;
- the seepage through the subsoil.

The bearing capacity of the subsoil is especially important with respect to the following items:

- the stability of the dam; under certain circumstances it may necessitate soil improvement below the dam.
- the deformation pattern of the dam.

Deformations occur:

* during construction;
* when the reservoir is filled for the first time;
* as a result of the level fluctuations in the reservoir.

57. The lining materials to be used must be able to withstand these deformations. If, however, this is impossible to achieve, measures must be taken to limit the deformations (soil improvement).

58. The seepage through the subsoil and the dam has the following important aspects:

- seepage means loss of water, and an ensuing loss of energy; the costs of any measures must be weighed against the capitalized energy losses;
- seepage through the dam causes an elevation in the water table of the dam; if this causes a decrease in dam stability, it will be necessary to take measures, which could include making the banks less steep;
- seepage through the bottom of the reservoir might penetrate the denser layers in the subsoil and cause an increase in local pressure; this could spread to outside the reservoir, where the upper layers may burst open;
- seepage through the subsoil may manifest itself below denser layers far outside the reservoir. The relatively salty groundwater in a coastal location that is carried off with it might cause an increase in the salt concentration in the adjoining more brackish but still arable areas.

59. The occurrence of seepage can be adequately prevented by means of the following measures:

- lining the banks of the dam on the side of the reservoir, if necessary in combination with drainage of the dam body;
- lining of the reservoir floor; in certain locations the present impermeable layers can be used to do so.

60. Whether or not these measures are necessary depends on the relationship between costs and benefits, and, of course, on the technical necessity. The chosen dam form is given in figure 8.

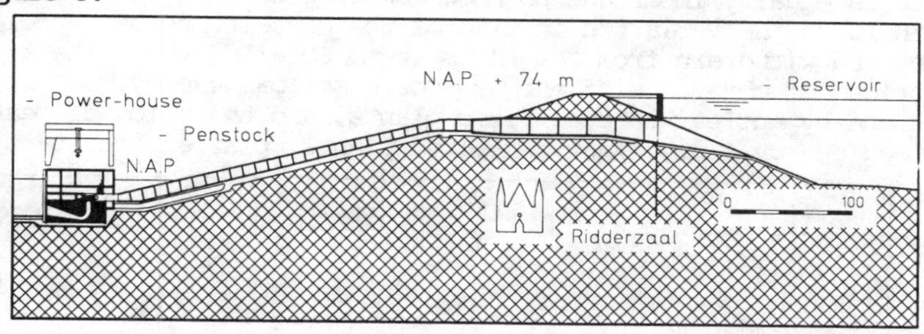

- Figure 8 Cross section of the dam profile

61. The costs of the dams are an important element of the total costs of a PAC. The total costs of pump turbines and their casings decrease as the head increases, whereas for dams this situation is reversed. This rate of cost increase is less rapid, however, than may be expected at first sight.

62. The volume of sand is proportional to the head, the surface area of the slope lining the reservoirs is constant and the surface area of the floor lining within the reservoir is inversely proportional to the square of the head.

63. On balance, however, there will be an upward tendency in the total earthwork costs, especially if the head is large. On account of the moderate character it is to be expected that the aggregate building costs will be at their minimum if the heads are reasonably high.

SAFETY, ENVIRONMENT AND PLANNING

64. In the history of large-scale hydraulic engineering projects in the Netherlands as well as abroad, the impact on waters and landscape used to be hardly, if at all, taken into account. The increasing attention for and appreciation of the environment also resulted in that, for larger projects, the effects on the environment and planning are investigated and considered more often when selecting alternatives and their design. This has even been laid down in legislation by way of the "Wet Milieu Effect Rapportage" (Environmental Impact Reporting Act, or MER). It will be obvious that the PAC study has also given ample attention to aspects such as flora, fauna, landscape, functionality of use, safety and to how these aspects can be optimally integrated in the system. The effects of the pump turbines on the fish stock and the influence of the reservoir on the quality of the water are particularly important points. A survey of the relationships between the PAC and these aspects have been studied are represented in a relational diagram (see figure 9).

OPTIMIZATION

65. Essential for dimensioning a PAC is the matter of optimal dimensions and optimal properties. Full optimization can take place only when the characteristics the PAC must meet have been determined. In addition to defining the potential energy contents of the reservoir and a guaranteed generating capacity, it is equally important to know how to use the system. This includes the variation in time of the capacity to be supplied by and withdrawn from the PAC system. Equally, the market prices of cheap - wind energy - or base-load energy, as produced by nuclear or coal-fired plants, and the price of peak-load energy from oil or gas-fired plants, play an important

Aspects	Environment					Planning			Technology	Safety		Economy
Constituent aspects / Characteristics	Landscape	Water quality	Fish	Birds	Flora	Water supply	Water sports	Infrastructure	Civil + mech. works	Working hypothesis	Risk analysis	Optimized building cost
I Construction phase												
Sand winning		□							■			■
Spoil disposal		+		+	+				■			+
Drainage						+						+
Soil conditions									■			■
II Operational phase												
Space occupied	□	□	□	□		□	□					□
Reservoir height	□						□		■		□	□
Distance to land	+			+	+	+	+	+				+
High voltage cable	+			+				+	+	+		+
Seepage		+			+	+		+	■		■	■
Flow rates		□	□	□					□			□
Level fluctuations				□		□	□				□	
Water depth		□	□									
Type of turbine		□							□			□
Runner diameter		□							□			□

□ relations important for preferential configuration
■ relations important for choice of location
+ relations important for both

- Figure 9 : Relations between PAC and environment

part in any calculation. Finally the choice of location determines many of the design considerations, such as the soil's bearing capacity and impermeability to seepage.

66. A number of potential locations was investigated on the basis of available geotechnical profiles, and the field narrowed down to two in the second phase of the study, one in the IJsselmeer, the other offshore. To this end, preliminary soil-investigations were carried out. Environmental, cost and safety considerations finally led to a preference for a PAC in the North Sea, in shallow water on the outside of the Brouwersdam in Zeeland province (figure 10).

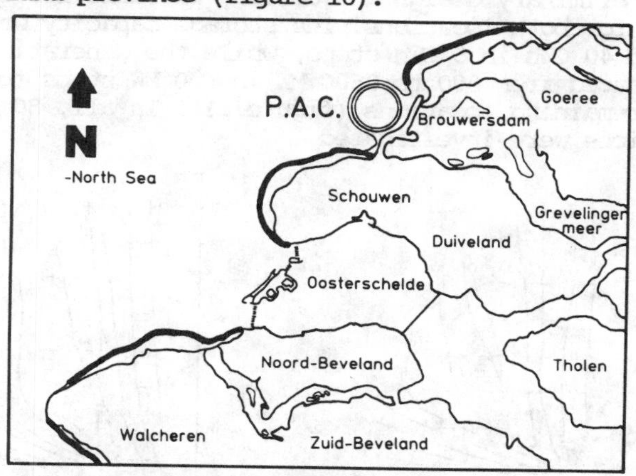

- Figure 10: PAC location off the Brouwersdam

67. Even then the number of possible combinations of reservoir height, reservoir surface area, water layer thickness to be used and number of pump turbines, with which the required capacity and energy supply can be met, is theoretically unlimited. The objective of optimization is to determine the combination that has the most favourable cost/benefit ratio related to, among other things, the energy costs. And although this is not necessarily the system with the lowest investment costs, a relatively low investment level is desirable.

68. In addition to the above-mentioned aspects of dimensioning and use, there are a number of important technical aspects that also play a role. For instance, the type and the runner diameter of the pump turbine and the number of pump turbines are decisive for the overall energy output of the system as a whole. Preference will usually be given to pump diameters with a large runner diameter, as the cost/benefit ratio of the system as a whole becomes more favourable as the diameter increases. However, the controllability of the system as a whole subsequently decreases. The technical feasibility contributes, of course, to the choice to be made, which obviously holds true for all component elements in a PAC system.

69. Other important optimization aspects are the seepage losses that occur and the ensuing energy losses in the system. Both have repercussions on the available and withdrawable storage capacity and can, therefore, be capitalized. In order

to keep these losses within reasonable and acceptable limits it will be necessary to make certain capital investments.

70. This will result in a possible increase in the total investments to be made, but in the end this will result in an optimal system with the most favourable cost/benefit ratio.

71. The optimization of a PAC system is a complex matter, due to the many related aspects. It was necessary, therefore, to use computer models when optimizing the system. As the dominant variables of storage capacity and generating capacity were also open to optimization in the second phase of the study, preliminary design and costing was done for each combination for both locations. The storage capacity was varied from 5 to 40 GWh in 5 GWh steps, while the generating capacity was increased from 500 to 2500 MW in 500 MW steps for each of the two remaining locations (figure 11). In all, 80 possible combinations were investigated.

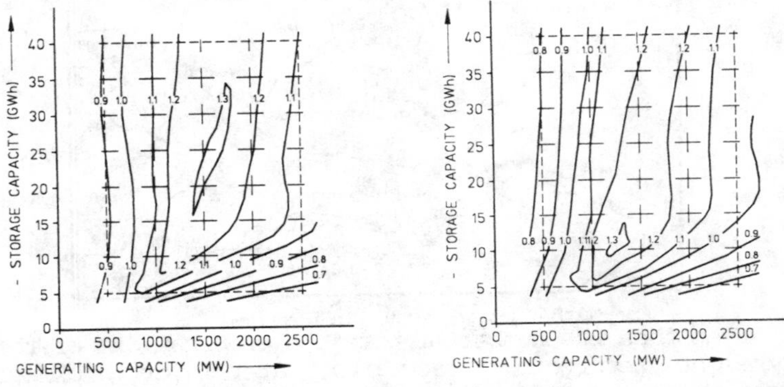

Figure 11: Lines of equal benefit/cost ratio for varying storage and generating capacities

72. These considerations led to a choice of optimal storage and generating capacities for the two locations of 30 GWh and 1500 MW respectively. With these parameters, see table 3, the preliminary designs were fleshed out into more detailed designs. The probable cost of the project was calculated at some Dfl. 3500 million ± 10%.

Parameters		IJsselmeer	Brouwersdam
Nett storage capacity	GWh	30	30
Generating capacity	MW	1500	1500
Maximum head	m	70	70
Minimum head	m	54.3	55.4
Reservoir diameter	m	4130	4220
Number of units	-	8	8
Maximum turbine capacity	MW	2150	2150
Average pumping capacity	MW	2150	2150
System efficiency	%	78.8	78.4

Table 3: Preliminary design parameters

CONCLUSION

73. We have given a brief outline of the many interesting civil - and financial - engineering problems occurring with a Pumped Storage Power Plant. The effects of the variations in energy prices, and accidents with nuclear reactors, have put the project on the back-burner for the time being. The time scale necessary for a project of this magnitude however calls for a long term energy strategy by the government, based not only on the cost/benfit ratio now, but also on the need for diversification in generating capacity in the years to come.

74. The realization of such a project will undoubtedly give a great impetus to the many sectors in the field of hydraulic engineering. It will also be a new dimension for Holland's engineering image, especially in connection with the export of knowledge and skill. It may also be emphasized here that the integration of civil engineering, electrotechnical engineering and hydraulic engineering technologies at the PAC must take place at a scale hitherto uncommon in the Netherlands. This poses an additional challenge for the future.

REFERENCES

1. Windenergie en Waterkracht, May 1981.
2. Appendix, Windenergie en Waterkracht, January 1983
3. Windenergie en opslag, July 1982
4. L.W. Lievense; Nieuw Plan Lievense voor Razende Bol bij Texel, De Ingenieur - no. 3, March 1985.
5. P. Henry and J.E. Graeser; Energy storage: developments in pumped-storage, Water Power & Dam Construction, June 1985.
6. S. Casacci, J. Bose, P. Hudon: Cylindrical protection gates for hydraulic turbine machinery, Neyrpic technical review.
7. A. Josserand, J. Delaroche; Neyrpic downstream gates, Neyrpic technical review no 1., 1982.
8. W.D. Geiseler and W. Kuhlmann; Asphaltic concrete linings for pumped-storage reservoirs, Water Power & Dam Construction, June 1985.
9. O. Hartman; Reliability improvements for hydro plants in the USA, Water Power & Dam Construction, January 1984.
10. W. Meier, J. Müller, H. Grein, M. Jaquet; Pump-turbines and Storage Pumps, Escher Wyss.
11. P. de Genst; The Plate Taille pumped-storage power station (Froidchapelle, Belgium), ACEC Review, no 3-4, 1979.
12. H. Grein, M. Jaquet; Operational flexibility of various designs of pumped storage plants, International symposium and workshop on the Dynamic benefits of energy storage plant operation, Wisconsin - Milwaukee, May 1984.
13. Masanori Nose; hydroelectric power Generation, Japan Society of Civil Engineers, 1983.
14. J. Parmakian; Flatiron Power and Pumping Plant, Mechanical Engineering, August 1955.
15. P. Dériaz; The Mixed-Flow Variable-Pitch Pump-Turbine, Water Power, February, 1960.

16. L. Gilde, J.A. de Ridder, A.L. van Schaik, C.J. Spaargaren and D.P. de Wilde; Een Pomp Accumulatie Centrale in Nederland, i^2, November 1985.

Pumped storage: the environmentally acceptable solution

R. WATTS, Balfour Beatty & Co. Ltd

SYNOPSIS There is seen to be a need in the industry for the provision of Peak and Spinning capacity after the end of this century and there is a need for this to be widely distributed around this country. This paper addresses the stage development of one proposal.

LOCATION

1. In searching for a suitable site, the following points have to be taken into account:
(a) suitable layout of the land
(b) suitable source of nuclear power to provide the pumping load at night
(c) as short a distance as possible from the main centre of load and pumping source.
(d) little environmental impact
(e) any other useful additions to the environmental equation

THE CHOICE OF SITE

2. The basis of this proposal is that a pump storage scheme be built on the coast in the South East near a suitable source of supply of cheap night time supply. Any location from Southampton to Ipswich could be considered. The design is for the Upper Reservoir to be in the sea and the Lower Reservoir and Power Station to be excavated in the carboniferous limestone strata.

3. The proposal of a site at Dungeness is based on the following points:
(a) A suitable source of nuclear power is on the site so that the transmission losses would be at a minimum for the greater pumping load.
(b) The power lines are already in position for the delivery of the generated load away from the station.
(c) The distance to the main centre of load in the South East of the country is far shorter than alternatives i.e. in North Wales or the Peak District (for Surface Schemes).
(d) The environmental impact will be minimal as the power station and the lower reservoir will be underground and all that will be above ground will be the transmission to the existing lines on the site, and some surface administration and back up.

(e) The nearness of the Trans-Channel cable crossing to France from Sellinge should bring in the possibility of buying electricity from nuclear stations in France.
(f) The lower reservoir and power station would be constructed in the carboniferous limestone which would give rise to a ready sale of crushed rock in the South East area of England.

THE REQUIRED ELEMENTS OF THE SCHEME

4. The Top Reservoir - This would be provided by the sea at a very cheap price and sea water would be used for the operation of the scheme.

5. The Shafts - The first shaft would have to be sunk by conventional means from the surface using mining type head gear and equipment. However, once the first shaft had been sunk the subsequent shafts could all be formed by means of Raise Bore equipment followed by enlargements as required.

6. The Power Station & Transformer Halls - These would be underground caverns formed in the carboniferous limestone by either a TBM or by conventional drill and blast techniques.

7. The Lower Reservoir - This would be the only item that would be relatively expensive compared with any conventional scheme, but here the cost can be taken to the lower limit by the use of a TBM of 10 to 12m diameter. This type of machine should be able to cut a tunnel some 150 to 200m long per week and so produce a volume of 20,000m^3 each week. It would be possible to design the underground tunnels so that they were independent and were able to be commissioned in stages. See Appendix IV for Hydraulic Layout.

THE ECONOMICS OF THE SCHEME

8. Electricity from nuclear and other base load stations can be supplied for pumping at night at low rates - 1.50p per KwH, or possibly lower. It may be bought from France at 1.00p per KwH. For instance the ordinary high voltage payment for peak loads in 1985 was at 6.05p in the winter for purchases in excess of 2 Megawatts, between 7.30 and 20.00 Monday to Friday. The possibility of spinning reserve for frequency control is important. The CEGB has been able to save at least £25m p.a. at Dinorwig from this facility alone.

9. The only cost comparison that can be made easily is with the Civil Engineering costs stated for the Dinorwig Scheme. It can be assumed that the Plant Cost will be higher and that the Transmission costs will be considerably less. The Plant Cost will be higher because of the higher head and the need to deal with sea water. (See Table I)

THE LAYOUT OF THE STATION

10. Three shafts would be required near the shore line by the Dungeness Nuclear Station.

11. One shaft would be required for the waterway from the sea to the turbines. The second large shaft would be for the access to the power station and the third shaft would be for the high tension cables and the ventilation.

TABLE I

COST COMPARISON USING 1983 PRICES FROM THE ICE PAPER

Items of Work	Dinorwig	Dungeness	Similar items proposed
Exploration & Access Tunnel	£ 3.8m	£ 1.0	SI only
Alternative Water Supply	£ 2.4m	Nil	Not required
Office & Miscellaneous (10 years)	£ 1.6m	£ 1.1m	Office etc. for 7 years
Vent Shaft 255m	£ 0.9m	£ 5.0m	Cable & Vent Shaft 1200m
Roads etc.	£ 5.0m	Nil	
Dam	£34.5m	Nil	
Main Works:			
LP Tunnel	£17.4m	£ 3.5m	Intake Structure
Penstock Tunnels 6 No. x 145m	£16.4m	£ 8.5m	Penstock Tunnels 6 No. x 75m
Plant Access	£ 3.1m	£26.0m	Main Access Shaft 1180m x 10m dia.
Personnel Tunnels	£ 0.9m	£ 0.9m	Similar layout
Diversion Tunnel	£ 9.1m	Nil	
Tail Race Tunnels	£10.5m	£ 2.4m	Six Tail Race Shafts @ £0.4m
HP shaft 450m x 10m dia.	£ 9.0m	£26.4m	HP Shaft 1200m x 10m dia.
Machine Hall, Trans. Hall, Valve Gal.	£34.5m	£30.0m	Layout as before but reduced for smaller high head turbines
Concrete Works approx. 0.5mm^3	£25.0m	£15.0m	Reduction in lining works
Lower Reservoir 4Mm^3 + Dam	£14.0m	£44.4m	TBM for 3Mm^3 £60m (less sale of stone £15.6m)
Sundry Items	£22.9m	£17.8m	Sundry Items
Total Cost of works in ICE Paper at £206m plus £6m to final account	£212.0m	£180.0m	Comparative estimate

12. An intake structure would be needed at the top of the water shaft able to keep the sea out during maintenance periods, but otherwise it would be open to the sea at all states of the tide.

13. The Power Station would be a cavern some 20m wide x 50m high and 150m long. This would be accompanied by several smaller chambers for transformers, valves, maintenance areas and the like.

14. The Lower Reservoir would be constructed by the driving of 10m to 12m diameter tunnels in the carboniferous limestone for a distance of about 2.25Km each, there being a need for two such tunnels for each of the six 300m MW sets.

15. All the above layout considerations are based on a duplication of the Dinorwig station layout and arrangements. In order to build the station in the sound carboniferous limestone it would be necessary to design for a larger head for the turbines. This, however, has the advantage of reducing the size of the Lower Reservoir, the Power Station, the civil cost, and the time for construction.

16. Excavation of the Lower Reservoir at a rate of 200m per week, will take three years to construct. (After the first shaft has been sunk and a method of muck raising has been installed.) See Appendix V and VI for Underground Layout.

PROBLEMS OF CONSTRUCTION AND THEIR SOLUTION

17. Site Investigation - So far the site investigation has been only a "Desk Study" of the Geology of the area and the collection of Bore Hole data from Public Geological sources (See Appendix I) Before the proposal is taken very far, a full geological investigation will need to be carried out to locate the carboniferous limestone over an area of 3km by 2km. The whole project depends, not only on the limestone being located at the depth anticipated but also being proved to be in first class conditions. It may well be that geological faults are located by the site investigation that will necessitate the redirection of the twelve tunnels proposed for the lower reservoir. The rock strength of carboniferous limestone is likely to be of the order of 200 MN/m^2 while the overburden stress is likely to be of the order of 30 MN/m^2 so no difficulty is to be expected from inherent instability of the rock.

18. Intake Structures - Many of these have been built on the coast and no difficulty is expected with a design for this type of structure to be built within a large coffer dam.

19. The First Sunk Shaft, for Cables & Ventilation - This shaft should be of the size most readily sunk using traditional methods of a Full Winding Frame and Shaft Sinking Gear. This is likely to be about five metres in diameter and sunk at a rate of 2.5m to 3m. per 24 hours. Thus with the set up time of two to four months it is likely to take about one and a half years to complete. There will be a need for a Main Kibble Winder, together with winders for the stage, and cables. Also as stand-by, a diesel operated hoist will be needed for emergencies. Concreting will be carried out on a

daily basis using a hanging form.

20. The Second and Third Shafts - 11m I.D. for Water & Access - These can be formed by Raise Boring methods being enlarged to 11m diameter after the raise boring has been completed and lined as the excavation proceeds. See Appendix VII for details of recent raise boring depths and accuracy. The water shaft will have to be lined for the full depth for structural and hydraulic reasons and will probably need a steel lining in the lower 100m and in the connections to the Main Inlet Valve Gallery.

21. Access and Safety - It is proposed that the installation would be served by sufficient lifts in the shafts for the whole of the work force to be able to evacuate the Underground Workings should this be necessary. There will be some saving in making the initial Shaft Sinking Gear in such a way that it could be adapted for the permanent installation of winding gear for the removal of personnel. The second safety feature that will need to be incorporated is the closure of the gates at the intake structure should there be any break in the Hydraulic Circuits below ground. The only high pressure waterways are between the bottom of the water shaft and the Machine Hall. These can be protected in the normal way by the installation of working valves backed up by emergency only valves.

22. The Machine Hall, Transformer Halls and Other Passages The techniques needed to excavate such caverns in sound rock are well established and need not cause any difficulties. Drill and blast using Mobile Drill Jumbos and Ramp designs are likely to be most successful, with the roof concreted in arches and walls sprayed and rock bolted.

23. The Lower Reservoir - Here the TBM is the best answer. Robins with their track record around the World (well tested and tried out in Norwegian granite) and specifically the experience on the construction of the Chicago Sewer Tunnels which have been taken out at 10.7m diameter in hard sandstone and with very good progress results. (545 lin m in 21 days.) See Appendix VIII. With full face tunnel boring machines the ground will be subject to the minimum of disturbance and normally will need no support but may need some occasionally on traversing any narrow faulted zones. These and any locked in stresses may be dealt with using rock bolts. These tunnels are only subject to atmospheric pressure and the pressure generated by filling them 95% full of water so the daily change in rock loading will be very small. They will all be connected by a ventilation tunnel running the full length of the reservoir area to the ventilation and cable shaft thus maintaining atmospheric pressure in the tunnels at all times.

24. Layout of Lower Reservoir - The layout of the Lower Reservoir has been made in the form of a series of parallel tunnels. This has been done in order to facilitate the quick transfer of the TBM from one drive to the next. After the first Lower Reservoir Tunnel has been constructed a cross

passage (big enough to take the TBM) will be driven across the end of the reservoir area. This will mean that at each end of the 2½Km drives there will be a cross passage through which the TBM may be translated quickly to the next drive for a further 2½Km of tunnel boring.

25. Ventilation - The ventilation of the project divides into several time periods:

(a)When the cable and ventilation shaft is being sunk the ventilation will be by ventilation tube down to the working level in the shaft.

(b)When the bottom of the shaft is being developed before the raise bores are complete the ventilation will continue under the same system.

(c)As soon as the second raise bore is complete it may be used for a return air shaft until the first raise bore has been completed whereupon the whole system can be put onto a two shaft system with the ventilation and cable shaft being used for the extraction of foul air and the access shaft being the main inlet for clean air.

(d)At an early date the connection from the lower reservoir tunnels to the ventilation shaft will need to be made so that good ventilation is available to the rock boring TBM which will product much dust.

26. Spoil Removal - Vertical - The vertical lifting of the spoil will present an unusual problem, but again turning to the Chicago Sewer project where the vertical steel reinforced pocketed conveyor belt was used with great success the design may be extended for this project. The Flexofast System will be capable of lifting 800 tonne/hour of rock in three vertical sections. Each section will be between 340 and 400m lift with transfer points at the pit bottom, twice in the shaft involving small enlargements and again at the top. The product of the TBM (which will be the greatest tonneage) will be 150mm down and so will be fed straight into the pocket conveyor system without any treatment. However any rock produced by drill and blast will need to be passed through a primary crusher before it is fed to the vertical conveyor system. (See Appendix IXA).

27. Spoil Removal - Horizontal - Here the idea would be to rely to a great extent on the use of conveyor systems (thus reducing ventilation problems). Static Conveyors would be used for the main runs about the Station and Lower Reservoir areas. A special conveyor would be needed to keep up behind the Tunnel Boring Machine. This conveyor would be equipped with a Belt Storage Bank able to hold 100m of Belt. The TBM would be equipped with a conveyor system which would overlap the main conveyor and give a shifts extension without any alteration, say 20m maximum. The conveyor structure would be in 3m lengths and at the end of each shift the appropriate number of 3m lengths would have to be installed to cover the shifts work. Up to four or five times a week a new length of belt would have to be installed. This again would be left to the end of the shift and would take a maximum of 1 hour non

productive time, which could well be used for maintenance on the TBM. The belt would be supplied already with a connector at each end so that it would be a matter of "unzipping and rezipping" which can be done in a matter of minutes.

28. The Use of Salt Water - Salt water in pump storage turbines has been used for a number of years on the Rance Tidal Scheme in Northern France. The problem has been faced and the solution found at fairly low heads. Some further work may need to be carried out to extend the durability and life of seals and other parts not normally subject to salt water at high pressure.

29. Stage Construction - As each turbine's lower reservoir is complete in its own right it would seem worth while to try to make the arrangements to keep these separate initially for a stage development of the station. At a later stage, it may be possible to connect up the reservoirs, so that any one turbine may make use of the whole lower reservoir capacity. This could be of great use in the provision of extra load when individual turbine sets were out of use for maintenance.

CONCLUSION

30. On the basis of the present investigation it appears that the project could be put forward with a comparative cost indication cheaper than Dinorwig. Proved and tried methods of construction are available to tackle the undoubted construction problems that a design of this kind produces. It would appear worthwhile therefore to investigate the proposal in greater detail.

APPENDIX Ia - GEOLOGY (JCM 6.4.84)

Dungeness:- hypothetical Strata thicknesses
based on CEGB Borehole, NCB Borehole, GS Memoir 289 305 306

Wealden	Beds.	38m
Wealden	Hastings Beds.	97m
Jurassic	Purbeck & Portland	-
	Kimmeridge Clay	64m
	U. Corallian	38m
	Corallian LST	38m
	L. Corallian	25m
	Oxford Clay	31m
	Kellaways Clay, Cornbrash & Forest Marble	12m
	Great Oolite, Inferior Oolite	23m
	U Lias	17m
	M Lias	5m
	L Lias	12m
Trias	SST & Marl	25m
Carboniferous	Coal Measures	735m
		1160m
Carboniferous	LST +/- 10%	120m

APPENDIX Ib - GEOLOGY

Adjacent borehole records of limestone:-

	Location	Depth M
1) Harmasole	1400 5270 TR	380 - 388
2) Elham	1970 4390 TR	697 - 716*
3) Adisham	2270 5380 TR	986 - 995
4) Bishopbourne	1890 - 5270 TR	966 - 986*
5) Fairlight	8592 1173 TQ	292 - 396
Further North other boreholes record limestone		
6) Trapham	-	845
7) Chislet	-	640
8) Stodmarsh	-	654 - 690*
9) Walmestone	-	694 - 697*
10) Littlebourne	-	796 - 798*

(The base was not reached in those marked *)

APPENDIX I (c)

MAP OF BOREHOLE LOCATIONS

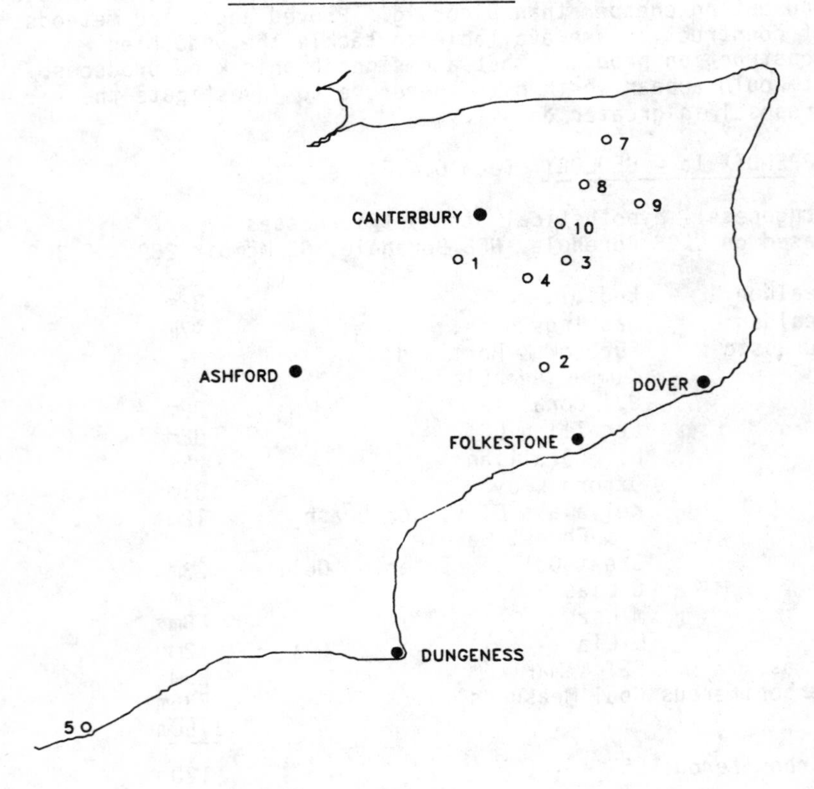

APPENDIX II - List of Recent High Head Pump Storage Schemes

COUNTRY	PUMPED STORAGE PLANT	NO. OF UNITS	UNIT RATING (MW)	HEAD (m)
Australia	Murray	10	119	521
Austria	Sellrain-Silz	2	220	1259
France	Emosson	3	64	750
	Bissorte	4	190	1194
Germany	Holzenwald	2	240	630
Italy	Chiotas-Piastra	8	150	1018
	Lago Delio	8	120	732
	Lete-Sava	2	52	624
	Roncovalgrande	7	121	746
Romania	Lotro	3	170	809
Switzerland	Hongrin	4	60	878
UK				
Wales	Dinorwic	6	300	530
USA	Blair Mountain	3	175	671
Yugoslavia	Bajina Basta	2	300	600

APPENDIX III - Cost of the Lower Reservoir at 1983 Price

This is based on the whole of the construction being carried out with the use of Large Diameter Tunnel Boring Machines of the type supplied by Robins for the Chicago Sewer Project and other Hard Rock Tunnel Projects.

	£/M^3
Cost of TBM £3.5M over 3M m^3	1.2
Spares @ 25%	.3
Power at 1000HP i.e. 746KW @ 3.05p = 5m^3/hr	4.6
Labour @ 0.34 MH/M^3 at £5/hr	1.7
Hoist & Travel, say	2.5
	10.5
Add 90% On cost	9.4
Projected Site Cost	£19.9
Say £20/M^3	

Allow in estimate £20/M^3 to cover for all extras

APPENDIX IV

HYDRAULIC SYSTEMS AT DINORWIG AND DUNGENESS COMPARED

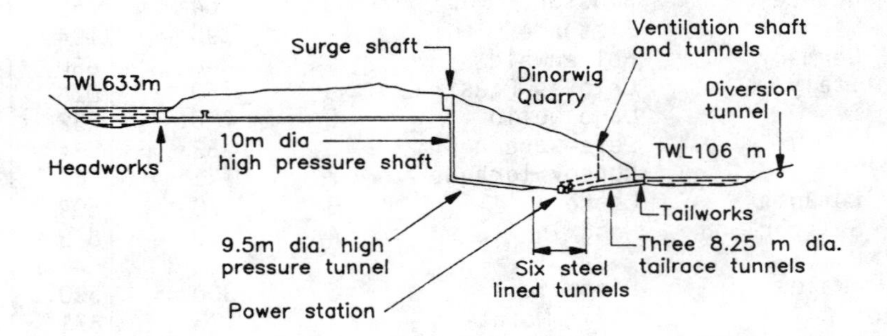

HYDRAULIC SYSTEM AT DINORWIG

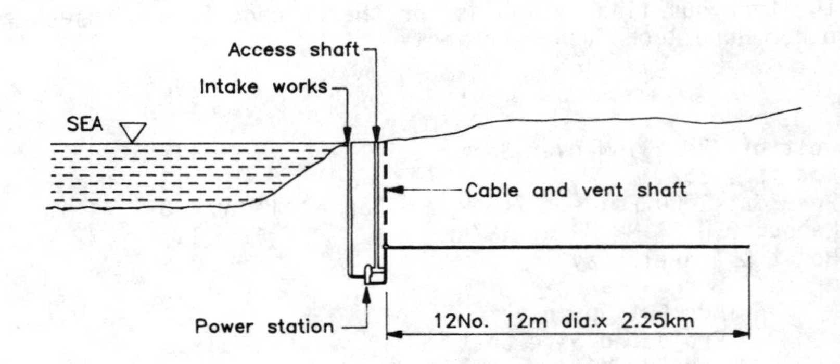

HYDRAULIC SYSTEM AT DUNGENESS

APPENDIX V

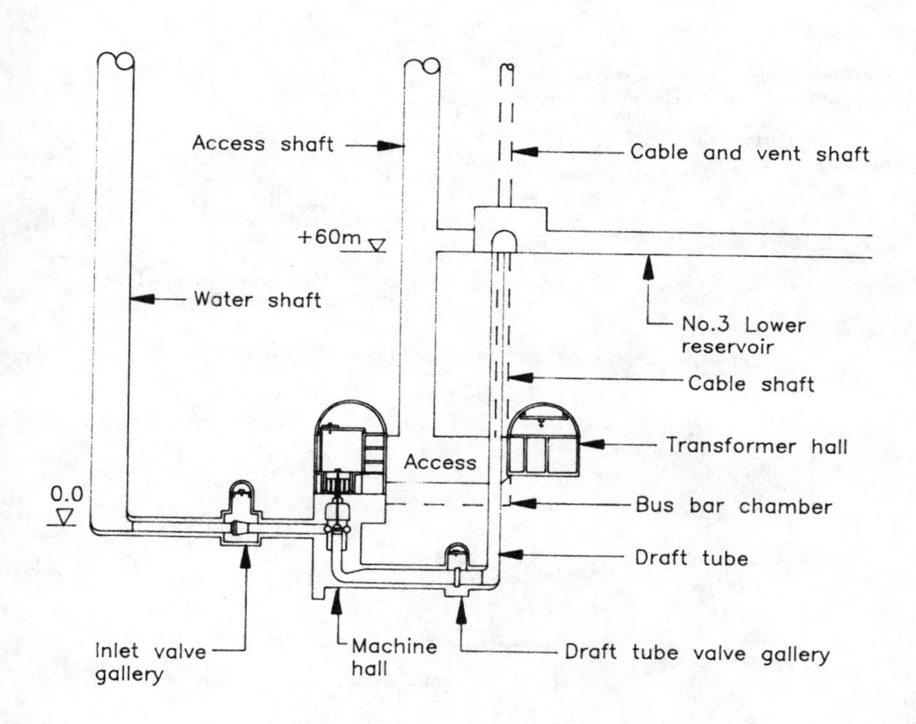

SECTION THROUGH POWER STATION

(MACHINE No.3)

APPENDIX VI

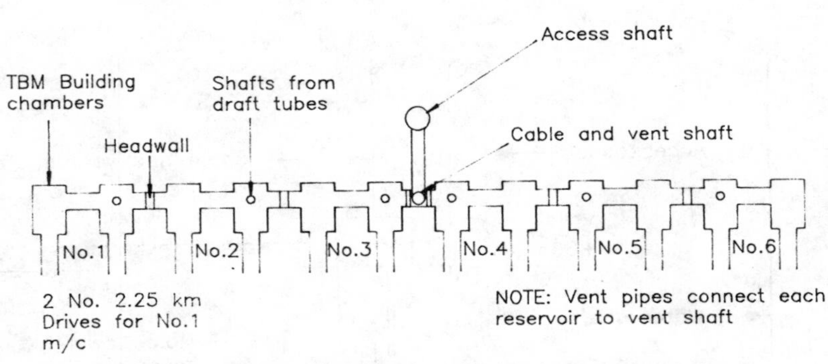

PLAN AT LOWER RESERVOIR LEVEL (APPROX.+60 LEVEL)

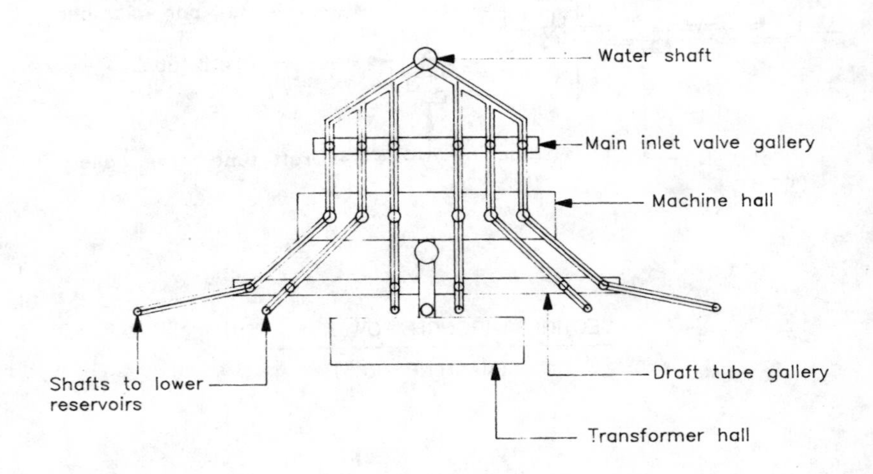

PLAN AT STATION LEVEL (+10m LEVEL)

(Turbines at 0.00 Level)

APPENDIX VII - RAISE BORING RECORDS

Raise boring for the muck shoot of the second and third shafts is quite possible. It is common nowdays to be able to drill 600m in depth and to ream out to 8m diameter. Greater depths have been taken out up to 900m. which may well be deep enough for this project. If it only proves possible to raise bore 600m. then the two longer shafts can be taken in two sections of 600m. each serviced by a small heading and chamber from the first shaft. Once the initial shaft has been drilled it can be enlarged with a raise bore and the final enlargement can be carried out with a V-Mole downwards. The current maximum diameter that has been constructed with this technique is 8.5m. but larger sizes have been designed up to 12m. ID and depths up to 965m.

APPENDIX VIII - EXPERIENCE OF ROBINS HARD ROCK BORING TBM's

Experience of hard rock boring machines on the Chicago Deep Tunnel Reservoir work before mid 1980 was that with six machines having an average diameter of 10m. they had been able to drive 21 kilometers of tunnel at an average machine efficiency of 42% and an average weekly output of 82m. or 247m. per month. The maximum average figures that they achieved were 247m. per week and 565 per month, with an average expenditure on labour of 0.343 man hours per m^3. TBM's working on the Channel Tunnel, have achieved consistantly outputs in excess of 200m./week. Currently, machines are planned up to 12m. in diameter with 480mm cutters and 6500 HP and prospects of 220m/wk. more than twice the experience at Chicago.

APPENDIX XI - The Value of Good Stone in the South East, 1989

The British Aggregate Industry (BACMI) have recently highlighted the need for more aggregate and especially in the South East. The situation in 1985 was that 35 million tonnes were produced in the region and 10 million tonnes dredged locally, the balance of 11 million tonnes was imported from Somerset and Leicestershire. The projection for the year 2006 is a similar quantity produced locally - 35 million tonnes, an increase in the dredged to 17 million tonnes and the remaining 25 million tonnes imported. Seven million tonnes being expected to come from Glensanda in Scotland, the only Scottish coast quarry currently able to export in bulk. The rest will come from Somerset and Leicestershire. There is a movement in both Somerset and Leicestershire to reduce the export of stone to the south east so these increased tonnages may not be met from these sources. Currently good stone is fetching a royalty of £2 per tonne and a sale price of £8-9 per tonne. With the excavation of at least 3.5 million cubic metres of rock (at 2.4 density) required for the project there could be an income of 3.5 x 2.4 x 8 = £67.2m., a not inconsiderable sum when compared with the Civil Cost (at 1983 prices) of Dinorwig of £180 million.

APPENDIX IX A

SPOIL REMOVAL–SHAFT BY VERTICAL CONVEYOR

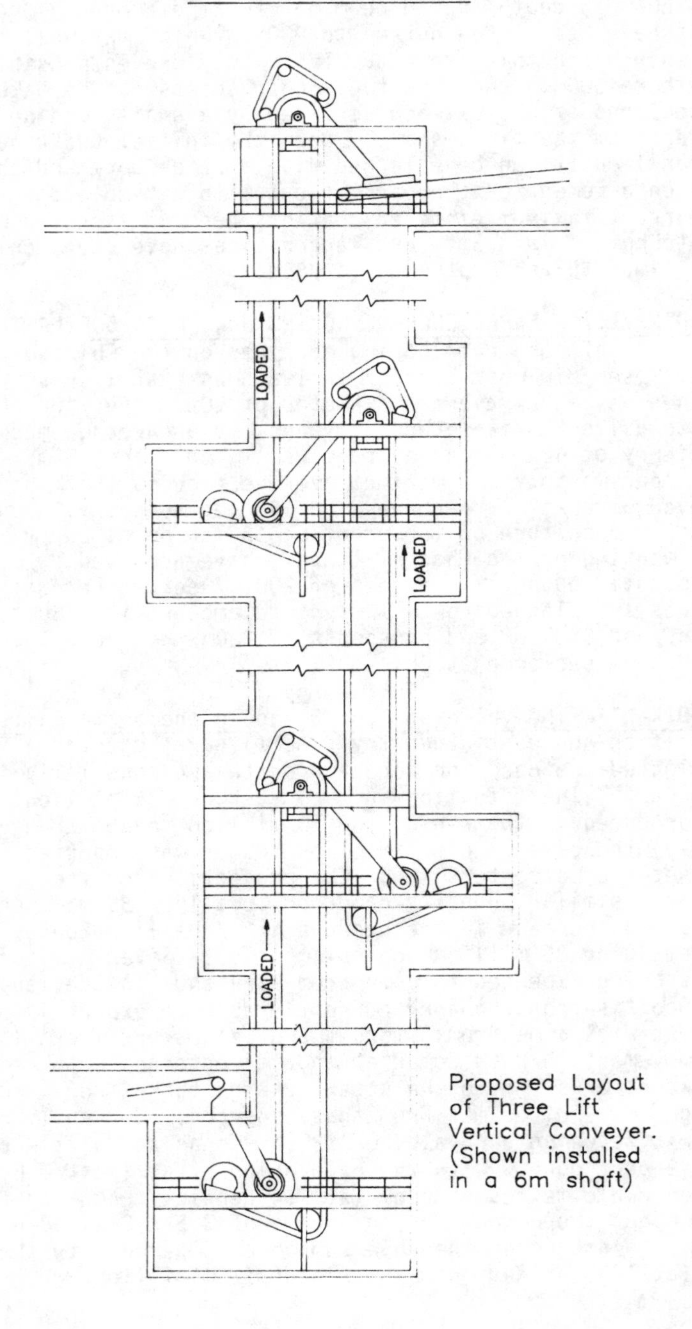

Proposed Layout of Three Lift Vertical Conveyer. (Shown installed in a 6m shaft)

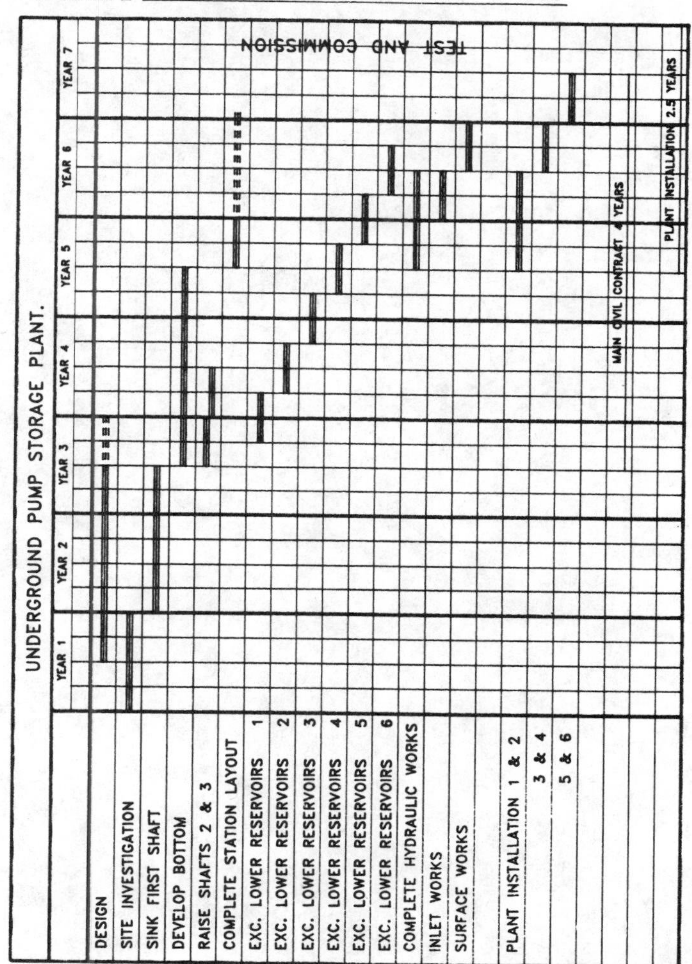

APPENDIX IXB - Spoil Removal Horizontally by Extending Conveyor

For the removal of rock from the TBM an extending conveyor can be used. This is envisaged as having a peak capacity of 800 tonnes per hour and the possibility of extending to a total length of 2500m. To do this it would be necessary to run a 600mm wide belt at a speed of 2.5m/second driven by two 150 HP motors. The extension each shift would take no more than a few minutes and the installation of a further 100m. of belt every third or fourth shift would only take about half an hour, so the disruption to production would be very small.

Pumped storage in the proposed Mersey tidal power project

E.T. HAWS, MA, FENG, FICE, FIPENZ, Rendel Parkman, London, and
E.A. WILSON, MA, FICE, and H.R. GIBSON,
BSc, MIMechE, Mersey Barrage Company

SYNOPSIS
Studies of a proposed 700MW tidal energy barrage across the River Mersey are well advanced. Pumping towards the end of the flood tide to augment water stored in the basin has been found highly advantageous to economics, shipping and the environment. This is a rare example of pumped storage giving a net energy gain.

INTRODUCTION AND HISTORY

1. Since 1983, a series of studies has been carried out on a potential project to develop tidal energy in the estuary of the River Mersey in north-west England (Fig 1), by means of a barrage (ref 1). To date, the studies have covered environmental impact, basic engineering, energy yield, economics and social/industrial aspects. Site investigations and hydraulic model studies have been carried out and the latest pre-feasibility study was presented in late 1988 (ref 2). This showed no fundamental impediment to the proposed barrage.

STATUS AND OUTLINE OF PROJECT

2. The Mersey Estuary is an attractive location for a tidal power barrage. The river discharges through a deep and narrow channel which could therefore be closed with a barrage of economical length. This would provide the water depths needed to set turbines and sluices at effective submergence levels and would also impound an upstream basin of large area. The mean spring tide range of 8.4m is favourable.

3. Two potential barrage sites have been studied in depth and of these Line 3 is shown on Fig 1. The barrage length at Line 3 would be about 1.8km. Strong tidal currents have scoured the estuary bed as deep as 17m below mean sea level and Bunter sandstone bedrock is overlain by several metres of sand, gravels and clays.

4. Fig 2 indicates the proposed layout for Line 3. 28 turbines of 8.0m diameter (25 MW) and 20 sluices 12m square are indicated as a maximum installation although it must be

Pumped storage. Thomas Telford, London, 1990.

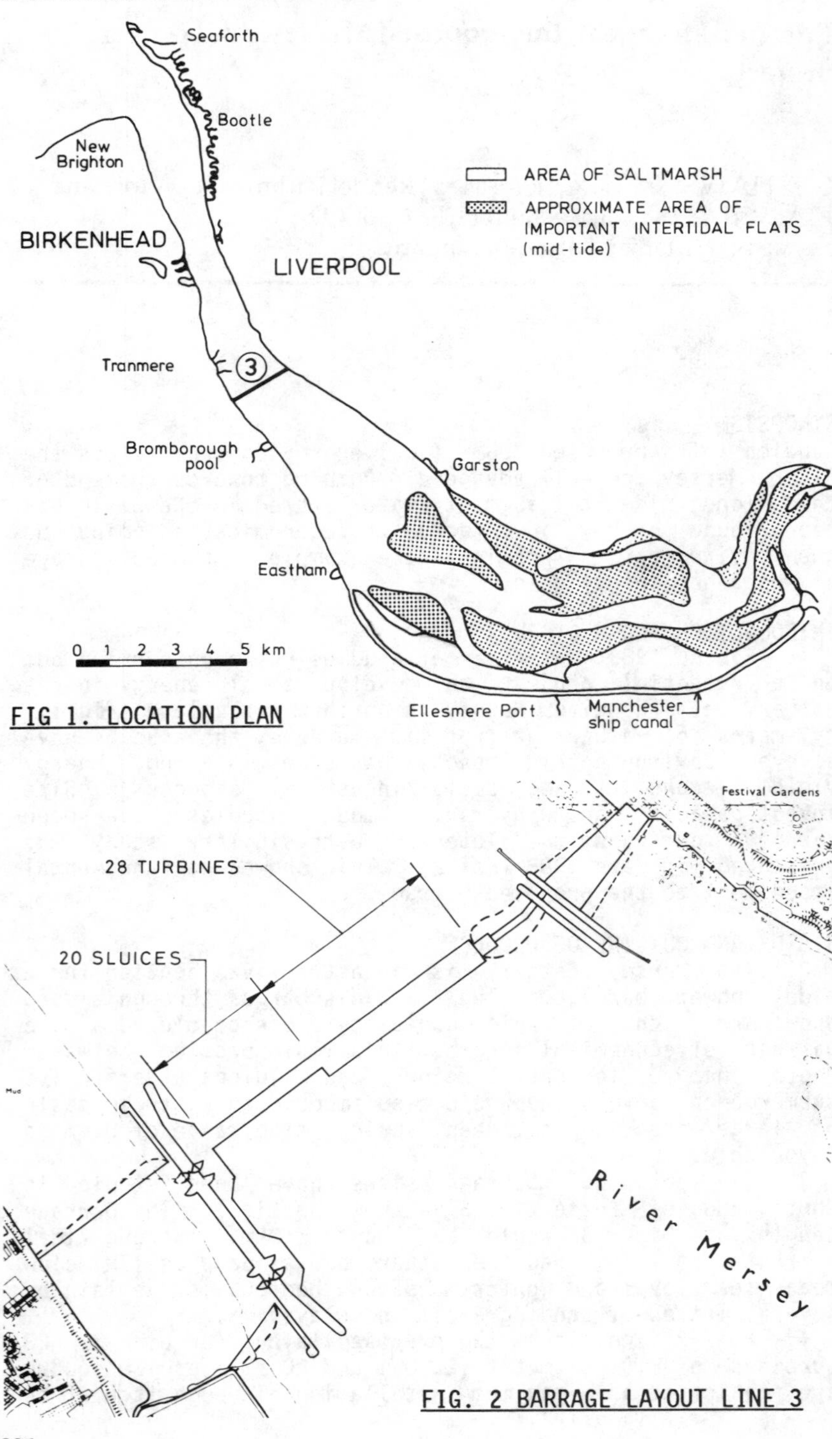

FIG 1. LOCATION PLAN

FIG. 2 BARRAGE LAYOUT LINE 3

emphasised that such a maximum is not yet proven as economically optimum. On this basis, the installed capacity would be 700 MW and the annual energy output from ebb generation would be of the order of 1.2 TWh. Fig 3 shows diagrammatically the change in water level during barrage operation over a single tidal cycle.

5. The River Mersey is a significant shipping channel and to permit continued navigation the barrage at Line 3 would include 2 large locks, 270m by 36m and 200m by 23m respectively.

6. Various construction approaches have been considered including both caisson and in situ methods.

7. The assessment of pumping within this paper is of a preliminary nature. The development of project concepts, including the pumping option, is the subject of on-going studies and designs, which may lead to significant variations.

FIG. 3 OPERATION DURING ONE TIDAL CYCLE OF EBB GENERATION BARRAGE

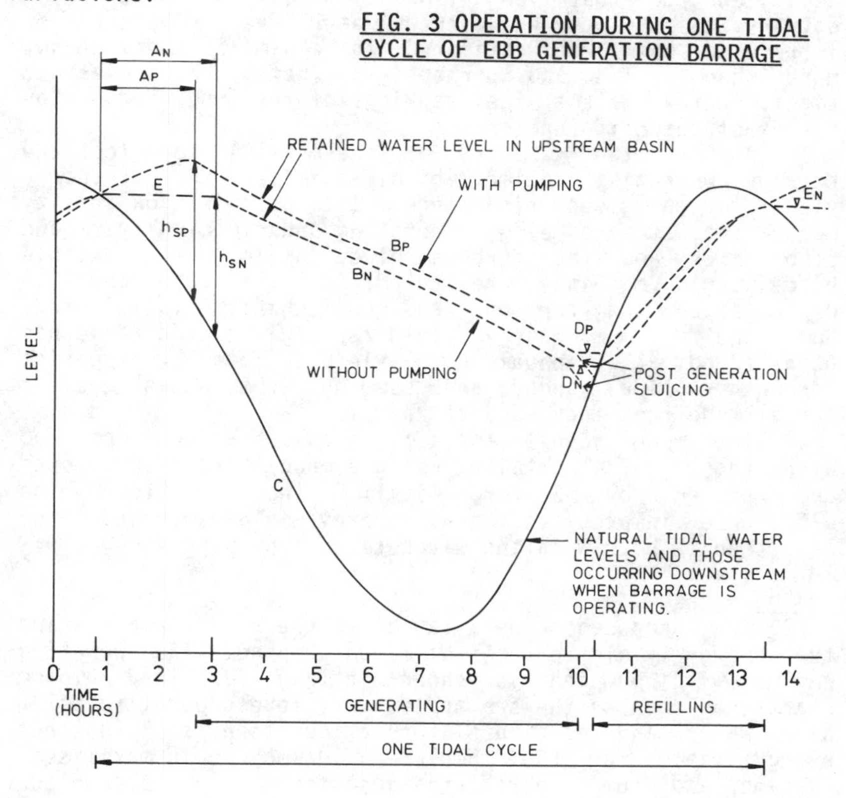

8 As a consequence of the gently shelving estuary banks of the Mersey Estuary the surface area of the reservoir impounded at MLWS is 20% of that impounded at MHWS. On this account only ebb generation is economically attractive and the proposed barrage operating modes are ebb generation with or without flood pumping.

ENERGY INCREMENTS OBTAINED

The model

9. A mathematical model of the water flows through the barrage was developed to predict the energy yield from operation of the barrage. This energy yield model is O-D because although the head losses local to the barrage are included, the dynamic hydraulics of the estuary are not simulated. Thus, the flow of water through the barrage is assumed to produce an instantaneous change in estuary level equivalent to that flow divided by the surface area of the estuary at the current estuary water level. The error so introduced has been estimated for a similar model at not more than ± 10% and perhaps ± 3% (ref 3). This does not significantly affect the rankings of the predicted energy yields of competing options.

10. An important ability of the model is to select the barrage operating regime which maximises the net energy yield for a given tide range. This optimisation ability means that the pumping, generating and sluicing start and stop times and the turbine blade angle (for a variable blade angle machine) are automatically selected and will depend upon the turbine (and pump) characteristics used. Once the optimum energy yield for a given tide range has been found, the annual energy yield is found by repeating for other tide ranges and summing from a histogram of annual tide frequency against range.

11. The model results for a given case have been compared with those (ref 4) obtained using a model already developed and proven by Salford Civil Engineering Ltd. The difference in predicted annual energy yield was found to be less than 1% both with and without flood pumping.

The base case

12. The proposed Line 3 layout is the base for assessing the energy yield effect of flood pumping. The operating regime with pumping is shown in Fig 3. Flood pumping starts as soon as the sea and estuary levels equalise after high water and continues for a duration such that net energy yield for the complete tidal cycle is maximised. Turbine and pump characteristics for a variable blade variable distributor (doubly regulated) bulb machine are used.

13. Using ebb generation with and without flood pumping the energy yields predicted for each tide range are shown in Table 1, and energy output is plotted against tidal amplitude in Fig 4.

Table 1: Base case energy yields for Mersey Line 3 Barrage

A	F	Yield N	Yield P	Pumping Energy Use	Net Yield P	Yield Gain %
2.25	1.0	0.12	0.42	0.23	0.19	61
3.25	22.5	0.36	0.79	0.26	0.53	49
4.25	79.5	0.70	1.21	0.29	0.92	30
5.25	124.0	1.07	1.92	0.50	1.42	32
6.25	145.0	1.60	2.32	0.39	1.93	20
7.25	173.0	2.12	2.72	0.29	2.43	15
8.25	113.5	2.61	3.07	0.22	2.85	9
9.25	41.5	3.04	3.40	0.21	3.19	5
10.25	5.0	3.33	3.57	0.11	3.46	4
Annual Total		1236	1667	233	1434	16

N = no pumping A = Amplitude (m) F = Annual frequency
P = pumping All yields and energy in GWh

Table 2: High, low and mean water Levels, tidal ranges and cubature in basin

Amplitude (m)	Natural Cubature (Mm^3)	Natural River		Without Pumping				With Pumping			
		HWL	LWL	HWL	LWL	MWL	Range	HWL	LWL	MWL	Range
2.25	93.7	1.32	-0.93	1.32	0.55	0.94	0.77	2.34	0.60	1.47	1.74
4.25	176.9	2.32	-1.93	2.31	-0.06	1.12	2.37	3.33	0.50	1.92	2.83
6.25	259.6	3.32	-2.93	3.27	-0.25	1.51	3.52	4.24	0.26	2.25	3.98
8.25	341.7	4.32	-3.93	4.11	-0.22	1.94	4.33	4.76	0.59	2.68	4.17
10.25	424.6	5.32	-4.93	4.86	0.83	2.85	4.03	5.27	1.19	3.23	4.08
Annual Cubature = 186.5 x 10^9				Mean		1.60	3.49	Mean		2.33	3.84
Cubature				Annual 126.3 x 10^9 Loss 32.8%				Annual 151. 4 x 10^9 Loss 18.9%			

Levels in m AOD, ranges in m, cubature Mm^3

FIG. 4 ENERGY OUTPUT v TIDAL AMPLITUDE (m)

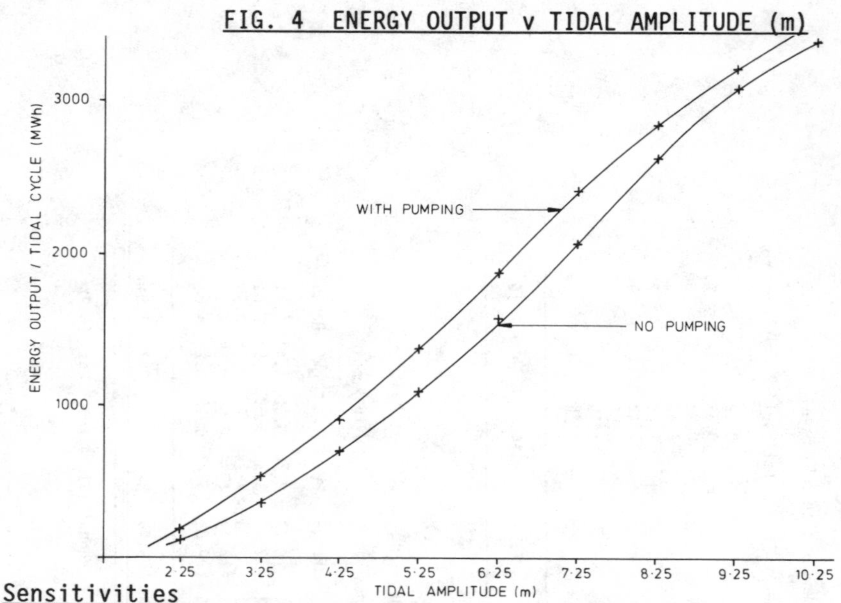

Sensitivities

14. The gain in net energy yield arising from the use of pumped storage on the Mersey Barrage is dependent upon many variables. Two of the most significant are the duration of flood pumping, and the machine characteristic as a pump. The economic gain is sensitive to the tariff structure. The effect of varying the duration of flood pumping is illustrated in Figs 5 and 6. The most frequent mid-range tides of amplitude from 5 metres to 8 metres yield the largest net gain per tide. The sensitivity of the annual energy yield to pumping duration is thus strongly influenced by the frequency of these tides.

FIG.5 NET GAIN PER TIDE v PUMPING DURATION

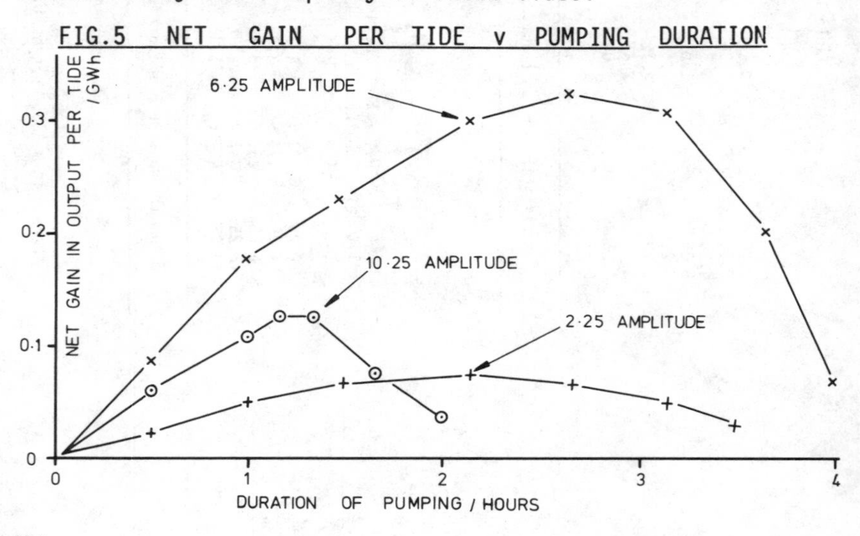

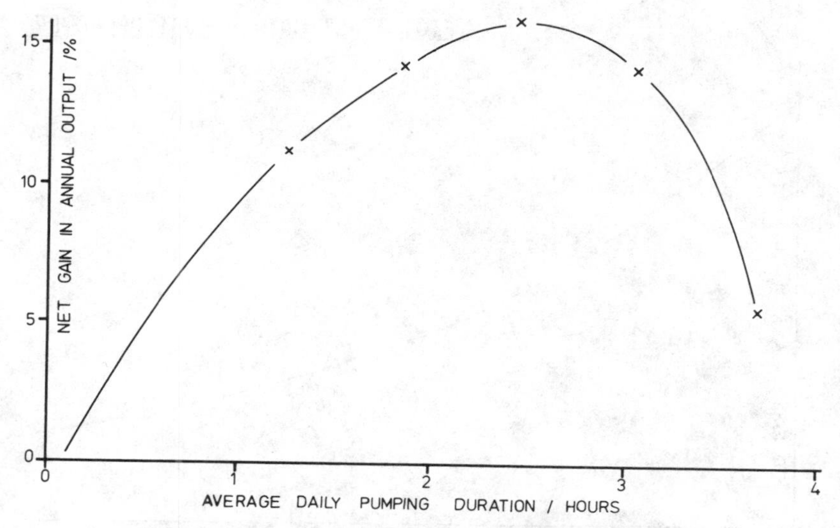

FIG 6. ANNUAL NET GAIN v PUMPING DURATION

15. The machine characteristic as a pump directly influences the benefit to be gained by flood pumping. However, the machine is primarily intended as a turbine and because the pumping is carried out at very low heads there is considerable uncertainty as to the correct characteristic values. From Table 1 it is seen that the net gain in annual output is 16% which is significant. The need for accurate determination of the pumping characteristic is apparent.

16. Only the net energy yield has been maximised. In practice it is the net value of the energy consumed and generated which should be maximised. To do this it is necessary to know the future energy purchasing and selling tariff structures but these are not presently available. The sensitivity of the gain arising from flood pumping to changes in tariff is illustrated by Fig 7 which shows the effect of changing the cost of power imported relative to the value of power exported.

EFFECT/VALUE OF EXTRA CUBATURE

Amenity

17. The Mersey Tourism Board published a paper jointly with the Mersey Barrage Company in March 1989 (ref 5) on the impact of a tidal barrage on tourism. The paper was enthusiastic and saw the Barrage becoming an integral part of Merseyside's tourism resource, to be promoted accordingly. There should be good accessibility to the Barrage and the riverside by public and private transport to maximise the potential of the impounded water.

FIG. 7 NET GAIN v COST OF POWER

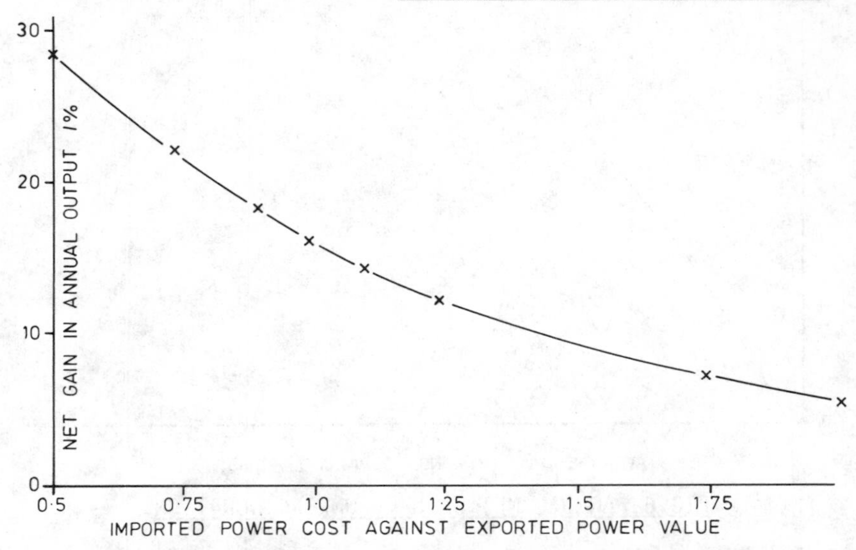

18. The Barrage, with a larger area of impounded water, would make more land available for the development of tourism attractions, pubs/restaurants and hotels. They would reap any benefits from higher land values arising from the presence of the Barrage and the impounded water. The longer the stretch of impounded water the more scope there would be for water-based activities, watersports and marinas to be developed and for those already in existence to expand.

19. There must be continued improvement of the water quality in the Mersey in order to maximise the full potential for water-based activities on the impounded water and to maximise any rise in land values which may occur.

20. On all these counts, pumping is advantageous because of its inherently greater retained water levels and cubatures. These result in larger water spreads and flushing capacities. Pumped and non-pumped cubature, with respective cubature loss on the natural tidal volume, are shown in Fig 8.

Shipping

21. Table 2 includes comparative highest retained water levels for a variety of tidal ranges, and also states an annualised average taking account of the frequency of the various tidal ranges throughout the year. The two barrage regimes, with and without pumping, are shown, as is the natural condition.

22. It can be seen that reservoir maximum levels with pumping are between 0.4 and 1.2m higher than without pumping. For all tidal ranges up to 9.25m, the reservoir maximum levels with pumping are higher than the natural

values, the greatest excess being approximately 1.2m at neaps. The greatest depression of maximum reservoir level below natural values is for the very highest tide and amounts to only 0.1m when the range is 10.25m. The implications of these facts for shipping are favourable and considerable. As compared with the no pumping case, there is no need to contemplate having to reduce the draft of the largest vessels using the up-river locks. The benefits of upstream low water being maintained close to existing mid-tide level remain and are indeed somewhat increased with pumping. All shipping will enjoy reservoir water depths greater than the natural condition over extended periods and over 95% of all tides. For the remaining 5% of tides, highest high water will be only marginally depressed, but the stand of high water will be extended for a much greater period (2 to 3 hours) than in nature.

23. The smaller vessels which form the greater number of the total movement of ships on the Mersey will have even longer periods of adequate water depths than in the no-pumping barrage regime, which themselves significantly exceed those in the natural case.

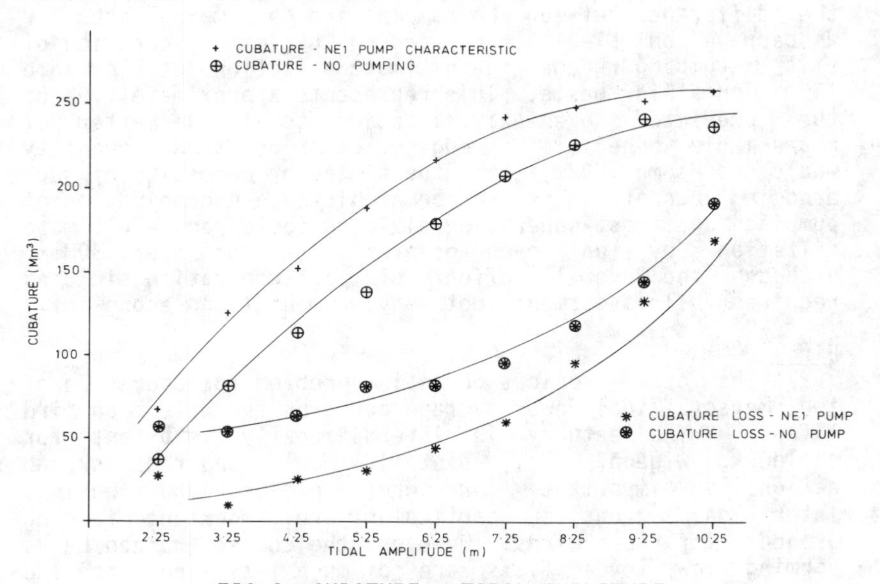

FIG 8. CUBATURE v TIDAL AMPLITUDE

Sedimentation

24. Studies by hydraulics Research Limited reported upon in July 1988 have shown that sedimentation is one of the most serious potential problems for operation of the Mersey Tidal Power Barrage, although not a fundamental impediment. It has been shown by MP O'Brien (ref 6) and Bruun and Gerritsen (ref 7) that there is a fundamental relationship between tidal prism (cubature) and cross-sectional area at mean tide level for any particular

estuary. This relationship for the Mersey was derived using bathymetric data from the former Mersey Docks and Harbour Board survey section lines between Rock Light and Widnes and tidal levels recorded during the 'Open River' runs on the physical model. The equation of the best fit-line, was:

$$Y = 665\, X^{1.38}$$

(or $Y = 747\, X^{1.36}$ excluding the inerodible narrows)

Where Y is the tidal prism in m^3 and X is the cross-sectional area below mean tide level in m^2.

25. The same relationship was shown to apply closely for each section up the estuary. The same relationship was also assumed to apply after barrage construction, permitting calculation of the new cross sectional areas for the reduced tidal prisms. Thus long term reductions in estuary volume were obtained giving an estimate of siltation during that period.

26. The same arguments and calculations were applied to the difference between the pumping and no pumping cases for a barrage on Line 3, concerning ultimate accretion of silt. Pumping reduces the ultimate siltation total by some $15Mm^3$ on this basis. This represents approximately 9% of the predicted eventual siltation total, a matter of appreciable benefit. Dredging costs for such a quantity would be some £22m, but the timing or necessity of such dredging cannot at present be predicted. A combination of pumping and post-generation sluicing could reduce ultimate siltation by an even greater total of over $30Mm^3$, although the overall effect of post-generation sluicing requires full assessment, both environmental and economic.

Birds

27. The other serious potential problem for operation of the Mersey Tidal Power Barrage concerns the effect on bird life. The estuary is internationally important for shelduck, wigeon, teal, pintail, dunlin and redshank and nationally important for grey plover and curlew. Inter-tidal areas in particular are important feeding grounds for these birds. However, the coarse grained banks forming low level flats are of much less importance and recent studies have concluded that permanent inundation of these areas would have little direct impact on birds. Most of the important feeding areas are at high and intermediate levels.

28. Looking now at the impact of pumping on the availability of prime feeding grounds for birds, Table 2 shows that pumping has the effect of increasing the mean basin range by 0.35 metres. Although the absolute levels are somewhat higher with pumping, the range effect is likely to be the most significant related to bird feeding.

Pumping should therefore be considered advantageous.

Other environmental effects

29. Aquifers The raising of low water and mid-tide levels will alter the hydraulic gradients between the estuary and the major sandstone aquifer, which exists in hydraulic continuity along the shoreline. This could lead to intrusion of saline water into the aquifer. Initial assessments (ref. 2) suggest that the effect on groundwater salinity is likely to be small but pumping will slightly exacerbate the situation.

30. Drainage The operation of the barrage would lead to a reduction of time during which water levels fall below natural mid-tide level. The implications of this change may be to cause sewers to surcharge, the banks of surface water courses to be overtopped and inefficient drainage of rivers discharging through inverted syphons. Corrections of these effects would probably require the installation of extra drainage pump capacity to aid discharge, upgrading of flood protection and the increased provision of flood storage capacity. Pumped storage would add to these effects marginally.

31. Red tides The high suspended sediment load of the Mersey imparts considerable turbidity to its water and prevents extensive algal growth. The barrage, by reducing the magnitude of tidal currents, will reduce the suspended load and decrease the turbidity of the water. Thus, there will exist a somewhat increased potential for algal growth. However, initial estimates suggest that 'red tides' are unlikely to occur as a result of the barrage and pumping is unlikely to influence this outcome.

32. Fish Fish populations are slowly increasing with improvements in water quality but such improvements need to go beyond those presently planned for a viable fishery to develop. The future development of a fishery could be impaired by a barrage acting as a physical barrier to migratory species, such as salmonids, but their reintroduction would probably only be through a deliberate stocking policy and can only be considered a long term prospect. Beneficial effects on fisheries may result from improvements of algal and invertebrate productivity, and greater depths of water at low tide. The latter would be enhanced by the pumping regime.

PLANT MODIFICATIONS

Mechanical and electrical

33. The introduction of flood pumping by operating the turbines as pumps and generators as synchronous motors necessitates reversing the direction of rotation of the generators and resetting the angle of the distributor and runner blades to the reverse direction of flow, ie: from sea to basin. Resetting both sets of blades would give the

best efficiency over the operating range of the pump. This range would depend upon the tidal amplitude and energy optimisation process. For pumping heads up to 4m the flow coefficient would change by about 15%, which means that the distributor and runner blades once reset could remain fixed for the pumping operation.

34. It is probably not feasible or necessary to optimise the turbine for a combination of turbine and pumping modes of operation because of:-

a) The low design head of the turbine, 4.4m to 4.8m, and the operating range as a pump up to about 4m.

b) The dual regulation of the distributor and runner blades.

c) The venturinated shape of the hydraulic passageway which in both directions gives a converging-diverging flow pattern.

The combination of these factors would give a peak efficiency of the order of 80% for the turbine operating as a pump making it necessary only to optimise the system (turbine and pumping) on the basis of turbine performance.

35. For turbines having a runner diameter of 8m, envisaged for the Mersey Barrage, the choice of turbine-generator will be of the bulb or pit type. To operate the generators as synchronous motors it is necessary to provide switches to reverse the direction of rotation of the generator rotor.

36. Two methods of starting can be considered:-

a) While the sea level is greater than the basin level, run the turbine in the reverse direction to achieve a percentage of synchronous speed and then start the generator as an induction motor. It is considered, however, that the head available would limit the speed to less than 50% of synchronous speed and would still generate high starting currents which would not be acceptable for regular operation.

b) Use variable frequency starting equipment. A variable frequency starting circuit is applied to the generator which acts as a synchronous motor and is brought up to speed and synchronised at the start of pumping.

<u>Cost</u>

37. Using a variable frequency starting method the additional cost of introducing pumping would be in providing switches and variable frequency starting equipment. If each piece of equipment starts say 8

turbine-generator units in sequence, then the additional cost would be of the order of 0.3% of the mechanical and electrical cost.

OPERATION AND MAINTENANCE IMPLICATIONS

38. When the turbine is operating in the pumping mode there is a significant change in the setting angle of the runner blades but once the distributor and runner blades are reset they could remain fixed for the pumping phase. However, it will be necessary during the design stage to prevent the possibility of flow induced vibrations being set up. Pumping then would simply include another phase in the operation of the barrage which would have to be included in the system control. Maintenance generated by the introduction of flood pumping would be confined to the additional electrical equipment which for correctly specified equipment would not be significant.

ECONOMIC ANALYSIS

39 It is apparent from the output and incremental cost figures that flood pumping is very substantially advantageous. Valuing the net additional energy produced at an arbitrary but realistic 5p per kWh yields the following net present values of the pumping addition, after full allowance for equipment costs and maintenance.

Table 3. Pumping Increment Economics (Net Energy Value 5p/kWh)

Discount rate	Net Present Value £M
4%	136.0
8%	71.9
12%	40.6
20%	16.3

40. Such figures are highly significant to overall project economics. The sensitivity of the energy gain to the machine characteristic as a pump, as referred to in Para 15, is again emphasised and additional work is required on this subject. However, Table 3 shows that even with a much less favourable characteristic the economic results would remain impressively favourable.

41. The final project economic evaluation will include optimisation both with and without pumping giving overall figures for comparison rather than these preliminary incremental figures.

CONCLUSIONS

42. pumping has a vital part to play in project economics for the proposed Mersey Tidal Power Project. In addition to the strong influence on project energy income, there are also considerable additional benefits from higher

reservoir levels advantageous to shipping and from reduced long term siltation.

Glossary

Bulb turbine generator	-	A type of water turbine generator particularly suitable for tidal energy applications. The water passage is straight and the generator is housed in a sealed steel bulb in the water passage
Ebb generation	-	A mode of tidal power generation in which water passes through the turbines in the same direction as the ebb tide, ie from the basin to the sea.
Pit turbine generator	-	A variant of the bulb turbine in which the bulb is extended vertically so that the generator is situated in an open pit directly assessable from upper floors of the power station. The water passages thus surround only 3 sides of the generator space.
Red tide	-	A discoloration of sea water produced by the presence of high concentrations of red coloured micro-organisms.
Saline intrusion	-	The intrusion of salt water into fresh water areas.
Stand	-	This refers to the extended period of high water on the upstream side of a barrage which is characteristic of tidal power projects. It lasts in the range of 2 hours (Springs) to 5 hours (Neaps)
Tidal prism or Cubature	-	The volume of water which flows past a given cross-section of an estuary as a result of the tidal cycle. It is approximately equal to the volume contained between the HW and LW marks upstream of the section in question.

Tidal range - Difference in water level between high water and low water.

REFERENCES

1. Rendel Parkman and Marinetech North West. Mersey Barrage : A Re-examination of the Economics, 1985
2. Mersey Barrage Co. Tidal power from the River Mersey; Feasibility Study, ETSU TID 4047, 1988.
3. DUFFETT G.L. and WARD G.B. Power generation studies of a barrage on the Severn. Paper 5, ICE Symposium on Tidal Power, 1986.
4. BALL M. and AUSTIN R.A. Mersey Barrage : Line 1 tidal energy computations ebb-generation and pumping. Salford Civil Engineering Ltd. Report reference 418/MB/1, September 1988.
5. GRIFFITHS C.W., Mersey Barrage Co and Merseyside Tourism Board. Tourism and the Mersey Barrage March 1989.
6. O'BRIEN M.P.Estuary tidal prisms related to entrance areas. Civil Engineering, 1931.
7. BRUUN P. and GERRITSEN F. Stability of coastal inlets. North Holland Publishing Co., Amsterdam 1960.

APPENDIX A: ESTIMATE OF REDUCTION IN SILTATION IN THE RESERVOIR BASIN DUE TO PUMPING

This estimate is approximate and is made on the basis of small increments/decrements in large quantities. The estimate has been made for an 8.8m amplitude tide, as adopted by HRL for the standard regression equation relating the volume of the tidal prism to the cross-sectional area below mean tide level, equation (1).

<u>8.8m tide:</u>

At Prince's Pier HWL = 9.30 mACD = 4.37 mAOD

LWL = 0.50 mACD = -4.43 mAOD

Tide levels and cubature volumes for an 8.8m tide at the Line 3 barrage are summarised below:

	With Pumping	Without Pumping
HWL (mAOD)	4.91	4.30
LWL (mAOD)	0.75	0.03
MWL (mAOD)	2.83	2.15
Cubature (Mm^3)	252.0	239.0

From HRL, the volume of the tidal prism (ie cubature), Y_p, can be written

$$Y_p = 665 \cdot X_p^{1.38} \qquad (1)$$

where X_p is the cross-sectional area below mean tide level.

Solving for the cross-sectional area,

$$X_p = \frac{Y_p^{0.725}}{111.1} \qquad (2)$$

To compare tidal volumes with and without pumping, consider hypothetical sluicing on the ebb cycle in pumped operation to make the mean tide level, behind the barrage, the same for each regime.

MWL (without pumping) = 2.15 mAOD
HWL (with pumping) = 4.91 mAOD

Thus for an equivalent MWL in the pumping regime, need

LWL (with pumping) = -0.61mAOD

Therefore, it would be necessary hypothetically to sluice at the end of power generation to reduce the reservoir level from 0.75 mAOD to -0.61 mAOD. From integration of the Salford surface area rating curve for the estuary, a reduction of the reservoir level over this range would necessitate sluicing of 55.4 Mm^3. This would lead to an increase in the pumped regime cubature:

$$\text{Cubature (with Pumping)} = 252.0 + 55.4 = 307 \text{ Mm}^3$$

From equation (2), the cross-sectional area below MWL without pumping, X_N,

$$X_N = \frac{(239 \times 10^6)^{0.725}}{111.1} = 10\,681 \text{ m}^2$$

For pumping and sluicing,

$$X_{P+S} = \frac{(307 \times 10^6)^{0.725}}{111.1} = 12\,807 \text{ m}^2$$

The volume at any station can be written

$$V = \frac{1}{k_h} \cdot A \cdot L \quad (3)$$

where V is the volume of the estuary above the station
k_h varies with level
A is the cross-sectional area
L is the length of the estuary above the station.

For Line 3, MWL = 2.15 mAOD:

k_h = 1.7
L = 20 km

Thus, the volume difference below mean tide level for non-pumped as against pumped and sluiced regimes can be written

$$dV = \frac{1}{1.7} \cdot dA \cdot 20{,}000 = 25.0 \text{ Mm}^3$$

of which, the amount reasonably considered to be due to pumping is

$$dV_p = \frac{22}{51} \cdot 25.0 = 10.8 \text{ Mm}^3$$

Hydraulics Research Ltd (HRL) quote total siltation is increased over the siltation below mean tide level by a factor of 160/115 = 1.39.

Therefore the saving of siltation in the estuary is approximately as follows:

Due to Pumping alone : 15.0 Mm^3

Due to Pumping and Sluicing (Hypothetical): 34.8 Mm^3

Afiamalu pump assisted hydro power project, Western Samoa

H. GUDGE, BSc, FICE, Hydro Consultant and A.B. HAWKINS DSc, FIMM, MICE, University of Bristol

SYNOPSIS
The paper discusses the concept of the scheme including how it enhanced the existing hydro provisions, providing greater flexibility and minimising the use of expensive diesel power.

INTRODUCTION

1. Situated at 172° west 14° south, Samoa (Fig. 1) consists of four main islands, Savai'i and Upolu (Western Samoa); Tutuila and Manu'a (American Samoa), which together with a number of smaller islands and sea mounts form a linear feature rising steeply from the floor of the Pacific Ocean.

2. At present 100,000 of the 160,000 inhabitants of Western Samoa live on the smaller island of Upolu; 35,000 in the capital Apia. The main occupation of the Polynesian indigenous population is subsistence farming and fishing.

3. A prominent feature of Upolu is the high central ridge which rises to a maximum of 1100 m ASL. On Upolu there is a marked difference between the high relief at the eastern end of the island and the flatter regions to the west.

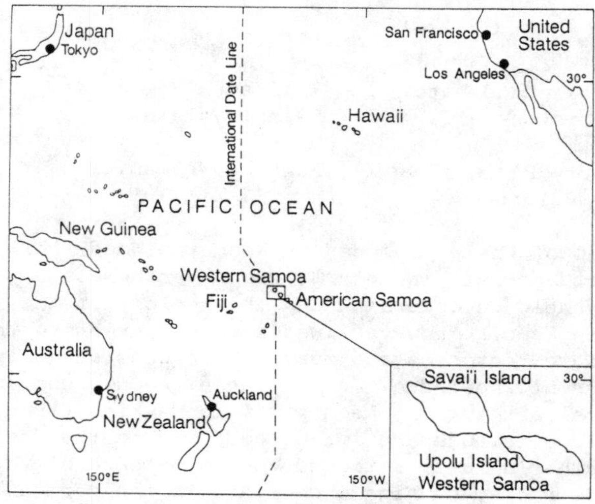

Fig. 1. Geographical position.

HISTORY OF WESTERN SAMOA

4. Prior to 1918 the islands were part of the German Empire. The League of Nations ceded the islands to the United Kingdom who passed them to the New Zealand Government for administration. Independence was granted in 1962.

5. The system of landownership is complex and difficult to establish. In addition to the limited amount of private ownership, land can be owned by the local, or "not-so-local", village or by the Government. Legally no compensation is payable when land is required for development of national importance and compulsory purchase is possible.

GEOLOGY

6. The islands of Western Samoa are composed almost entirely of basic volcanic rocks with olivine basalt forming the main suite. Like Hawaii it is believed that the initial island was a shield volcano which originated above a hot spot in the lithosphere. Later volcanic eruptions in Upolu were related to over forty minor cones along a central fissure belt; itself caused by fractures resulting from the down-drag of the Pacific Plate into the Kermadec-Tonga Trench.

7. On behalf of the New Zealand Government, Kear and Wood prepared geological maps (1958) and a Memoir NZGS 63 (1959) for both Savai'i and Upolu. This work was based mainly on the interpretation of aerial photographs and the 1:20,000 topographic maps, with only a limited amount of fieldwork. Their studies resulted in a proposed stratigraphic sequence as shown in Table I.

Table I. Stratigraphy of the volcanic rocks of Upolu (after Kear and Wood, 1958/9)

Holocene	Aopo Volcanics Puapua Volcanics Lefaga Volcanics
Middle and Late Pleistocene	Mulifanua Volcanics Salani Volcanics
Early Pleistocene or Pliocene	Fagaloa Volcanics

8. The map of Upolu produced by Kear and Wood, whilst obviously not accurate in detail, was nevertheless of considerable help in understanding the geology of the islands. The Memoir, however, was prepared prior to the theory of plate tectonics and must now be considered in the light of current knowledge. A more detailed discussion of the geological setting has been given in Hawkins (1988).

9. The older islands are in the east. Here the deep tropical weathering has destroyed the competence of the older basic igneous rocks, causing a decrease in the permeability of the older lavas.

METEOROLOGY

10. In Western Samoa the daily and seasonal temperature variation is only slight; the mean daytime temperature being between 27° and 30° C, although from July to September night temperatures may descend to 18 to 20°C.

11. The islands receive considerable rainfall throughout the year. Historically the wetter half is from November to April but heavy rain can be experienced at any time. The precipitation is partly orographic in character, the average annual rainfall of approximately 3 m at sea level rising to over 5 m in some higher regions of the islands. Fig. 2 shows the long term rainfall distribution for two gauging stations, Mulinu'u at 2 m elevation and Afiamalu at 735 m ASL. The relative humidity varies between 80.4 and 84.8%.

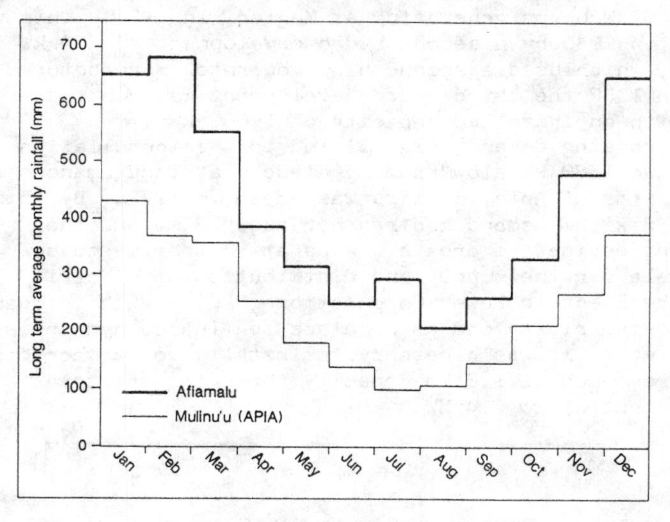

Fig 2. Comparative rainfall figures (mm) for Afiamalu (735 m ASL) and Mulinu'u (2 m ASL)

HYDROLOGY

12. Typical river lengths are less than 15 kms hence with the high relief the individual river catchments are very small. The effect of a heavy rainstorm is almost immediate, producing high river flows which return to pre-rainfall levels only a few hours after the cessation of the storm. River flows are also influenced by the high permeability of the post Fagaloa rocks. In Savai'i where the volcanic rocks are young and less weathered, the permeability is so great that there are virtually no perennial streams and the ground water level is close to sea level. At the western end of Upolu a similar situation occurs where during times of low flow streams may go underground, sometimes resurfacing some distance downstream. In the central and eastern part of Upolu where the older rocks outcrop, more perennial streams are found.

13. Despite the heavy rainfall in the central ridge area, most of the catchments produce average annual flows of only about 1 m^3/sec. Any loss of flow therefore becomes of considerable significance and has resulted in only small run-of-river type hydro electric schemes using small storage ponds with impermeable membranes.

SYSTEM DEVELOPMENT

14. The initial small electricity supply organisation set up as part of the Public Works Department (PWD) of the Government appears to have been successfully managed. Under the auspices of the PWD, development was confined to the Apia area using both diesel and hydroelectric power. The largest diesel generator was of 450 kW and the first hydro development of 80 kW began generating at Magiagi in 1928. This was followed in 1949 by a second hydro development of 230 kW at Fuluasou, which used a second-hand generator manufactured in 1934. In 1957 the third hydro development was constructed at Alaoa, with an installed capacity of 1,000 kW .

15. Increasing demand (Fig. 3) led to the installation in 1966 of two 1680 kW slow running diesels at Tanugamanono on the outskirts of Apia; a third was added in 1972. By this time the maximum demand had reached about 3 MW and the Government decided to create a separate authority to be responsible for the supply and distribution of electric power, the Electric Power Corporation (EPC). With increased usage of electricity and as no plans to finance hydropower had been made, it was necessary to install two further diesel generators, each of 1.8 MW capacity, bringing the total diesel potential to 6.4 MW.

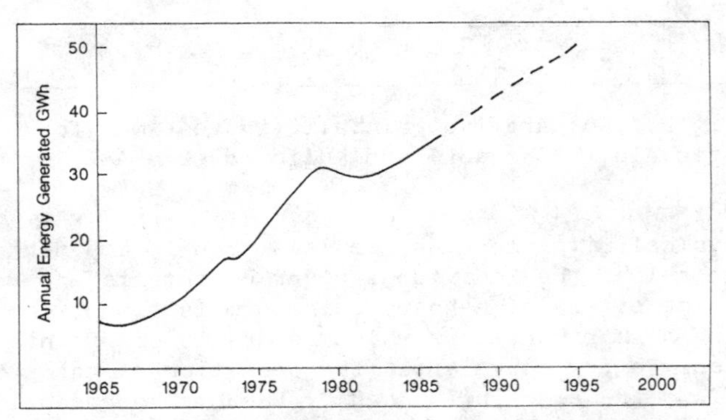

Fig. 3. System demand for Uppolu, projected to a 4% forecast.

16. In 1977, as a result of a grant from the European Development Fund (EDF), MRM Partnership were appointed to produce a Master Plan for the development of a generating capacity to meet forecast loads up to the year 2000. This Master Plan recommended a programme of hydro development should be undertaken. The recommendations were adopted and with the aid of financing from a European Development Grant and the Asian

Development Bank (ADB) by 1985 three run-of-river hydro-electric developments had been constructed. These were at Samasoni, Fale o'le Fee and Sauniatu, with two 800 kW, one 1600 kW and two 1750 kW generating sets respectively.

17. When these schemes were in commission, not only could the wet season demand be met but there was a considerable surplus for six months of the year and at night. In the other six months, however, there was a large shortfall and it was still necessary to use expensive diesel generation. The ADB therefore financed a feasibility study into the development of a storage scheme in the Afulilo area of Upolu where it was possible that suitable sub-soil conditions may exist to allow the creation of a reservoir. The initial study confirmed that the area warranted more detailed investigation, but as it was somewhat remote from Apia a completely new 66 kV transmission line over 50 km long would be required to convey the energy to the main load centre. The additional cost of the transmission line made the scheme uneconomic at that stage, particularly as it coincided with a considerable downturn in the economy of the country and a corresponding decrease in the current and forecast energy demand. Nevertheless the large shortfall in the dry season still had to be accommodated and hence diesel back-up was necessary for some 80% of the demand.

19. During the field studies it had been noted that at one of the higher central regions of the island there was a flattish area where it would be possible to construct an artificial reservoir. The original concept was for a pipeline from the Samasoni development in the north to a reservoir at Afiamalu; the water being pumped when surplus hydro power was available and the storage then used to generate when there was a shortfall. The proposal was abandoned when closer consideration drew attention to two important factors: that the pumping costs would make the scheme a net consumer of energy and that the only feasible route for the pipeline was through the gardens of several of the best properties on the island, mainly occupied by High Commissioners.

20. It was then suggested that the headwaters of small streams on either side of the watershed could be impounded and pumped to the reservoir via smaller and shorter pipelines. As seen in Fig. 4 there were several small streams within five kilometres which could be pumped into the reservoir, the average pumping head being less than 260 m. As the reservoir would be constructed some 725-737.5 m ASL with a coastal power station the full head could be utilised.

21. Although clearly the main demand was in the north, study of the aerial photographs showed that there was a shorter and better penstock route to the south coast. The pre-feasibilty report therefore suggested that this proposal warranted further examination. This was accepted by ADB who agreed a feasibility study into a pumped storage scheme at Afiamalu should be financed.

THE PROPOSED DEVELOPMENT.

22. To study the effects of constructing the previous three hydroelectric schemes a sophisticated computer programme had been developed to monitor changes as each new development was introduced to the system. This was relatively easily adapted to determine the amount of energy which would be absorbed by pumping and hence the nett addition of energy which would be produced. The analyses showed that the effect of introducing pumped storage into the system would be to reduce the amount of thermal generation by 10 GWh per annum. This was lower than the 12 GWh anticipated by the earlier studies which had been based on a more optimistic view of the hydrology than was proved by the flow measurements and detailed rainfall analysis undertaken during the feasibility study.

DESCRIPTION OF PROPOSED DEVELOPMENT

23. General The proposed development was based on the concept of abstracting water from streams in their upper reaches and pumping it up to a storage reservoir (Fig. 4); the stored water being drawn as required to produce energy at a power station located near sea level.

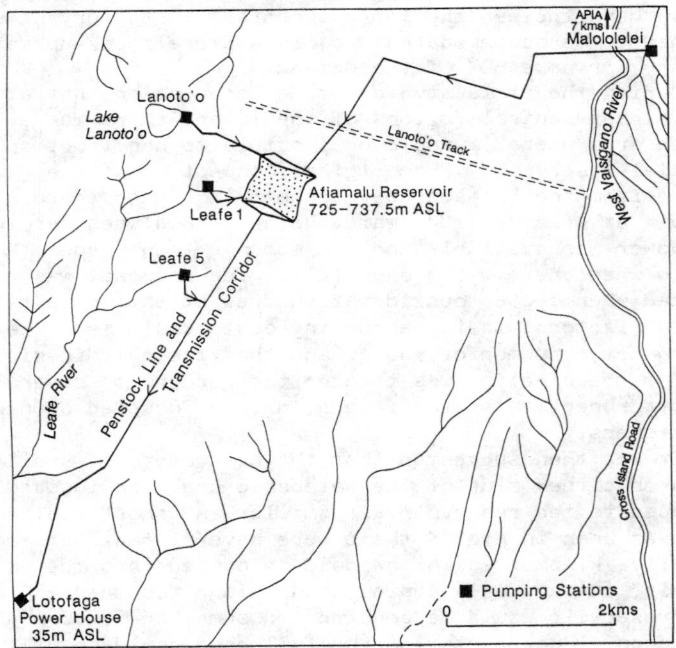

Fig. 4. Layout of the various components of the pump storage scheme.

24. The main elements of the Afiamalu development are four abstraction points, with pumping stations near Malololelei on the West Vaisigano at 500 m ASL, at Lake Lanoto'o at 702 m ASL and at two points on the Leafe streams at 675 m and 453 m ASL, Figs 4 and 5.

25. The first three sites pump the water via their own pumping mains directly into the proposed reservoir at Afiamalu with an initial top water level at an elevation of 737.5 m ASL, whilst the fourth pumps directly into the 6.3 km penstock. The powerhouse was to be located on the Lotofaga stream at an elevation of 35 m ASL and would contain a 5 MW Pelton turbine driven alternator.

26. The power produced would be transmitted to Apia via a 17 km long transmission line, in part following the penstock route, either using two lines at 22 kV or one line at 66 kV.

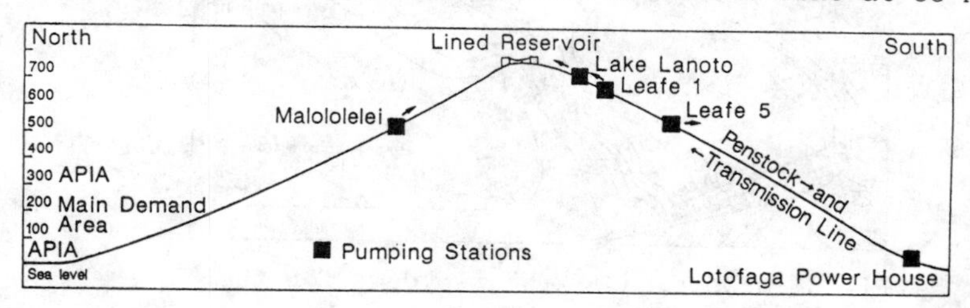

Fig. 5. Generalised section through the island to show the relative positions of the components.

27. <u>Reservoir</u>. The reservoir storage required to obtain the best use from available water resources is about 3.5 million m^3. A suitable site, in the saddle to the west of the Cross Island road at Afiamalu, would require the construction of two embankments some 16 m high to enclose the area on the north and south sides. Possible locations had been identified during preliminary work but the subsurface exploration carried out during the feasibility study soon revealed the foundation material under the initial site for the northern embankment would not be suitable to support the proposed structure, Fig. 6.

28. Fortunately it proved possible to relocate the embankment within the planned reservoir area where it could be of a shorter length and with foundations at a higher elevation. However to provide the required storage it was necessary to raise the water level from 733 to 737.5 m ASL.

29. The reservoir would be formed by stripping the soft overburden and infilling the lowest area to a finished level of 725.5 ASL. This stripping would expose suitable rockfill, which would be excavated by ripping and blasting and then placed in two closure bunds with side slopes of 1 vertical to 2 horizontal and a 3 m wide crest. The southern bund was to be 18 m high, 700 m long and contain 180,000 m^3 of fill, the northern 16 m high, 640 m long with 155,000 m^3 of fill.

30. On the downstream side the bunds would be faced with topsoil and covered with vegetation, while the slopes on the upstream faces of both bunds would be blinded with fine aggregate to produce a smooth face free of sharp protusions ready to receive the impermeable membrane. The remainder of the reservoir area would be graded to produce a suitable bedding for the membrane.

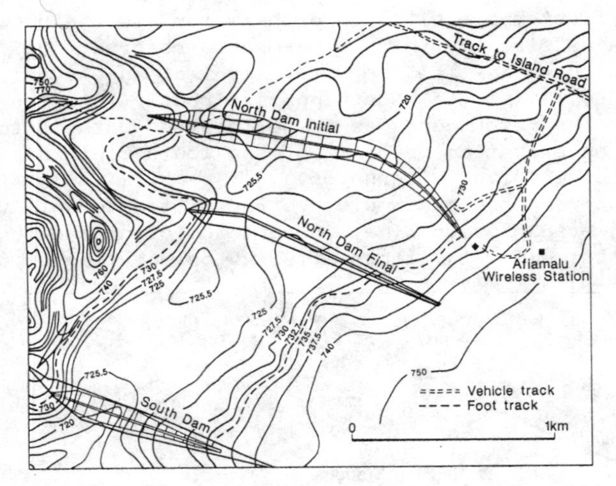

Fig 6. Position of the two dams for the Afiamalu reservoir.

31. Considerable attention was given to the most appropriate material for the membrane. To be suitable it should be flexible; easily but efficiently jointed; immune as far as possible to outside conditions, such as heat, uv radiation etc; tough and resistant not only to natural attack but also to vandalism; easily repairable, and in view of the requirement for some 300,000 m^2, as cheap as practicable.

32. Attention was focused on three possibilities:-

a) high density polyethelene (HDPE);
b) low density polyethelene (LDPE), and
c) a polymerised oil product reinforced with a geotextile.

33. HDPE was used with reasonable success in the Sauniatu headpond with dimensions of approximately 110 m x 90 m and side slopes of 1:1, giving a storage capacity of 23,000 m^3. As the Sauniatu excavation was to be open for about 12 months before lining, the side slopes were blinded with concrete to eliminate erosion but the bottom left as a coral sand blanket.

34. The 2.5 mm thick material was delivered in 10 m wide rolls weighing about 5 tonnes. Only one expatriate laid and jointed the lining; the work being done quickly and successfully even though the shape of the headpond was quite complicated. A small amount of vandalism with machete cuts was easily repaired by non-specialist staff.

35. The price of HDPE sheeting is almost directly proportional to its thickness hence the 2.5 mm grade, the thickest commercially available, is also the most expensive and was considered too costly for the Afiamalu reservoir. The only advantage of the additional thickness is its increased resistance to natural and human attack, hence other more economic means of obtaining such benefits were considered. It was appreciated that if the bed was well prepared and overlaid with a geotextile material of felt-like constitu-

tion a 1 mm material could be used. To obtain protection against malicious damage it was considered necessary to cover the material with between 50 and 100 mm of natural fine grained material. A net-like plastic sheeting would be required on the steeper slopes to ensure the soil cover was retained.

36. LDPE has similar characteristics to HDPE except that its modulus of plasticity is greater, hence if subject to angular penetration the material reforms over the projection instead of being punctured. Because of its composition it is also cheaper than HDPE.

37. Polymerised oil product is applied in-situ onto a suitable open weave geotextile. The ground is first levelled and the geotextile then laid directly on the soil. The sealant, which is a two pack pre-mixed liquid, is sprayed onto the fabric to produce a build-up of some 1 to 1.5 mm. The material has the advantage of being free from distinct joints but the disadvantage that it must be very carefully applied to ensure a uniform thickness. As the cost estimates were considerably higher than prefabricated sheeting, its use was not justified.

38. It was proposed to line the whole of the area below 740 m ASL with 0.75 mm LDPE laid on a geotextile and covered with a minimum 150 mm of fine grained material. At the top the LDPE would be returned into a trench 300 mm deep, refilled with fine grained material.

39. The characteristics of the four proposed pumping stations are as follows:

Location	Pump Capacity (m^3/sec)	Pumping Head (m)	Pumping Length (m)	Mains Diameter (m)
Malololelei	0.75	260	5200	0.76
Lake Lanoto'o	0.06	35	1350	0.20
Leafe 1	0.10	72	365	0.25
Leafe 5	0.25	300	760	0.50

Malololelei Pumping Station, Diversion Weir and Pipeline.

40. General. The location of the pumping station on the West Vaisigano had to meet two contradictory requirements: to obtain as large as possible a base flow yet restrict the pumping head to an acceptable and economic level.

41. The selected site was located between two waterfalls. If the site was moved downstream the available flow increased but the head became unacceptable, ie the energy used in pumping became equal to or greater than the energy obtained from generation. If the pumping energy was solely excess hydro power this might be reasonable but it would virtually exclude the option of using thermal power for this purpose. If a site upstream of the upper falls was adopted the diminution in flow would be greater than the effect of the decrease in head so again less economic than the chosen site. The best size and number of pump units has emerged as eight units,

each discharging 100 l/sec with a power requirement of 370 kW.

42. Power Supply. Two alternative voltage levels for the supply at Malololelei were examined, 22 kV and 66 kV primary voltage. In addition two pump motor voltages were considered, 415 V and 3.3 kV, but in both cases the rating of the switchgear at the top of the river gorge was 3.3 kV. With 415 V pump motors the method of supply would be to feed each motor on a unit basis, the 3.3 kV cable terminating in a cubicle that contains a 400 kVA, 3.3kV/415 V transformer, a 415 V, 300 kVAr capacitor bank and a 400 kW soft starter with bypass contactor. With motors rated at 3.3 kV the Star/Delta contactors, together 3.3 kV/110 V control transformers would be located at the top of the gorge with 2 x 3 more 3.3 kV cables per motor laid to the pump house. Motors would be CACW cooled to prevent large temperature rises within the building and avoid large section air ducts within the pump house. Intermittent operation of these motors in the high humidity environment in Western Samoa is also a factor in this choice of cooling method, and motors would be fitted with internal heaters to prevent condensation.

43. Malololelei Pumping Route. This pumping route climbs directly up the 75 m side of the Vaisigano gorge and traverses westwards mainly through pasture areas, Fig. 4. The maximum quantity to be lifted to the reservoir would be 0.75 m^3/sec and the head when the reservoir is full would be 240 m. To reduce transportation costs a series of diameters was proposed. The very steep valley section with the highest head utilised the smallest diameter, 711.2 mm od. This would be an all welded length until it reached the flatter ground in a thrust block. At that point the head was reduced by 60 m hence a diameter of 762 mm od was chosen for the next 2500 m followed by 864 mm od for the remaining 2500 m. Flexible couplings would be used, minimising the amount of foundation preparation necessary. Where possible to reduce land problems, the pipeline would be buried with a 100 mm concrete surround.

Lanoto'o Pumping Station and Pipeline

44. The abstraction from the Lanoto'o crater is very small but being only 1350 m from the reservoir and requiring a pumping head of only 35 m its contribution would be very economic. As the lake is designated a future National Park it was intended to limit any reduction in water level to 2 m. A small single pump station would supply the 200 mm diameter steel pipeline leading to the reservoir, Fig. 4.

45. The electrical requirements of this station are unlikely to be greater than 50 kVA and would be supplied by a short spur from the proposed 22 kV line. Because of the terrain and the fact that Lake Lanoto'o is a potential National Park the 22 kV pole mounted transformer would be sited on the reverse slope of the crater 150 - 200 m from the lake and power at 400 V would be taken to the pumping station

by underground cable installed at the same time as the pumping main.

Leafe 1 Pumping Station and Pipeline

46. Again the 0.1 m^3/sec abstraction would be small but its 365 m distance makes it extremely economic. A small submersible pump would be placed behind a concrete weir with a reinforced concrete pump well on the left abutment. The power supply would again be a short spur off the 22 kV transmission line protected and isolated by drop out fuses to a 200 kVA 22/.400 kV pole mounted transformer adjacent to the pump well.

Leafe 5 Pumping Station and Pumping Main

47. This abstraction site is located considerably further downstream and on a different tributary of the Leafe River (Fig. 4). Although the abstraction is only 0.25 m^3/sec with a 300 m pumping head it is economic as the pumping main is reduced in length as it discharges directly into the penstock. A small diversion weir with three submersible pumps, similar to that proposed for Malololelei, would discharge into the 760 m long 500 mm diameter main connecting to the penstock. While the power rating of these pumps is almost identical to those at Malololelei, the duty is very different in that each pump is rated at 83 l/s at a discharge head of 300 m. The duty head means that 1500 rpm nominal pumps could not meet the required discharge hence the pumps would be 2900 rpm nominal submersible pump/motor units rated at 415 V and the supply arrangements similar to the 415 V alternative at Malololelei. Both 22 kV and 66 kV supply voltage levels were considered.

Penstock

48. The route selected for the pipeline from the southern end of the reservoir to the Lotofaga Power House is fairly direct along the ridge between two of the southern flowing river valleys (Fig. 4). The overall length of the route is 6.3 km and the difference in elevations 700 m. In the previous hydro developments the pipes were supplied from Europe and it proved most cost-effective to design the line with three to five different pipe diameters so that they could be inserted into each other for transport. In the past shipping was charged on bulk rather than weight but it is likely that in future pipes will be charged at dead weight rather than by bulk. Nevertheless in view of the protection this method affords it is still considered preferable.

49. The 700 m head on this pump storage was far higher than on previous projects hence for the first time it proved beneficial to opt for thinner high tensile steel for the pipes in the lower areas. If shipping is charged by dead weight, the cost benefits are two-fold: the reduced thickness more than compensates for the increased price of the high tensile steel and the transport cost is reduced.

50. As a result it was proposed that the penstock should be constructed with:

2.1 km of 864 mm od 6.5 mm thick A43 steel
2.1 km of 762 mm od 8.0 mm thick X52 steel
2.1 km of 711.2mm od 13.0 mm thick X52 steel

51. Generally the main penstock would be erected using couplings of the Viking Johnson type, with welded connections at difficult sections and on the last steep incline to the power house.

Lotofaga Power House

52. This power station was to be located on the Lotofaga river at an elevation of 35 m and about 1.7 km upstream of the south coast road. At this point the gradient of the river reduces considerably and no further head can be economically developed.

53. Two arrangements were examined: two 2.5 MW Pelton Turbine generators or a single 5 MW machine. As the efficiency curve of a Pelton type turbine is very flat between 20% and full load there is little penalty in opting for one machine. The number of occasions when the load on the station would be less than 1 MW would be minimal.

COST AND RATE OF RETURN

54. The estimated cost of the scheme was 18 million US$ in 1985. This produced a financial internal rate of return of 8.6% with an equalising discount rate against a diesel generating alternative of 9%.

POSTSCRIPT

55. On the completion of the Sauniatu Development a 22 kV line existed from the load centre to Sauniatu, only some 8 km short of the site of the Fagaloa power station selected for the development of the natural storage basin at Afulilo. Studies by the EPC showed that the line had been overinsulated hence it was economically feasible to double the line's carrying capacity. By uprating the transmission line to operate at 33 kV and extending it to the proposed power house it would be possible to carry the proposed 5 MW output to the load centre in the Apia area. This clearly made the Afulilo scheme more economic and as a result it was given preference and is about to be constructed. It is considered, however, that rather than cancelling the implementation of the Afiamalu pumped storage scheme, it has merely postponed its construction.

ACKNOWLEDGEMENTS

The study reported was undertaken as part of an investigation by the MRM Partnership and the work in Samoa was financed by the European Development Fund and the Asian Development Bank. The EPC, particularly the general manager J R Worrall, gave continued assistance and encouragement throughout the work.

Discussion

F.G. JOHNSON, ***formerly of North of Scotland H.E. Board***

It is interesting to note that a micro pumped storage scheme has recently been constructed in the Island of Foula, Shetland. It comprises aerogenerator reservoir and pump generators and is a community project. Formerly the islanders used small diesel generators mainly on a 'house by house' basis.

T.J. PATERSON, ***James Williamson & Partners, Glasgow***

The proposed conveyor hoisting system is unproven at the 900-1000 m hoisting depth proposed. Would the well-proven mine winder system, consisting of winder, headframe and ships, not provide a better alternative?

The excavation of the underground reservoir by TBM is a very interesting concept. Where large volumes of underground excavation are required in hard rock, excavation by drill and blast of the largest possible cavern cross section has proved the most economical excavation method in the past. Can the reasons for using TBM excavation be explained in more detail?

The unit cost of £20/m^3 for TBM excavation of the lower reservoir appears optimistic. A figure of US$45/$m^3$ was quoted in 1975 for the TARP project in Chicago, which used TBMs of similar diameter in limestone/dolomite. As the viability of any mined underground pumped storage is dependent very largely on the underground excavation cost, can the figure of £20/m^3 be justified?

E. GOLDWAG, ***GEC Alstrhom Power Plants, Manchester***

If I understand Mr Haws correctly he intimated that the amount of additional energy due to pumping based on the use of a 2-D model was marginally lower than that given in the Paper and based on an 0-D model.

In view of the above, could Mr Haws give us some indication regarding the magnitude of the change.

F.J. BINDON, *Institution of Electrical Engineers*

Could the Authors say whether line 3 on Fig. 2 is now the final location for the barrage or will line 1A previously examined in Stages I and II of the feasibility studies be considered. What are the main reasons for selecting line 3?

R. WATTS, *Author*

In reply to T.J. Paterson, the selection of vertical conveyors is made in the paper as a result of studying the successful use of such conveyors on the TARP project in Chicago, and after discussions with the manufacturers.

They have installed over 40 000 conveyors and have taken over 800 t/h up over 100 m lift on many occasions. They were unwilling to contemplate a lift of 950 m but were quite able to consider lifts of 350 m in each of three successive units. The use of a continuous process is preferred, when taking the produce of a continuous process, to an intermittent method such as winder, headframe and skips. The use of a TBM for the excavation of large volumes is preferred in this case for the following reasons:

(a) There are no blast fumes to be removed.
(b) The rock arisings are all to a size.
(c) The work is continuous not cyclical.
(d) The remaining structure is not damaged in the same way as the perimeter of a cavern is damaged by blasting.
(e) The proposal envisages a stage development and blasting near to turbines that have been set running is to be avoided.
(f) TBM excavation has taken many strides forward in recent years (and is still improving in its capacity) so that it is a direct competitor with drill and blast in a number of cases nowadays.

The figure of £20/m^3 for TBM excavation is a bit of a hybrid figure, based on 1990 technology and production projected back to 1979 to compare with the figures from Dinorwig.

E.T. HAWS, E.A. WILSON and H.R. GIBSON, *Authors*

In reply to E. Goldwag, the energy assessed by 2-D modelling is generally slightly lower than that assessed by 0-D modelling; the difference is not limited to the pumping mode.

Significant reasons for the difference appear to be the hydraulic losses in the seaward channel when large turbine discharges continue at late stages of the ebb and early stages of the flood, and draw down towards the machines in the headpond. Both effects continue to be investigated with finer grid 2-D modelling, and the quotation of numerical differences should await the outcome of this later work. A fuller response to Mr Goldwag's question would therefore be available by the end of 1990.

In reply to F.J. Bindon, line 3 is now the firm general location proposed by the Mersey Barrage Company. The decision is based on slightly better economics and has slightly less impact on shipping. The line does, however, pose greater foundation problems which are now being investigated by further fieldwork. This could lead to fine tuning of position fixing.

How the Hungarian energy system can be rationalized by pumped storage

I. SZEREDI, PhD, Pumped Storage Division, Viziterv

INTRODUCTION

Hungary is a small country with 10 millions inhabitants and an area of 93 000 sq.km; it is situated in the center of Europe. The first hungarian electric light bringtened in the 1888 year and the first power station - an small hydro power station - was taken into operation.

Since Hungary has almost no own resources it was obliged to trade with foreign countries at all time. Due to the development of the high voltage grid in Hungary and foreign countries one large common network for the whole eastern Europe was created. Today Hungary has a very important position it the European networks. It is connected to the eastern and western countries with 11 lines, operated with 120, 220, 400 and 750 kV.

The country's total consumption of electricity (including self consumption of producers) in the year 1988 amounted 40,5 TWh. The electric power companies supplied 26,6 TWh and the imported electric energy was 11,3 TWh. Last year the nuclear power generation 13,4 TWh. The share of the different ways of electricity production in the 1988 year

(a)	nuclear power station /fuel purchased/	34,0 %
(b)	coal fired power station	21,0 %
(c)	oil fired stations /fuel purchased/	4,4 %
(d)	natural gas for electricity production /partly purchased/	11,6 %
(e)	hydro power stations	0,4 %
(f)	import electric energy	28,6 %

The peak load of last year 6 523 MW. To meet of peak load and daily cycle utilized mainly the electric energy import changing between 1950-500 MW limits.

As a result of the energetic crises which arose during the past 15 years bitter experiences were collected and the price-explosions afflicted strongly hungarian systems which were depending on the import of fossile, first of all hydrocarbon sources. The aspiration on security necessitated actions in the technical, economical and political fields.

The satisfaction of the growing electric energy demands, the changes in the structure of consumption, the replacement of the gradually aging power plants needs the evaluation of possibilities on wide scale simultaneously with the technical - economical tendencies concerning their application.

THE PRESENT OPERATION CONDITIONS

The present operation of the hungarian energy system is defined considerably by the power plants, supplied with hydrocarbons. The total capacity is about 3000 MW. Since the power plants are becoming obsolete, the steadily reducing domain of load changing makes the system more rigid and following the time schedule of operation is getting more and more costly. The experiences of gas turbine operation in the system, especially in the case of Inota power plant can't be already regarded in technical sense as real successful solutions, because their costs of energy production didn't proved to be optimal.

The sucessful starting of operation at the Paks Nuclear Power Plant gave an impetus to the development of nuclear power production. At the same time the planned 1000 MW block sizes are calling the attention in an increased degree to the need for rapidly utilizable reserves of suitable capacity in the energetic system.

The deficiency for a few hours in the 2000 MW peak capacity energy import, as well as that in the operation of the recently planned 1000 MW nuclear blocks or large capacity coal fired blocks causes restrictions on the country's level, when rapidly obtaiable reserves are not for disposal.

The security of parallelly working electric energy systems and as a consequence, the security of imported energy was not sufficient during the last decade. Disturbances in operation occured 50-100 times annually, thus effecting a reduction of 600-800 MW in the import, which caused also at many events disturbances in frequency and reduction in consumption.

The amount of the aviable high-load power for import is going to decrese, and concerning exchange on basis of trade conditions going to large variation of prices, aviability and safety. Growing the prices possible in high-load period to 5,0-8,0 US cents/kWh.

One potential path of the East-West and North-South international electric energy transportation leads through Hungary. The direct current connection needed for the energy transport and the 400 kV transmission line will be realized in the following years. At the same time the Hungarian system can participate only on a restricted level and with low transit fees in absence of suitable security of operation.

The creation of elasticity in the Hungarian power system seems indispensable. The elasticity has in this case twofold meaning. It means from one hand the development of a system being able to follow the changes of world market prices in fuel consumption and by the other hand the significant improvement in security and in the quality of energy supply.

THE POSSIBILITIES OF DEVELOPMENT

Forecasting the electric energy requirements and maximum power demand serve as the basis of the long-term electric power development plan. Growth of the consumption in last years was lov

(a) 1980/1981 + 3,1 %
(b) 1981/1982 + 3,1 %
(c) 1982/1983 + 3,9 %
(d) 1983/1984 + 5,0 %
(e) 1984/1985 + 3,0 %
(f) 1985/1986 + 2,4 %
(g) 1986/1987 + 4,7 %
(h) 1987/1988 + 0,4 %

The mean annual increase in aggregate national power requirements during the ten years perion in the last development plans forecasted at pessimistic 1,5-2 %.

The programme of electric power development for the satisfaction of perspective electric energy and output demands can be based by part on basic power plants securing the elesticity of the electric energy system. Increasing the capacity of the basic power plants, first of all the by nuclear considered. Regarding the increasing rate of demands at present, around year 2000, further basic power plants must be put into operation.

Based on the proved level, referencies, technical and economical parameters of the possible solutions, there are from the point of view of energetic policy in the fuel utilization modifications in the adverse direction possible. These are as follows

1. To increase the oil and natural gas consumption of the electric system, by constructing new gas turbine power plants, and increases the dependence of the system on hydrocarbons.

2. To equalize the load of the energy system, to reduce the utilization of hydrocarbons and simultaneously rationalizing the structure of power sources

by the construction of a pumped storage scheme, securing elasticity for the system on a suitable level.

The two directions of development are neither technically and economically, nor operationally and environmentally equal.

THE FUEL STRUCTURE RATIONALIZATION

The rationalization of the structure of energy sources in the electric energy system investigated on the base of cheaper fuel utilization. According to this the replacement of hydrocarbons, among them first of all oil and natural gas secures the greatest economic preferencies as a result of structure modification. The reasons of this can be cleared by relating the costs of various fuels, utilized in electric energy production. Thus the growing fuel costs of energy production in coal fired power plants 5 times, in oil power plants 7,5 times, in gas turbines 21 times more, comparing to that in nuclear power plants. The fuel cost of hungarian gas turbines about 10-11 US cent/kWh.

The proportions in production costs of electric energy are adverse to the previous conslusions because of other factors considered. Production cost of energy produced with gas turbines is ten times greater than in nuclear power plants.

The utilization of the advantages in the price system of imported energy is of great importance, because the cost of the energy, imported in the daily hours is 1,7 times, that imported in the peak period 2,7 times higher than the energy imported at night.

The load equalizing effect of the pumped storage lifts the major part of time schedule constraints and enables a brand new strategy of operation. For this reason the effect of pumped storage can be cleared on the base evaluation of fuel consumption balance.

TRENDS IN FUEL BALANCE

The investigation on the conditions of satisfaction of electric energy and power demands, was performed for the parallel pumped storage and gas turbine development, regarding 2,0-2,5 and 3 % increase in output demands, as well as for cases with identical and different increases in energy demands, for the period 1988-2012 years.

The parallel investigations of fuel balance - for pumped storage and for gas turbines - belonging to various measures of growth justifies that the most important tendencies are in every case indentical. These can be summarized as follows

1. There is a definite connection between the changes in the energy balance and the planned basic power stations entering into service. The capacity of the new nuclear or coal fired power plant, intended in perspective project can be applied to the system only in cooperation with pumped storage.

2. In the case of gas turbine development, the consumption of coal will be supplanted from the energy balance rapidly, while in case of pumped storage, the level of consumption, as well as the subsequent gradual decrease can also be secured.

3. The hydrocarbon consumption of the system can be kept on level, or gradually decreased by the pumped storage, while in the case of gas turbine application, the utilization of hydrocarbon sources increases gradually.

4. The exploitation of advantages prevailing in the price structure of electric energy import would be significant in the all planned period.

TRENDS IN FUEL COSTS

From the evaluation of fuel cost changes the following can be concluded

1. In case of pumped storage, the fuel costs in the system can be maintained approximately on level, the occuring differences are caused by change in the proportion of nuclear plants. In case of gas turbines the fuel cost on system level increases continously.

2. The system fuel cost in the case of pumped storage is in every period and every investigated case, lower than that of the development with gas turbine driven plant.

The savings, following the commencement of pumped storage can be characterized with the fuel cost difference of gas turbine development and pumped storage. On the base of the savings' evaluation, the following most essential tendencies can be determined

3. The fuel cost's savings of the period before starting of new basic power station are less than Ft 2-2,5 milliard annually.

4. The savings of the period after putting into operation of new basic station can be characterized with the annual value of Ft 3,5-4 milliard. The magnitude of annual cost savings increase.

The changes in the specific fuel costs of the electric energy system for 2% increases in demands are shown on Fig. 1, The tendencies of specific costs are among others, the followings

5. The development of specific costs is determined definitely by the structure of energy sources, so by proportional share of nuclear power.

6. In case of hydrocarbon based gas turbine development, the specific costs quickly increases, while the pumped storage, common with the nuclear power plant development causes a monotonuous decrease.

7. The sudden changes of specific cost differences are also underlining the importance of separate investigations for the periods, beginning with starting of new basic energy blocks.

TRENDS IN OIL PRICE DEPENDENCY

The most substantial tendencies of changes in hydrocarbon utilization on system level, are indentical with the conclusions from the fuel balance as follows:

1. The strategy depending on gas turbine development increases the hydrocarbon consumption, as well as the dependence on hydrocarbons. In case of energy storage, the hydrocarbon utilization can be maintained on level.

2. The hydrocarbon fuel saving with pumped storage after putting into operation the new basic power station can be characterized with the annual value 30-150 PJ, depencing from increasing of energy consumption.

3. For the 4x320 MW pumped energy storage, there is, in case of 2,5 % increase in demands after 2008-09 and in case of 3 % increase after 2007-08, a supplementary gas turbine capacity needed. This points to the fact that the regulating capacity in the system will be restricted.

4. The investment costs of the gas turbine capacities are surpassing already in the period of 1999-2004, the investment costs of all investigated possibilities of pumped storage in Hungary. The increase of gas turbines investment cost are contained in Fig. 2,

THE OPTIMAL CAPACITY OF PUMPED STORAGE

The optimal built in capacity of the pumped storage, can be determined by the common investigation of the total regulating power plant investment costs and of the effects arising in the system. The most important tendencies in the investigated domain of pumped storage capacity can be characterized with the followings

1. The tendencies of fuel costs and savings on system level are showing a pumped storage capacity, changing in time, the most effective.

2. In the period extending until putting into operation the new basic power station ped storage capacity over 600 MW, can't be demonstrated.

3. In the periods from starting of new basic power station to 2005-2007, the increasing of efficacy can't be guaranteed above an tuilizable pumped storage capacity of 1100-1200 MW.

4. The installed capacity of pumped storage can be determined primarily by the utilizable capacity of the pumping operation. In period 2000 needed 4x320 MW pumping capacity.

Based on the investigation of energetic parameters, in the period extending to 2010, the demands can be satisfied with a daily 5,1 GWh turbine operating capacity of upper reserfoir on the days with the highest energy turnover. The water stored in the upper reservoir of the pumped storage system, operated in daily cycle, enables the scheduled operation. In case of greater breakdowns in the power plant system, there is most likely that the capacity is needed for a longer period as usual within the day. For this reason the needed capacity reserves estimated with probability calculations is about 2,0 GWh. For the scheduled regulation of the system, a storage capacity of 0,9 GWh, however for the reserves a further storage capacity of 1,6 GWh was allocated.

THE STARTING TERM OF PUMPED STORAGE

The most expedient term of starting can be determined by minimizing the needed power sources of national economy. The two most definite components of the required sources of national economy are the followings

(a) the means of investment needed for the development of the system's production capacity,
(b) the costs of produciton, or obtaining the fuel needed for the continuous satisfaction of demands.

The losses of national economy originating from surplus cost allocations in consequence delayed of starting Fig. 3, shows. On the base of national economic allocation survey, the most significant conclusions are as follows

1. The minimum national economic cost allocations are only in the case of starting in 1998-2000, together with the new basic power plant.

2. The national economic loss in consequence of the delayed starting, surpasses the value of investment cost allocations of pumped storage in year 2001 and these expenses are increasing exponentially after this period.

RECOVERING OF PUMPED STORAGE CONSTRUCTION EXPENDITURES

The actual recovering of pumped storage construction expenditures for the cases of investigated increase in energy demands and for the different credit constructions were surveyed on base price level as well as for predicted prices for various terms of starting.

The most essential tendencies are the followings

1. In case of higher increasing of emands, the recovering of investment costs proceeds more rapidly.

2. Each of the credit constructions resulted a shorted recovering, compared to the Hungarian investment practice, estimated with an interest of 18-24 %.

3. The results of the investigations, performed with the cost savings on system level, are referring in every case to the possibility of effective pumped storage realization.

SUMMARY

The results of the investigations and economical evaluations can be summarized in short as follows

1. After putting the pumped storage into operation, the function of the Hungarian electric power system would be equalized significantly, the structure of the utilized energy resources would be rationalized to a great extent and its rapid regulating capacity would facilitate the scheduling and avoiding the operation failures of the system.

2. The results of economic evaluations and the possible annual savings in hydrocarbons, as well as the problems in the operation of the electric power system are proving unanimously the necessity of construction. The economic parameters of the work are favourable, compared to those of the alternatively investigated gas turbine investments. The low investment cost of the pumped storage enables the forming of an effective development strategy with minimum investment cost demands. Further, the electric power supply can be made significantly independent from the not predictable consequences of hydrocarbon priceexplosions.

3. On the base of minimizing the employment of national economic sources, the year 1998-2000, seems to be the most favourable to put the scheme into operation. The optimal capacity of the plant is 4x320 MW, with a reservoir of 8,23 million cu.m volume. In this case, the plant can perform for 10 years, the regulation of the system without a supplementary peak power plant.

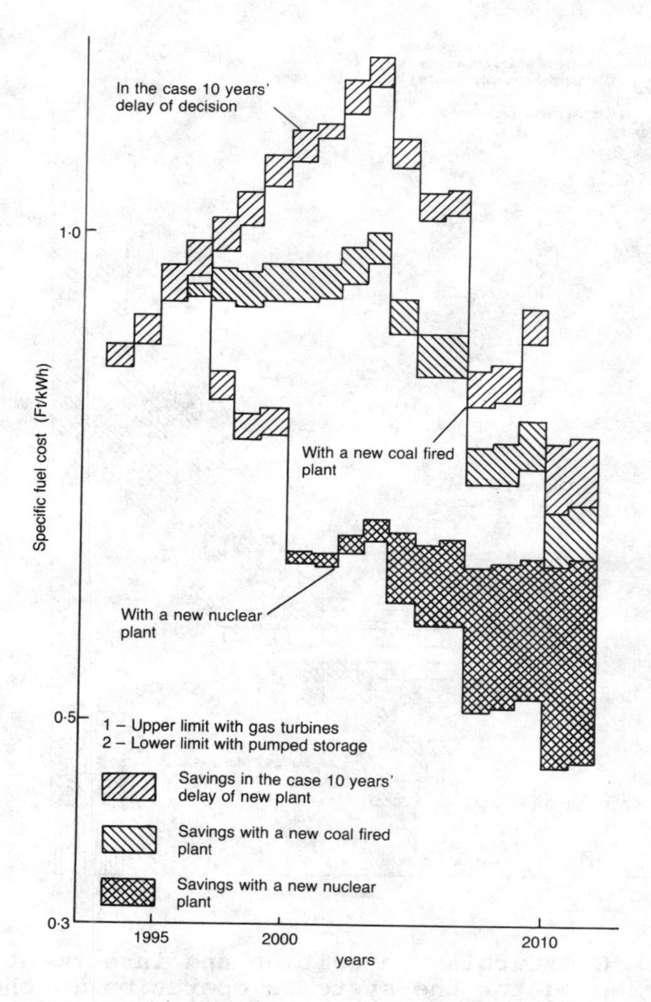

Fig. 1 Specific fuel cost of energy production for two per cent increase of demand

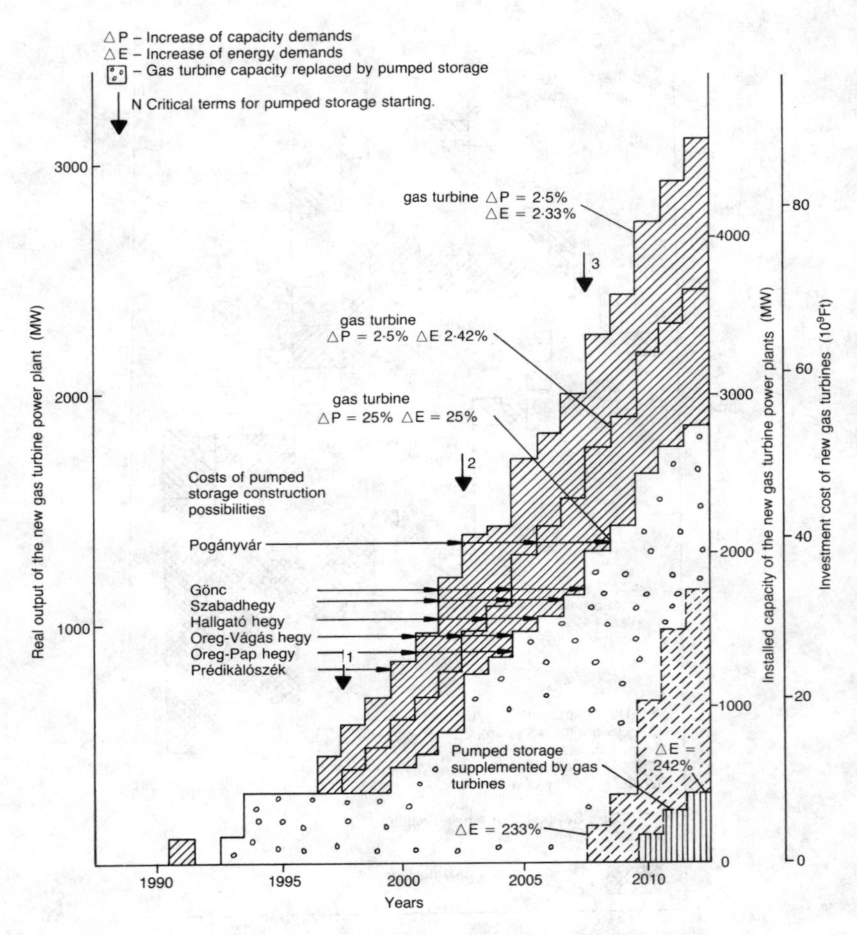

Fig. 2 Gas turbine capacities and investment costs needed for the system's operation in the period 1988-2012 at 2.5 per cent capacity demand increase on the 1985 price level

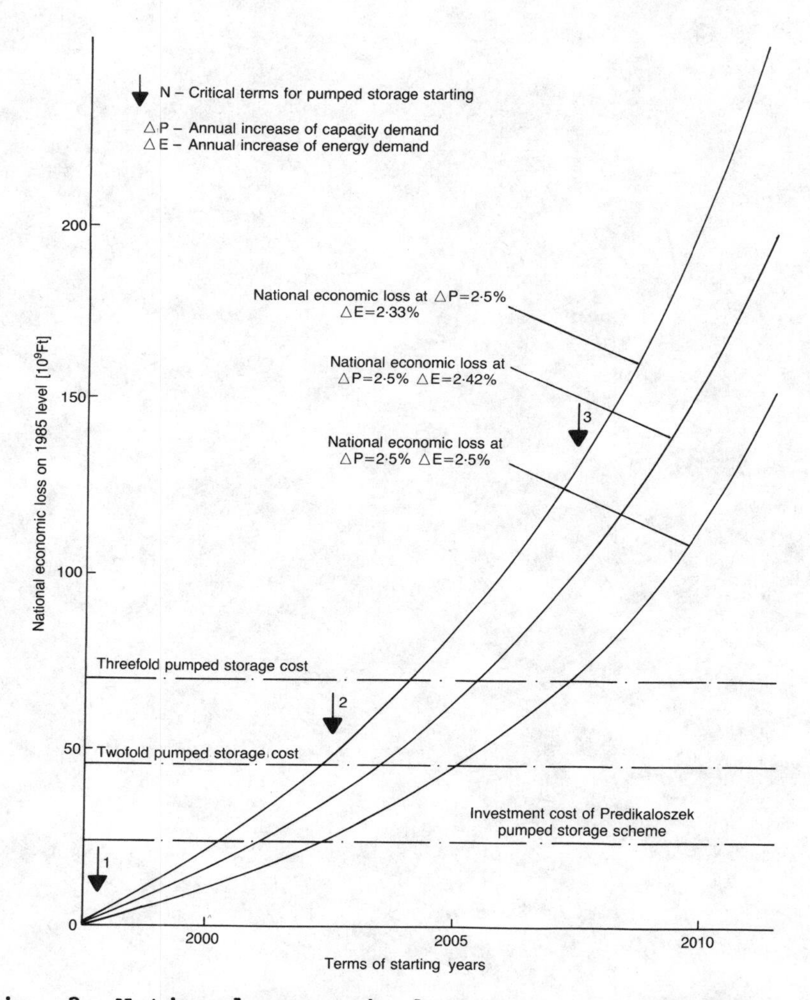

Fig. 3 National economic losses in consequence of delayed starting of pumped storage on the 1985 price level at 2.5 per cent increase in demand

Some recent pumped storage studies in the UK

J.G. COWIE, FICE, MConsE, T.H. DOUGLAS, FICE, FIStructE, FIWEM, FGS, MConsE, and T.J.M. PATERSON, FICE, MIMM, FGS, James Williamson and Partners

ABSTRACT

With the commissioning of Dinorwig, the pumped storage needs of the three mainland Electricity Boards - the NSHEB and SSEB in Scotland and the CEGB in England and Wales - were satisfied for the foreseeable future. Since then, a number of private studies have been undertaken and, in addition, the Boards maintain their own lists of sites so that, when system demand and the economic climate dictate, potential sites can be selected and promoted. Particular reference is made to studies, with which the Authors have been associated, for underground pumped storage and of means of minimizing the construction duration for pumped storage schemes.

INTRODUCTION

Four pumped storage power stations (Fig. 1) have been built in the UK: Ffestiniog, 360 MW (1963); Cruachan, 400 MW (1966); Foyers, 300 MW (1974); and Dinorwig, 1800 MW (1981).

Since the promotion, construction and commissioning of Dinorwig by the CEGB in 1981, only limited studies have been undertaken into new schemes within the UK. The CEGB maintain a list of possible sites throughout England and Wales and the potential integration into the Grid of some of these has been studied in more detail for comparison purposes during promotion of other power projects such as gas turbine, coal-fired or nuclear power stations. In addition, some other pumped storage schemes have been prepared in outline and one of these is referred to in more detail later in this paper.

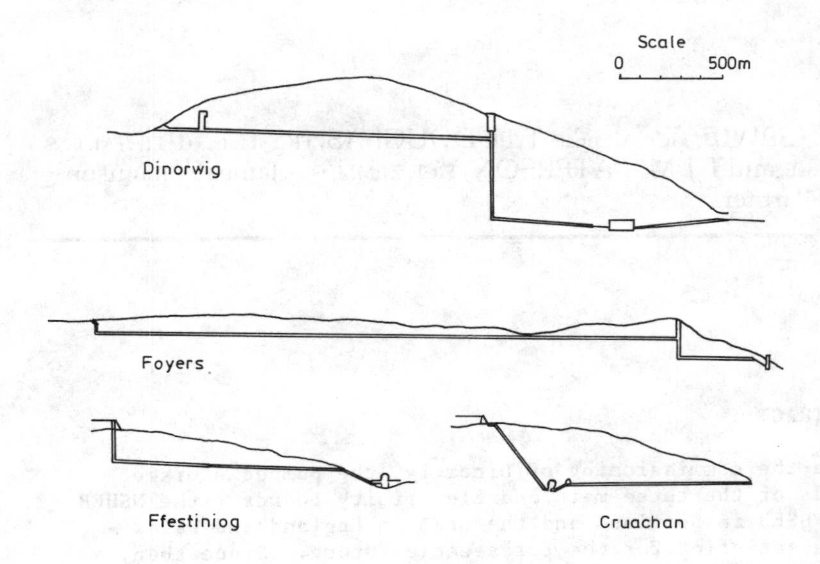

Fig.1 Arrangements of UK Pumped Storage Stations

Sites for conventional pumped storage schemes do exist in the area of the South of Scotland Electricity Board, but these would be restricted in scale and in available head because of the lesser height of the hills in comparison to those in the North of Scotland. There many excellent higher head sites have been identified, including the Craigroyston scheme already investigated in some detail by NSHEB.

In England and Wales, with 1800 MW of pumped storage at Dinorwig and 360 MW at Ffestiniog, the proportion of pumped storage capacity within the system is less than 4% of the total. Studies in a number of countries suggest that the optimum installed capacity of pumped storage in a predominantly thermal system would be rather higher than this, with further increases required as the proportion of nuclear generated power and the size of individual generating units increased. Benefit could, therefore, be obtained from additional pumped storage capacity within the system, probably in the range 600 - 1000 MW, to reduce the economic penalties and operational problems associated with overnight shut-down and start-up, load cycling and load following with low merit thermal plant, including gas turbines. To maximise benefit, such a scheme would have to be located in South-East England, necessitating resort to underground storage.

In addition to schemes promoted and designed by the Electricity Boards, a number of special or dual purpose schemes have been built or are being considered. The Carsington Water Supply Scheme in Derbyshire, for example, is designed to pump water from the River Derwent to a reservoir which is currently under reconstruction. Also, a series of studies have been undertaken in the U.K. on estuarial barrages and the proposal for the construction of one on the Mersey incorporates a pumping facility to enhance the economics of the project (Ref. 10).

Continuous search must be made for more economic forms of development - each scheme being tailored to suit particular topographical and geological conditions and to meet specific requirements associated with the operation of the Grid.

UNDERGROUND STORAGE

The idea of locating the lower reservoir of a pumped storage scheme underground has been current since the 1960's,(see Ref. 1 for literature up to 1976). More recent papers include two in this symposium (Ref. 2 & 3). So far, no such scheme has actually been constructed anywhere in the world.

Underground air storage schemes have, however, been developed in conjunction with gas turbine power stations (Ref. 4 & 5) and this type of development is in direct competition with pumped storage. Sites so far developed or promoted have utilised existing underground caverns or used 'solution excavation' of salt deposits. Many such sites in U.K. have already been used or reserved by British Gas for gas storage (Ref. 6) and by ICI for storage of gases and liquids. This type of development seems unlikely to have a future in the U.K., although possible off-shore storage sites in salt deposits are available and these could be developed in conjunction with wave or wind power generation.

Viable sites for underground reservoir pumped storage are relatively few due to the constraints inherent in this type of scheme. It is desirable for the caverns to be located in 'basement' rock and for the shafts to avoid aquifers as far as possible. For economic viability, such a scheme should be located reasonably close (say within 100 km) of the principal area of demand. The addition of environmental aspects relating to the upper reservoir further reduces the possible choice of site unless the use of sea water is contemplated.

In 1979, consideration was given by the CEGB to investigating the technical feasibility and economic viability of the underground pumped storage concept. Submissions for the necessary study were made (e.g. Ref 7), but the investigations were not continued. A private, unpublished desk study (Ref. 8) was however made by James Williamson & Partners in 1984/5 to examine the idea in more detail and further reference to this is made later in this paper.

Underground pumped storage uses an upper reservoir which provides water storage capacity at ground level, and a lower reservoir excavated in rock at a depth underground to give a suitable head. The power station itself, accommodating appropriate pump-turbine units and associated plant, is also underground. A typical layout is shown in Figs 2 and 3.

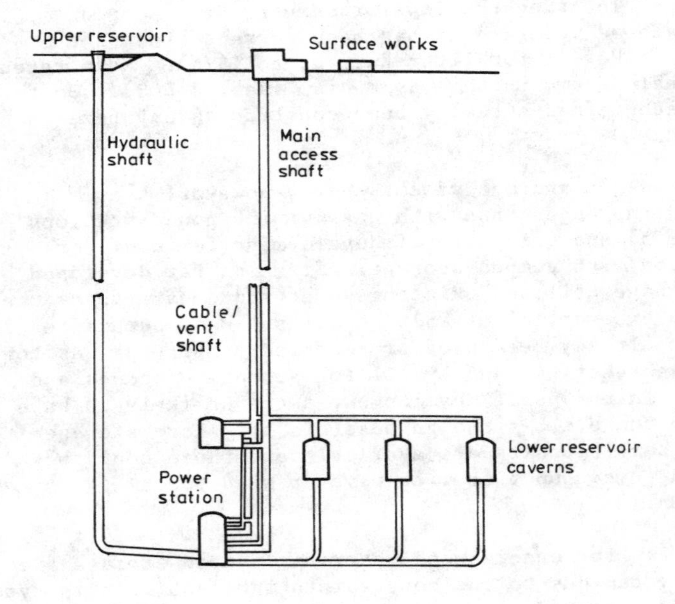

Figure 2 Proposed Layout of Underground Pumped Storage Scheme

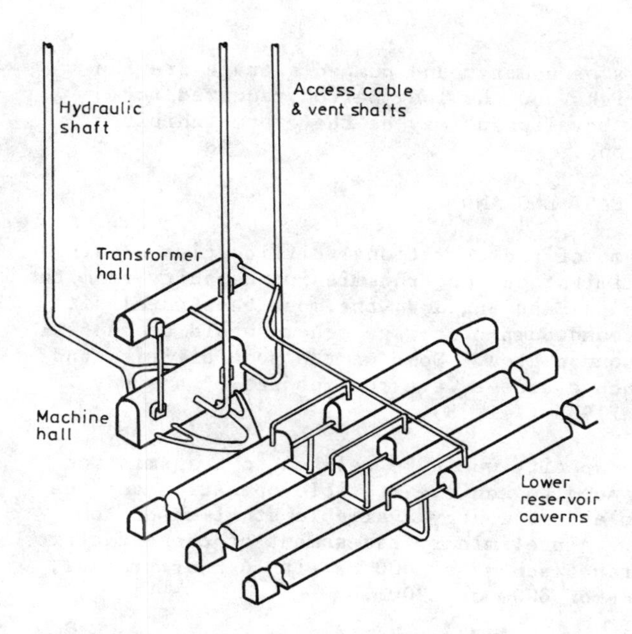

Figure 3 Underground Pumped Storage Scheme
Isometric view of lower works

A maximum head of about 800 m is the present limit of single stage pump/turbine technology and, as increasing head reduces the volume to be excavated to form the lower reservoir for a given energy storage, this head is indicated as the optimum for this type of plant at present. The alternatives of two UPS stations in cascade or multiple stage pump/turbines or separate pump and turbine units could increase operating heads to 1200 m to 1600 m or more with consequent reduction in lower reservoir volumes.

The principal advantage of underground pumped storage is that suitable upland topography is not required. Such a scheme has been investigated in some detail in Holland (Ref. 9). In the U.K. context, this allows a scheme to be sited in an area near a major load centre, such as London or S.E. England, requiring only supply or storage of water at ground surface and suitable geological conditions at the lower reservoir and power station levels. In addition, the environmental impact could be minimised by using a redundant industrial facility for the site of the surface works. Alternatively, an estuary or a river of adequate flow could be used as an upper reservoir.

Major drawbacks of underground pumped storage are the geotechnical risks and the long period required for construction. Development may be therefore inhibited for these reasons.

UGPS IN SOUTH-EAST ENGLAND

To overcome some of the operational difficulties arising from current limitations on transmission capacity from the Midlands to South-East England, the most beneficial site for an underground pumped storage scheme would be in or close to the London area. Some of the possibilities and problems of such a scheme were the subject of a study carried out in 1985 (Ref. 8).

The installed capacity would be related to transmission capacity, the need to achieve flexible operation and the requirement to absorb energy available at times of low system demand. A preliminary assessment of these factors led to a reference scheme of 800 MW with 6 hour storage, utilising heads of 800m or 1200m.

Five principal areas of potential benefits were identified:

- load following to allow low merit thermal generation to be significantly reduced,
- savings due to removal of limitations in grid transmission capacity from major generation areas such as the East Midlands,
- savings arising from deferring construction of new thermal plant and the closure of older, uneconomic stations,
- savings due to central location, reducing need to import energy, improving system loss load factor and voltage control,
- additional dynamic benefits from rapid response time and spinning reserve.

Over a 30 year life at 5% discount rate, these savings were estimated to be in the range £500 - £1,000 million (1985 prices).

Additional benefits may accrue from peripheral activities such as the sale of spoil for aggregates or land reclamation or the use of the site for some secondary activity (e.g. recreation or commercial) after completion of construction. These savings are obviously site specific and have not been included in the above assessment.

Within the London area, there are a number of possible options for the development of sites for underground pumped storage schemes. There are a number of existing, disused thermal power stations which have good transport access and are, by definition, close to the distribution network. Some of these are either too small or are being developed for other purposes, but a number of sites within this category may be worth considering further. If the problems of using saline water can be overcome, such a scheme could draw direct from the Thames estuary with adequate provision to preclude silt, eliminating the need for an upper reservoir.

A second category would require either an existing reservoir of suitable capacity or an undeveloped site of about 40 hectare for a freshwater upper reservoir. An adjacent area of up to 10 ha would be required for temporary works and permanent structures. The proximity of existing transmission lines and of rail or water borne transport is highly desirable. It is unlikely that any site meeting these criteria could be found within 20 km of Central London due to the high density of urban development.

The third possibility is to adopt a hybrid approach where a balancing reservoir of limited area is provided, interconnected with the Thames to reduce maximum flows abstracted from and returned to the River. Sites of 2 to 30 ha would be feasible with possible use being made of redundant dock sites or the sites of disused power stations.

The sites considered were all within the syncline which forms the London Basin. The shafts would pass through the London Clay; the Upper, Middle and Lower Chalks; and the Gault Clay; together with sands and gravels. The power station and lower reservoir could be located in underlying Palaeozoic strata.

The Tertiary sands and the Cretaceous chalks constitute a major aquifer system which extends over most of the London Basin. The Lower Greensand is also an important aquifer but is not present north east of London. Between 25% and 50% of the strata through which the shafts would be excavated might be aquifers and a small proportion of these are likely to be artesian. Clearly there would be major problems to be overcome in sinking and lining three or more shafts, but orthodox construction methods could be adopted, with ground freezing or grouting used to control water inflows.

The underlying Palaeozoic formations are structurally complex; the upper surface of these forms a flat surface dipping gently to the south between 150 and 450 m below sea level. These formations comprise mudstones, sandstones and shales together with some conglomerate and intruded dolerite dykes. Ideally, both underground caverns and reservoirs should be located in the Cambrian basement rock, the depth of which below ground is typically of the order of 950 m.

The primary structure at ground level would be either the upper reservoir or a river intake. An upper reservoir would most likely be formed using orthodox dam construction techniques with an intake/outfall structure incorporated into the bed of the reservoir. An alternative possibility is that an existing reservoir could be used,subject to checking that its integrity can be maintained under the onerous conditions of rapid and frequent changes in water level. With either form of upper reservoir, a suitable source of water would be required for filling the system initially and maintaining the system volume together with provision for disposal of surplus water.

If the feasibility of using an existing site beside the River Thames can be demonstrated, then an intake/outfall structure would be required along the river bank. This would have to be an extensive structure to limit peak velocities and avoid disturbance to adjacent properties, to shipping and to the river bed regime. It would require to function over the whole range of tides and flows in the river and adequate provision would require to be made to avoid the entry of fish, floating debris and silt into the power station.

Information on the rock conditions at the power station depth is limited but the data available suggests that conditions would be suitable for construction of the caverns using methods similar to those used at Dinorwig.

The lower reservoir would take the form of a large excavation or series of excavations located at a higher level than the Machine Hall to maintain minimum submergence of the pump-turbine runners in all conditions. Construction would probably be by orthodox drill-and-blast methods rather than by tunnelling machines.

The completed station should have little adverse effect on the environment as surface works will be relatively small and there is no pollution. During construction, the principal problems would arise from the need to remove around 3,000,000 m^3 m of spoil from the underground works by rail or water transport.

The principal disadvantage of this proposal is that the period from inception to completion is unlikely to be achieved in less than 8 to 10 years. This arises partly from the period required for research and investigation prior to design and partly from the period needed to excavate shafts to a depth of perhaps 1200 m before construction of major underground openings can commence.

The capital costs of such a project, estimated to be in the range £810 to £980/kW (1985) may be compared to the estimated benefits of £660 to £1250/kW. At the time of the study, therefore, the economic justification of the scheme was uncertain. As underground construction techniques improve, the demand for electricity becomes greater and the constraints on electricity supply increase, it is perceived that the economic viability of a scheme of this nature will increase steadily with time.

REDUCTION IN PROGRAMME TIME

In addition to studies of specific sites for pumped storage development in the UK, more general investigations have been carried out to achieve designs which can be built in the shortest possible time and at the least construction cost. These studies can be summarised under the following headings.

1. Hydraulic System Layout

The decision to opt for a single hydraulic conduit between the upper reservoir and the penstock bifurcations in the Dinorwig Scheme has proved successful. Unless geological and geotechnical conditions preclude the use of large diameter hydraulic conduits, a largely single conduit hydraulic system is likely to be preferred in future projects.

Vertical pressure shafts have so far been favoured in the UK except at Cruachan, but world-wide the use of inclined shafts is common. The benefits of reduced hydraulic losses in an inclined pressure shaft system are partly offset by the quicker, cheaper and safer construction of a vertical shaft. In addition, the extent of steel linings on the upstream side of the powerhouse may be substantially less in a vertical pressure shaft system because of the greater rock cover. The vertical pressure shaft also offers advantages in access for inspection and maintenance (Fig. 4).

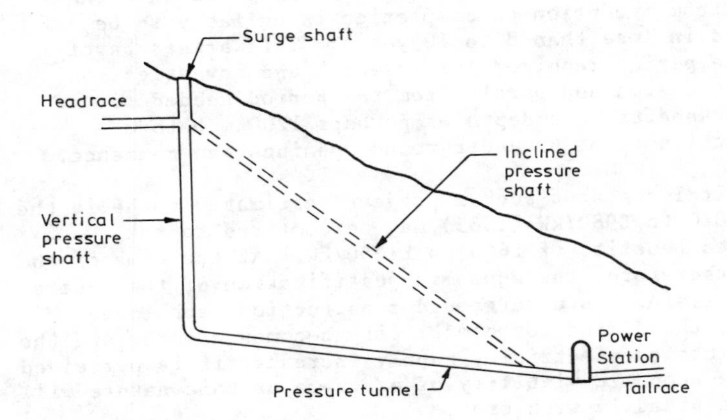

Figure 4 Alternative Configurations of Pressure Shafts

2. Underground Powerhouse Layout

The Dinorwig layout of 9 caverns in the underground powerhouse area was very complex. It was a solution to the problems of that project, in particular the requirement for rapid response to changes in load.

Several significant changes from the Dinorwig layout would merit serious consideration for future projects including:-

i) Reduction of span of machine hall to the minimum required to accommodate hydraulic plant.

ii) Incorporation of main inlet valves in machine hall, with penstocks entering the machine hall on a line skewed at about 60° to the machine hall axis, to minimise cavern width.

iii) Accommodation of ancillary plant in an extended machine hall rather than in separate caverns.

iv) Incorporation of draft tube valves below transformer hall.

By these means it should be possible to limit the number of major power station caverns to two, machine hall and transformer hall (Fig. 5).

Figure 6 shows a typical programme for a new station and illustrates the opportunity of obtaining earlier commissioning by overlapping activities.

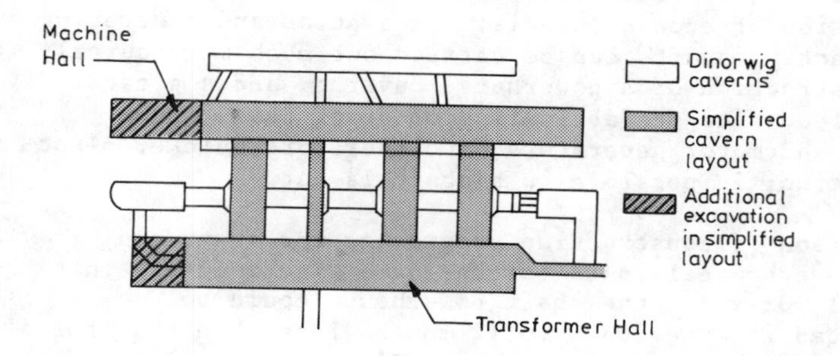

Figure 5 - Simplified Layout of Power Station

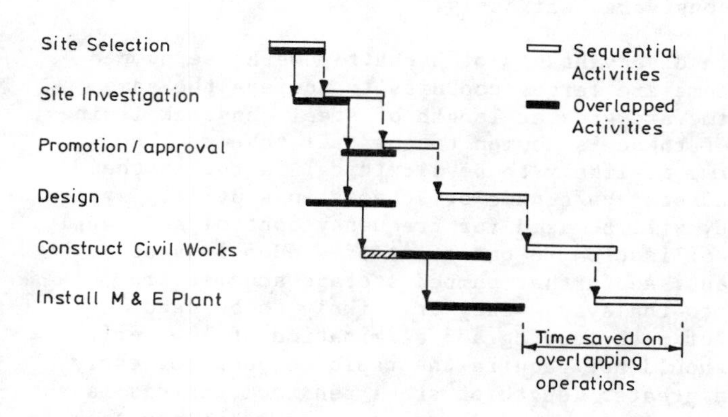

Figure 6 - Alternative Programmes

3. Powerhouse Location

There has been an increasing tendency over the last twenty years to locate the powerhouses of pumped storage schemes in underground caverns. This allows great flexibility in the layout and operating characteristics of the scheme scheme and may be essential for geological or seismic reasons. However, where a tail (downstream) location for the powerhouse is possible, the alternative of a non-cavern (shaft) location for the powerhouse warrants serious consideration.

Table 1 compares some of the advantages and disadvantages of cavern and shaft powerhouse locations. The key factors allowing more rapid construction of a shaft powerhouse are that construction of the powerhouse is not delayed by the construction of access tunnels; excavation and concreting of the machine shafts can be carried out much more quickly than construction of a powerhouse cavern; and a greater degree of on site prefabrication of draft tubes, pump turbines and motor generators, and therefore quicker plant installation, is possible in a shaft layout.

A comparison of construction programmes for shaft and cavern powerhouse layouts for the same site suggests that the civil works for the shaft powerhouse could be constructed in a period 9 to 12 months less than that for a cavern powerhouse. The consequent financial benefits of lower financing costs during construction plus quicker return on capital investment, in addition to lower construction costs, require that the shaft powerhouse layout be considered seriously.

The two main disadvantages of a shaft powerhouse pumped storage scheme are larger conduits to achieve the same response time,and greater length of steel penstock linings. The first of these is common to any tail scheme. The response time is likely to be a critical factor in the first pumped storage scheme or schemes in a utility system because they will be used for frequency control and supply network stabilisation in the event of sudden loss of thermal plant. As further pumped storage schemes are introduced to the system they are likely to be used primarily for peak lopping and elimination of low merit plant and should not require the rapid response of early units. The greater length of steel penstock linings is an inevitable result of the reduced cover over the penstock in a shaft powerhouse scheme compared with a cavern powerhouse scheme.

Although a shaft powerhouse will often have to be constructed on a site which is less favourable in terms of geology (e.g. Foyers) than the site for an underground powerhouse,the geology of the shaft powerhouse site is likely to have been much more thoroughly investigated in advance of construction. Techniques are available for the construction of deep, large diameter shafts even on unfavourable sites, as was demonstrated at Presenzano in Italy (Ref. 11).

TABLE 1

COMPARISON OF CAVERN AND SILO POWERHOUSES

Item	Cavern Powerhouse	Shaft Powerhouse
Geology	Fairly wide choice of cavern location to find best geology	Limited choice of site, at base of slope and edge of reservoir
Site Investigation	Limited unless access tunnels driven in advance of decision to proceed	Full SI possible for machine shafts
Access tunnels	Required	Not required
Steel lined tunnels	Length may be short in suitable geology	Longer tunnels required in general
Surge chamber (downstream)	Required in head scheme & some mid schemes	Not required
Plant installation in large units	Limited by access tunnel size, crane capacity and erection area in powerhouse cavern	Installation in larger units possible with gantry or mobile craneage
Construction cost		Generally lower
Construction programme		Generally shorter, irrespective of geology
Interest during construction (IDC)		Lower
Response time	Downstream surge chamber can reduce effective conduit length and hence response time	Generally slower than for cavern powerhouse layout because of greater upstream conduit length

4. Unlined Hydraulic Conduit

Unlined pressure tunnels and shafts, commonly excavated by full face tunnelling machine, are coming into common use in hydro schemes located in hard, competent rock, particularly in Norway. Such unlined conduits offer the possibility of shorter construction durations. In a cavern layout such a saving may not be important because, in general, the underground powerhouse is on the critical path. However in a shaft layout the hydraulic conduit system may be the critical activity in the construction programme.

The feasibility of such unlined conduits in pumped storage power plants is considered doubtful because of the much more onerous operating conditions to which those plants are subjected due to the frequent load changes and load reversals. In the early pumped storage schemes constructed in the U.K. a considerable amount of remedial work was required to repair concrete lined conduits which had been damaged at construction joints,etc - much more than expected from experience of earlier conventional hydro tunnels. Problems with unlined conduits in pumped storage schemes are likely to occur where support to fault zones, thrust zones, or zones of closely jointed rock has had to be provided by the use of shotcrete. Such shotcrete linings may well remain in position long term in a hydro scheme with infrequent load changes but may have only a limited life in a pumped storage scheme.

5. Colliery Shafts in relation to UGPS

As the very lengthy construction period is a factor weighing against any underground storage scheme, the possibility of expediting construction by utilising existing underground workings has to be considered. Another symposium paper (ref. 2) describes a scheme in USA based on an abandoned limestone mine. No similar large scale workings exist in UK, but abandoned colliery workings are available in abundance.

The geology of a coalfield is generally unfavourable to a pumped storage development, but a review of the British coalfields identified South Leicestershire as possibly offering reasonably suitable conditions. Here faulting has brought up the pre-Cambrian basement rock to abut directly on the Coal Measures, and to outcrop nearby. However, existing colliery shafts are only some 300m deep and would not make a very significant contribution to the economics of a pumped storage scheme. This fact together with the distance from London (150 km) discouraged pursuit of a site in this area.

More conveniently situated is the Kent coalfield, 100 km from London and only 40 km from Dungeness and the cross-channel transmission link. The five collieries in this group all ceased production between 1969 and 1989, leaving a number of shafts from 450 to 900m deep. Palaeozoic formations within which it should be possible to find a suitable location for a competent and watertight lower reservoir, lie at depths between 600 and 1200m. Problems with water ingress to shafts through the younger strata could however be serious, as at least one of the pits had to deal with considerable inflows of highly saline water. A favourable point, however, would be that the provision of a 40 ha surface reservoir should not present undue environmental problems in an area already scarred by colliery waste.

No detailed study has, to our knowledge, been made of the potential value (if any) of such colliery works in the future development of UGPS schemes, and it may be that further research into this aspect is justified.

FUTURE REQUIREMENTS

Before the next pumped storage scheme is promoted in the U.K. it is essential first to establish the need for such an installation in relation to the projected demand and associated system control and other requirements. When the need is established, a wide selection of potential sites is available near load centres or in remoter areas. The selection requires to be made not simply on previous experience but also taking account of the latest techniques which have continued to be developed overseas during the current period of inactivity in the U.K., e.g.

- techniques for engineering assessment of rock conditions underground at depth,
- more economic use of plant in gaining early access underground (e.g. improvements in shaft sinking),
- extended use of TBM's

With the ever increasing experience in the development of pumped storage at home and abroad, it is apparent that early commissioning of pumped storage stations is not only highly desirable but quite practical. Further studies along the lines suggested in the paper are required to demonstrate this and to impress on electricity undertakings the full potential of pumped storage in their systems.

REFERENCES

1 ELECTRIC POWER RESEARCH INSTITUTE, CALIFORNIA. Underground pumped storage - research priorities, April 1976.

2 WILLETT D. Summit energy storage project. Int. Conf. on Pumped Storage, London 1990, (In draft).

3 WATTS R. Proposal for construction of pumped storage station on coast of South-East England. Int. Conf. on Pumped Storage, London 1990, (In draft).

4 HILL & McMILLAN. Compressed air storage systems for power generation. World Energy Conference, 1977.

5 MASS P. & STYS Z.S. Operating experience with Huntorf, 290 MW: World's first air storage system transfer plant. Proc. Am. Power Conf, 1980.

6 LISTER P., BANNISTER W.E., and DUNN T.H. Salt cavity storage - the first 25 years. North of England Gas Association, March 1985.

7 JAMES WILLIAMSON & PARTNERS. Proposal for pre-feasibility study for Underground Reservoir Pumped Storage (Submission to CEGB), October 1979.

8 JAMES WILLIAMSON & PARTNERS. Underground Pumped Storage in the London Area. Unpublished Report, March 1985.

9 UNDERGROUND PUMPED STORAGE IN THE NETHERLANDS. Pilot Study, Dec 1980.

10 HAWS E.T., WILSON E.A., and GIBSON A.I.R. Pumped Storage in the proposed Mersey Tidal Power Project. Int. Conf. on Pumped Storage, London 1990.

11 BETERO M., PAVIANI A. & VANNI D. Deep consolidation using the jet grouting system for the tunnels & bottom plug of the Presenzano hydro-electric power plant - execution & checks. Foundations & Tunnels - 87, 1987.

Planning of the Guangzhou pumped storage power station: the first high water head large capacity pumped storage power station in mainland China

Y.Z. CAI, Guandong General Power Company

Summary
In this paper, an outline is given of the Guangzhou Pumped Storage Power-Station (GPS): the necessity and reasonableness of constructing this power-station; the options of a lower reservoir site and of penstocks; the selection of electrical and mechanical equipment; financial aspects, etc. In the latter part of this article, the prospect of constructing a second-stage pumped storage power-station, with the installed capacity of 1200 MW, will be discussed.

Outline of GPS project
The GPS is located at Lu-Tian Conghua county, Guangdong province. The distance from Guangzhou to the station is 90 km (Fig. 1). The project is composed mainly of dams of upper and lower reservoirs, delivery water systems, and underground power house, etc. The upper and lower reservoirs

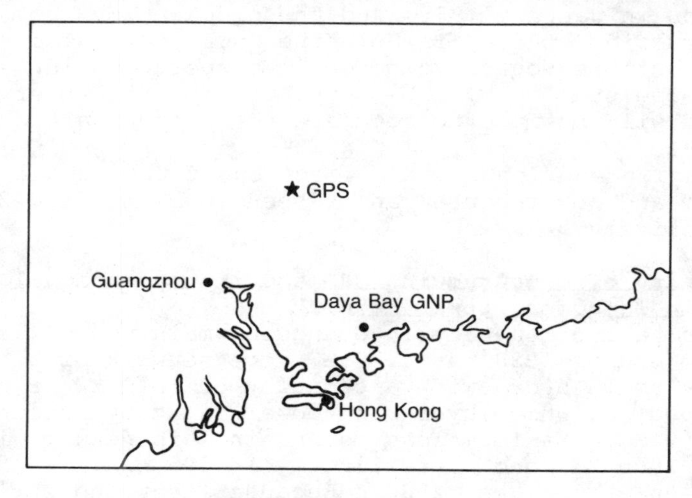

Fig. 1 Location of project

are situated separately on the secondary tributaries of the Liu-xi River, where there is a beautiful scenic spot with plentiful water resources and surrounded by high mountains. The reservoirs are in good natural basins; the total capacities are 17 million m^3 and 17.5 million m^3 respectively. The maximum heights of the dams are 60 m and 37 m respectively. The lasher foundation of the upper and lower reservoirs is formed of granite with single rock properties. It is small in size for the fracturing structure, thin in bed alluvial layer, and slight in weathering. No seepage problem exists at the reservoirs.

The total length of the penstock and tailrace tunnel is 3751 m. The dimensions of the underground power house are: width 21 m; length of 145 m; height of 45.6 m. Both the delivery system and the underground power-house are arranged in the thick rock mass of granite. Except for the alteration of varying degree which occurs in the rock mass around the structure fissure, rock properties are comparatively uniform and integral.

The installed capacity of the station at the first stage is 1200 MW, which consists of four single-stage reversible pumped storage units of 300 MW each. The maximum water head is 535 m. The daily average pumping time is 6-7 hours and the generating time 4-5 hours. The overall efficiency of the station is 75 percent. After its completion, the station will be linked separately to the Guangdong electric network by a 500 kV transmission line with a total length of 225 km.

The central government gave approval to build the station in January 1988, and the preparation for construction started in July the same year. The first unit is scheduled to go into operation in 1993. The station will be completed in 1994. Every effort will be made to complete the project ahead of schedule.

The plan and longitudinal section of the tunnel system and power-house can be seen in Figs 2 and 3 respectively.

Necessity of constructing GPS and its operational role in Guangdong power network

The installed capacity in Guangdong main network up to the end of 1988 was 4230 MW (the network of mainly small hydro with a capacity of 1810 MW is not included) of which hydraulic power capacity accounts for 990 MW. Maximum load demand in this network is about 3800 MW, and approximately 15 000 MW is foreseen at the year 2000. Because light industries are the main consumers in the power network,

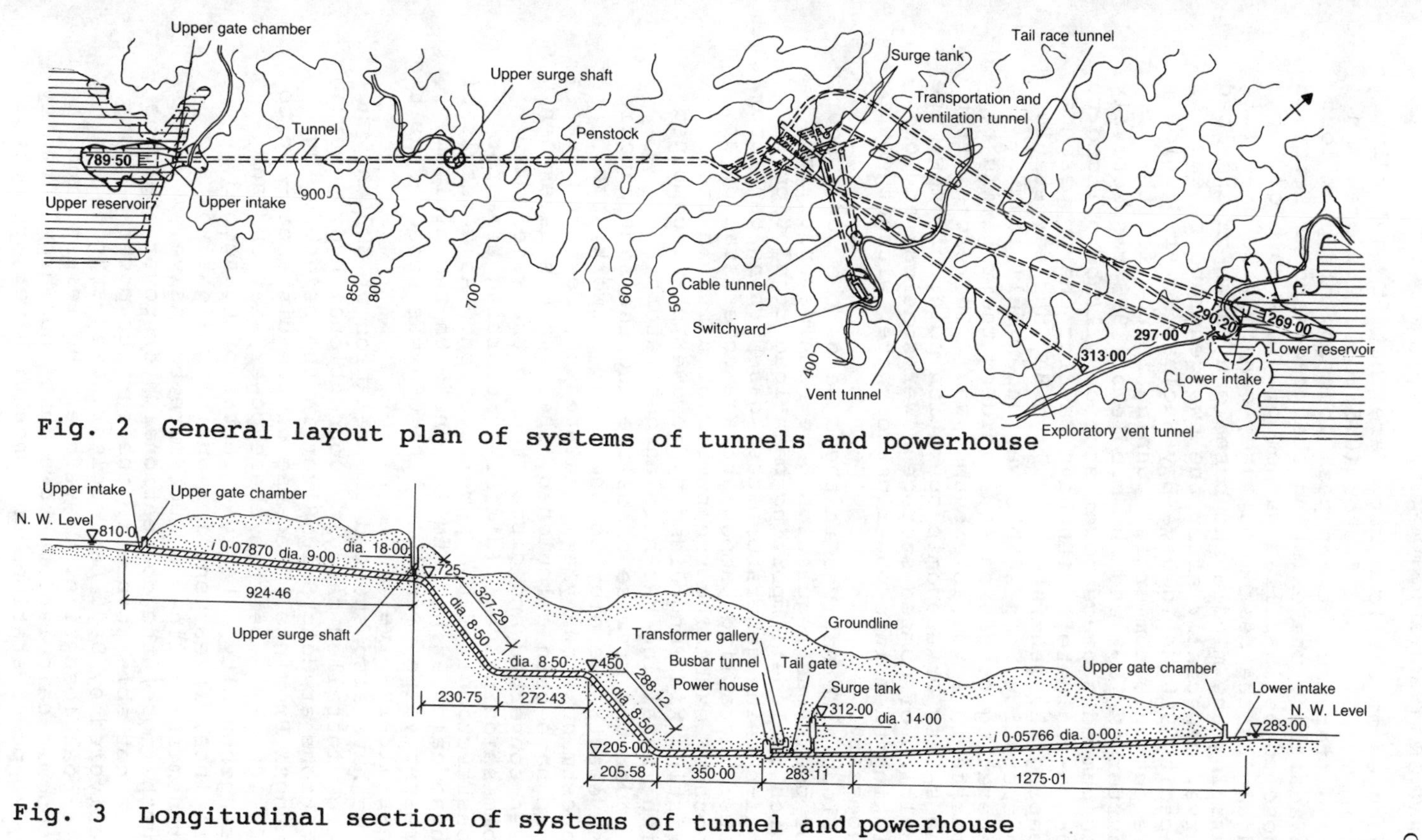

Fig. 2 General layout plan of systems of tunnels and powerhouse

Fig. 3 Longitudinal section of systems of tunnel and powerhouse

features of the load curve are that it is low in minimum daily load factor (0.5 to 0.57), great in the difference between peak load and off-peak load. In addition, the existing thermo-generator sets and those to be installed cannot play the role of adjusting the peak load; they can only be used for carrying the middle and base load shown on the load curve, especially after the two generators, each of capacity 900 MW in Daya Bay Nuclear Power Station, are put into commercial running in the fourth quarter of 1992 and the middle of 1993 respectively. The pumped storage power station will be obliged to run co-ordinately with Daya Bay for the purpose of keeping the generators in Daya Bay nuclear station in operation in a stable way under full load generating condition and avoiding their having to operate in a variable output way. Therefore, from the safety or economic operation point of view, the building of pumped storage power stations and of linking them into power network are, without doubt, necessary.

Eight alternatives in the case of the power source for peak load regulation have been demonstrated, which include: importing peak-load regulation coal-burning generator sets; gas-burning generator sets; installing pumped storage power station at Shezhen; extending the number of units in the existing Xin Fung Jiang hydropower plant combined with installing gas-burning generators.

Under the premise of delivering the same capacity and electric energy to the power network, the investment flow, yearly maintenance cost and fuel cost have been calculated. These values have then been converted and discounted to the same commissioning year (different starting time for construction) of the power station. As a result of these calculations and in line with the minimum present value principle, the GPS was selected as the best alternative.

We have also studied the relation of GPS and the large installed capacity conventional hydropower stations at Hong Shiu River in the neighbouring Guangxi Province. Because of the distance - up to 700-1000 km - between these power stations and Guangzhou city, the load centre of Guangdong Province, investment for building light voltage transmission line is much more expensive. Furthermore, the conventional hydropower stations are not able to absorb the surplus power output delivered by Daya Bay nuclear power-station at off-peak load time. Therefore, the result from the economic calculation shows that the large capacity hydro power-stations on Hongshui River are better to

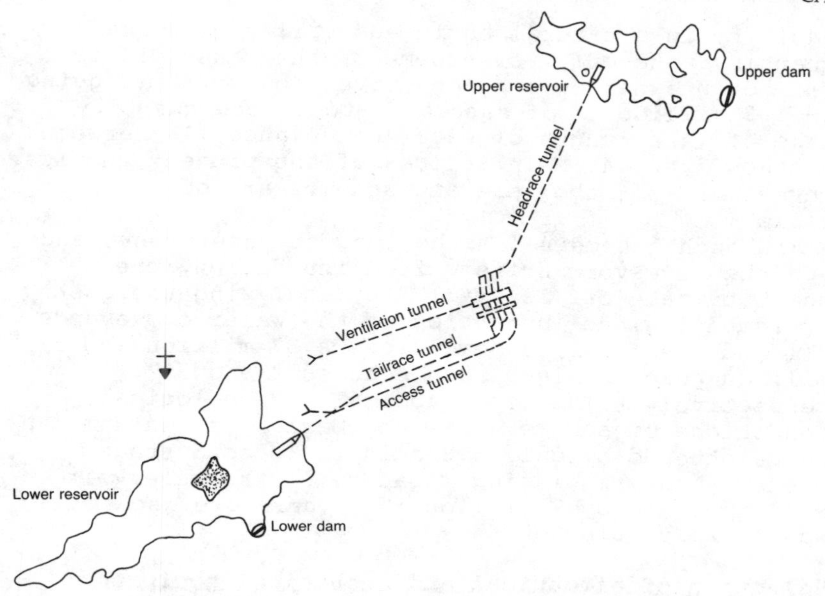

Fig. 4 Layout

supply power for the middle part of the load demand curve of Guangdong Province.

Summarizing this analysis, the GPS will play important regulating peak and filling off-peak roles in the Guangdong power network. Furthermore, in the design of the reservoir, a stand-by reservoir capacity for storing water to run two turbines for two hours with full load output has been included, so that, if necessary, the GPS can act as a short duration emergency stand-by power station. Meanwhile, the GPS can also be used for condenser running and power network frequency regulation.

Lower reservoir sites and selection of water conveyance

On the basis of the topographic and geological conditions indicated in Fig. 4, we have studied four alternative schemes for lower reservoir sites and their relevant water conveyance system.

For scheme 1, since the dam base and karstification are well developed in the reservoir site, leaking and foundation settlement are difficult to deal with. It was given up after preliminary work. The other schemes have different characteristics. Scheme 2 has the shortest water conveyance; it is proposed to locate the storage facility within an encircling embankment on hill slopes, with the most quantity take-off. Scheme 4 has the longest water conveyance and it is proposed to locate the storage facility in a natural basin and valley with least quantity take-off. For scheme 3, the reservoir is formed in a narrow and long river course, by digging up 1.8 million m^3 of sand and stone; the quantity take-off and length of water conveyance lie between schemes 2 and 4. Main indices of the project and its investment for the relevant schemes are given in table 1.

Although scheme 4 has the longest water conveyance and the reservoir suffers from inundation, the quantity take-off is small. By analyzing the cost of remedying the inundation of the water conveyance, dam and reservoir, scheme 4 costs 37 million and 39 million yuan RMB less than schemes 2 and 3 respectively. The topographical and geological conditions of scheme 4 are the best. If the dam can be heightened by only 6.8 m in the second stage project, the regulating capacity of the reservoir will increase by 7 million m^3. Therefore, scheme 4 was finally selected.

Selection of electrical and mechanical equipment

Because the water-head of the GPS is high, up to 535 m, a single stage pump turbine unit with an output of 300 MW and rotating speed of 500 rpm is adopted. The manufacture of units with such output and rotating speed involves the world's most advanced techniques applied in the production of similar machine sets; as such, they cannot yet be built in China and have to be purchased from abroad. However, in order to reduce the financial burden incurred during the project construction period, a favourable treasury mixed loan, including a loan from friendly countries has been considered for this project. According to international practice the equipment must be purchased from the country which gives the favourable loan. One disadvantage of such a system is that prices of equipment will not be competitive. As a result of this, the loan country and the manufacture of the equipment is finalized by way of combining equipment quality, equipment prices and loan conditions at the very beginning, thus creating a bid package open to competitive tender.

In line with the technical requirements set out in the 'GPS letter of inquiry of electrical and mechanical equipment' which we have submitted, seven Consortia from five countries entered the competition. After more than one year's discussion of technical problems, price comparison, and bidding negotiation, France was selected as the loan country in a deal finalized with the CGEE ALSTHOM Consortium of France. The GPS electrical and mechanical contract was signed in January 1989, and the

TABLE1

Designation	Unit	Alternative 2	Alternative 3	Alternative 4
1. Overall capacity of reservoir	$10^2 m^3$	713	788	1750
2. Total length of water conveyance system	m	2217	3069	3751
3. L/H		3.3	5.1	6.9
4. Dam height/Dam length	m	43.2/1970	41.6/606	33.6/129
5. Variation in water level of lower reservoir	m	13.6	17.5	5.7
6. Farm land inundation	Mu	159	261	1056
7. People to be settled		160	233	943
8. Investment difference of dam and water conveyance system	10^4 yuan	3700	3920	0
9. Geological condition at dam and reservoir site		Located at karstification area, with complex geological condition	Great variation in water level; not good for the stability	Geological condition good
10. Extension possibility of reservoir		Limited	Limited	Possible

treasury soft loan contract and the Buyer's credit contract were signed at the end of April and beginning of May, respectively, in the same year. The prices of equipment are reasonable: the cost per kilowatt of main equipment and its auxiliaries is 120 US dollars.

With regard to the performance comparison of the main machine, the pump turbine unit, we studied and repeatedly compared the rotating speed of the unit, i.e. 428 rpm and 500 rpm. The 500 rpm alternative is most favourable for the following reasons: comprehensive efficiency will be around 1% higher; it is small in unit dimension; 13% less cost for the unit as well as 2.2 million yuan RMB less for the civil works of the power house. Even though the immersed depth of the turbine runner of the 500 rpm unit is 13 m deeper than that of the 428.6 rpm unit - i.e. the depth will be 70 m - it will have little effect on the civil work quantity of the underground power house. One aspect is worth noting. For the 500 rpm alternative, the specific speed Nsp for the pump turbine proper should be higher: i.e. its Nsp will be up to around 35, which means that this pump turbine will bear more serious working stress. However, this problem can be solved with up-to-date mechanical manufacturing techniques. Therefore, all the manufacturers who joined the competition suggested the 500 rpm rotating speed alternative as the rated speed of the unit. The diameter of the runner is 3.9 m. The runner technical data will be finalized by model test.

Prospect of second-stage pumped storage project

The GPS is only 90 km from the load centre of the power network, namely Guangzhou city. It processes good topographical, geological and water source conditions. On the basis of the experience gained from carrying out the first-stage project, it is obvious that an extension of the second-stage project would be both economical and reasonable.

The necessity of the extension of the second-stage project is

on account mainly of the following: the quick increase of load demand; the absolute value of the difference between peak load and off-peak load becoming greater and greater; as well as the Daya Bay nuclear power station, it is possible to build another nuclear power station at Guangdong province, and as a result, more pumped storage machine sets will be needed to operate in co-ordination with it. Therefore, it is considered that, before completion of the GPS first-stage project, advantage can be taken of the existing tentative construction devices and, by the way of bid invitation, to construct a second-stage project.

The installed capacity of the second-stage project is considered to be 4x300 MW at the preliminary stage. The following could be undertaken without affecting the construction and running of the first-stage project; proper heightening of the dams of the upper and lower reservoirs; the building of a new penstock, new tailrace tunnel as well as of an underground power-house. The purchasing of the electrical and mechanical equipment will also be undertaken by means of international bid negotiation and competition. At present the feasibility study is progressing; the preliminary design will be carried out in 1991; by 1992, it is hoped to have government approval for construction to begin.

Discussion

F.G. JOHNSON, *formerly of North of Scotland H.E. Board*

In assessing sites for potential pumped storage schemes, it is not adequate to investigate only hydraulics, geology, engineering and economic aspects. An increasingly important field which must be addressed in parallel is the environmental and promotional situation. In studies undertaken by the Hydro Board over the last two or three years, parallel studies have been carried out to provide preliminary environmental assessments not only of the potential pumped storage scheme sites, but particularly the transmission route corridors which are essential to convey the power between sites and the entry point into the grid, including strengthening works on existing lines. I consider these environmental aspects will have a predominant influence in determining the acceptability and promotabilty of future pumped storage schemes.

Similar preliminary environmental assessments have also been carried out in appraising sites for conventional hydro schemes.

T.H. DOUGLAS, *Author*

In reply to F.G. Johnson, while our paper covered technical matters, we agree with Mr Johnson's comments on environmental assessment, but would like to add two specific comments. First, it is now mandatory for all future major schemes in the UK to be supported by an environmental assessment and this is also confirmed under recent European directives. Secondly, it is well within the competence, ingenuity and skill of engineers to design the works - even the transmission routes - so as to minimize their effects on the environment. Mr David Jefferies in his opening remarks to the conference pointed out how successful Dinorwig had been and we heartily endorse this approach.

T. H. DOUGLAS, *Author*

When embarking on a study of a proposed pumped storage project it is considered essential as a first step to investigate and establish the electrical system requirements. In the UK, for example, peak load generation, particularly with the National Grid covering England and Wales, does not necessarily have the same value in economic and functional terms as spinning reserve and frequency regulation. Fig. A shows the principal grid lines within mainland Britain.

The second step is to determine the type of scheme

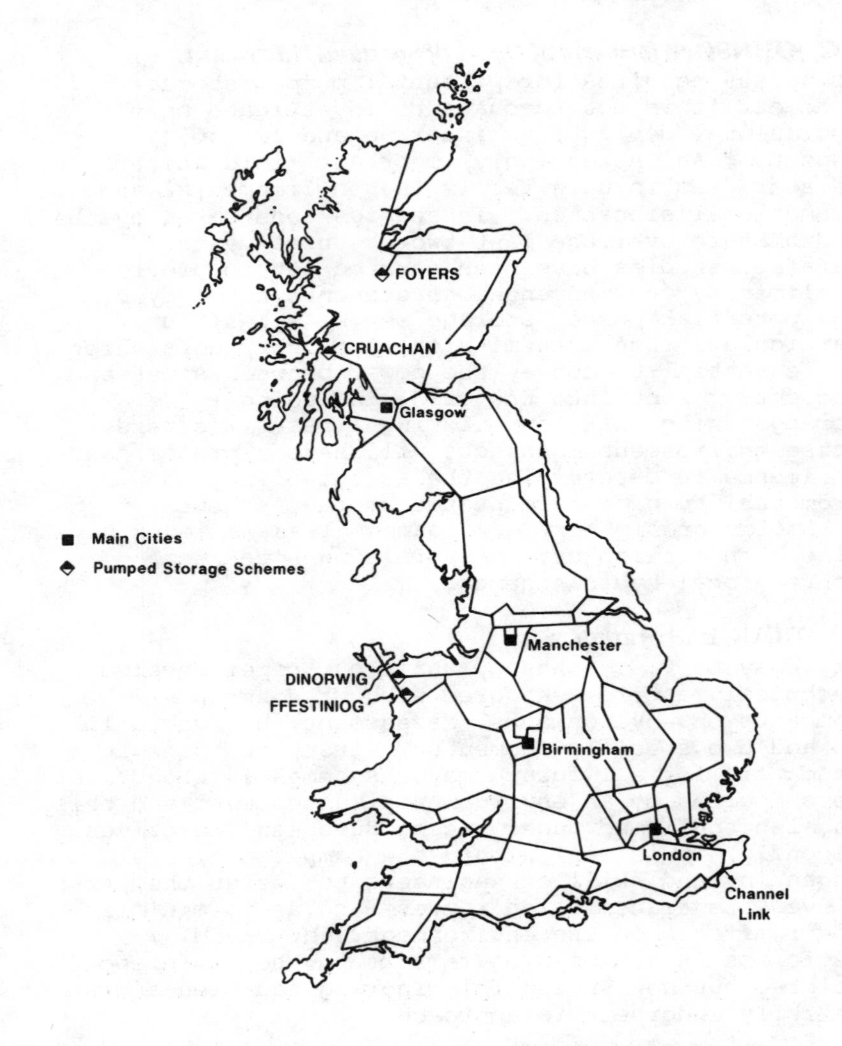

Figure A - PRINCIPAL GRID LINES IN MAINLAND U.K.

which will most closely fulfil these requirements. For example, should it be a single large scheme with a system control function or a series of more modest or smaller capacity schemes with local control functions. Again, economic conditions may favour a multi-purpose scheme. As part of this stage, the choice has to be made as to whether the storage element is to be designed to operate on a weekly

Figure B - TOPOGRAPHY OF U.K. MAINLAND

cycle or the much smaller storage associated with daily cycle. Perhaps economic conditions may favour even shorter term storage options.

The third step is to identify potential sites. A first assessment would be to consider a conventional scheme in mountainous terrain. Fig. B shows land

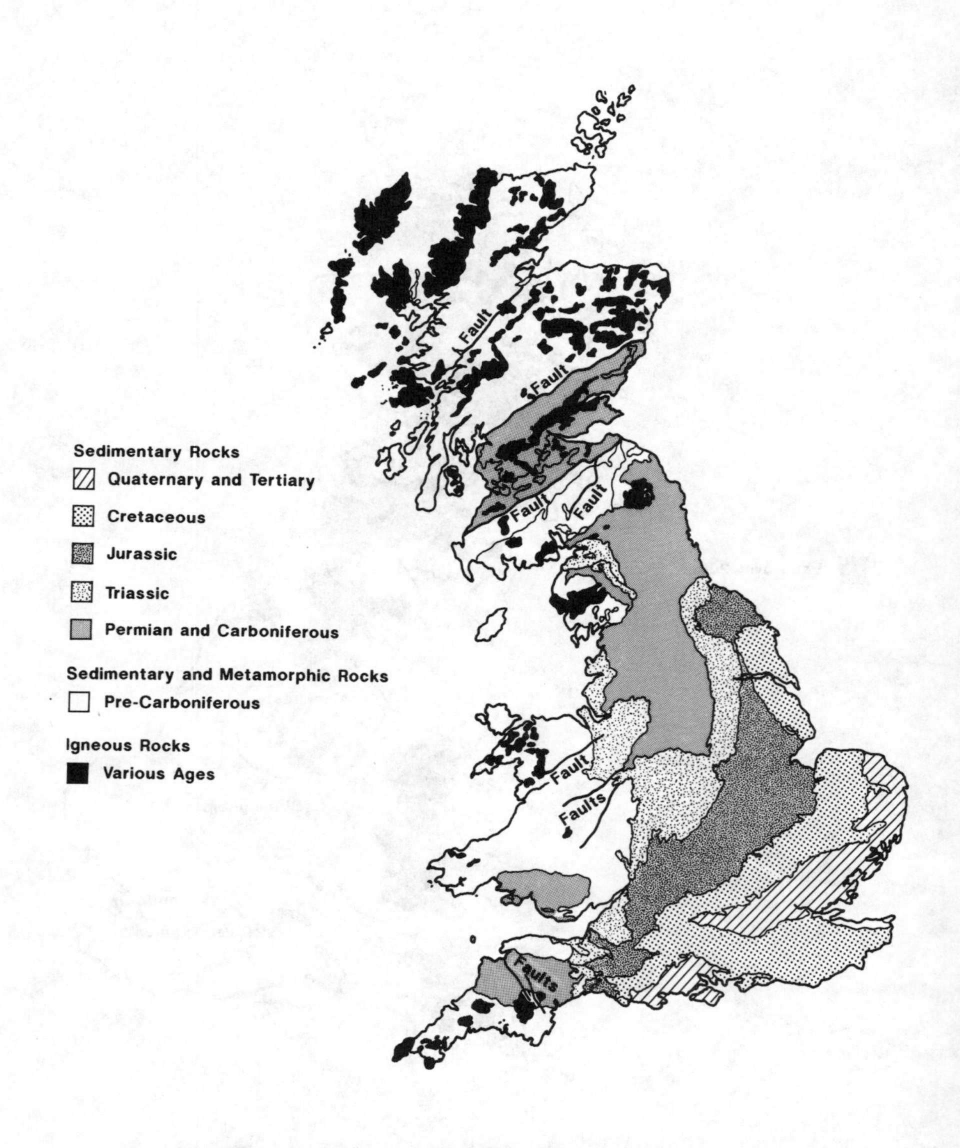

Figure C - GEOLOGY OF MAINLAND U.K.

over 200 m above sea level in the UK mainland, although such investigations today are more likely to favour level differences of at least 500 m which would sensibly limit initial studies to a lesser number of locations, principally along the west coast. It may be practical to utilize existing underground space. In the UK the largest excavations have been developed for the extraction of coal but these are generally unstable and collapse in a relatively short time. New excavations are dependent on the soundness of the rock formation for ease of construction and initial assessments would be centred on favourable rock types. Fig. C shows the base geology of mainland Britain.

Finally, the economic options should include various scheme configurations and a review of improvements in shaft technology, such as blind drilling and raise boring techniques. These and other techniques, such as overlapping activities (shown in Fig. 6 in the paper) will influence considerations of alternative programmes. In the final analysis, the most economical option will be based on the least sum of the particular site development plus the transmission costs.

R.W. BUCHANAN, ***Author***

In his paper Dr Szeredi brought out the advantages of investment in pumped storage against the alternative of gas turbines and other thermal plant. He recommended a pumped storage facility of 1280 MW capacity to be commissioned by 1997. Since the paper was written, the scheme has been postponed and this timetable is unlikely to be achieved. Hungary is heavily dependent on nuclear generation and imports from the USSR.

Consideration of transient phenomena from load rejection in the waterway of Shimogo pumped-storage power station

M. HORI and M. KASHIWAYANAGI, Electric Power Development Co., Ltd, Tokyo, Japan

SYNOPSIS. Transient phenomena of internal water pressure induced by full load rejection have been observed. Particularly, water-hammer pressure in the headrace tunnel was investigated with relation of the surging pressure.

SHIMOGO POWER STATION.

Shimogo power station is a pumped-storage power station of maximum output 1,000 MW, comprising from four 250 MW reversible sets under an effective head of 387 m. There are two independent waterway systems, each serving two reversible sets. In each system, it consists of about 2.2 km long headrace tunnel, restricted orifice type surge tank and penstock of approximately 700 m in length.

MEASUREMENT OF DYNAMIC INTERNAL WATER PRESSURE.

Direct measurement of water-hammer has hardly ever been made for the reason because it is not desirable in operation of power generating facilities to install measurement devices inside a waterway. Accordingly, a verfication of an analysis method for water-hammer has hardly ever been attempted. Furthermore, it is not clear how to conditions of reflection of water-hammer pressure in a restricted orifice type surge tank should be assumed.

To accomplish the purpose, a pressure transducer has been embedded in the concrete lining at the section of the headrace tunnel which is located in 72 m upstream from the surge tank. In the surge tank, a pressure transducer was also set at a position of approximately 20 m above the orifice to monitor a surging water level or pressure. In addition, a monitor of dynamic pressure was also performed at the end of the penstock by transducer.

MEASUREMENT RESULT OF WATER-HAMMER AND SURGING

Figure 1 shows the pressure variations at the three points after full load rejection by simultaneous closing of two 250 MW units. The maximum of water-hammer pressure at the end of penstock arised at approximately 10 sec after occurrence, and the water head at this time was approximately 570 m. When considered as head variation taking the initial head into account, the maximum head by water-hammer was approximately 125 m, and it was about 28% of the initial static head.

Meanwhile, with regard to surging, it is noticed that there was a small water-head variation at the initial stage of surging. The peak of this minute variation occurred at approximately 20 sec after start of surging with the influence of water-hammer pressure being superposed on surging. It can be seen that the same trend appeared in the water pressure variation at the measurement section in the headrace tunnel. In this case, the peak of the pressure due to water-hammer arised at approximately 14 sec after occurence, the value being approximately 3.3 kgf/cm^2. The subsequent pressure variation was superposed on surging water-head variation, with the maximum value occurring at the time of maximum water level rise on surging. The surging pressure on this time was approximately 3.0 kgf/cm^2.

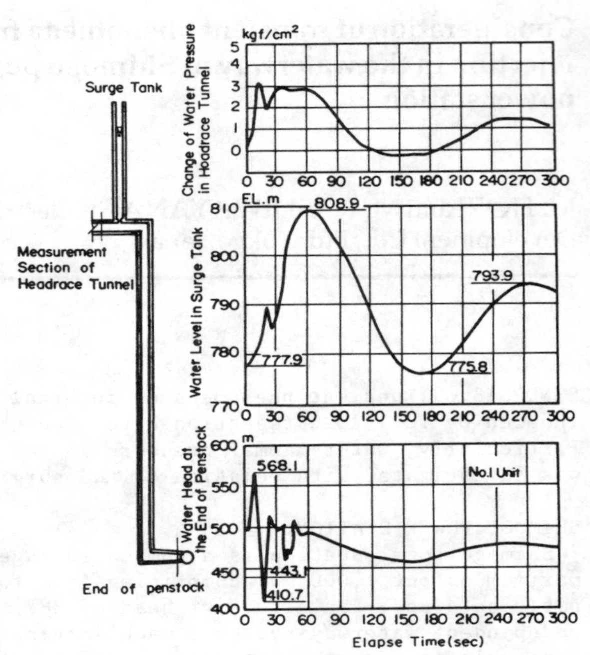

Figure 1 Transient Water Pressure in Waterway at Full Load Rejection

It may be concluded from the above resluts that water-hammer pressure appears prominently in surge tank water-head variation and at the upstream side in the headrace while producing changes in its waveform. The pressure variation measured inside the headrace tunnel is in a form of water-hammer pressure superposed on pressure due to water-head variation of surging. According to this, it may be said that maximum value of water-hammer pressure at the headrace measurement section become approximately 2.7 kgf/cm^2. Since the static water pressure at this section in operation was 7.4 kgf/cm^2, the water-hammer pressure was approximately 36% of this, and in comparison with the maximum water-hammer pressure of 12.5 kgf/cm^2 at the end of the penstock, approximately 22% had been propagated.

Workshop summary

C. P. STRONGMAN, Merz and McLellan

C.P. STRONGMAN, *Merz and McLellan*

The session was chaired by Dr H.L. Grein of Sulzer Escher Wyss, and attended by about 12 delegates.

The session opened with a summary of the paper: 'Consideration of the transient phenomena from load rejection in the waterway of Shimogo pumped storage power station' by M. Hori and M. Kashiwayanagi of EPDC Japan, given by the latter co-author. The Shimogo station contains four 250 MW reversible sets under a head of 387 m and its hydraulic system comprises a 2.2 km long headrace tunnel, restricted orifice surge chamber and 700 m long penstock.

The discussion was opened by C.P. Strongman who expressed interest in the site measurements and particularly in the observed discontinuity in the surge chamber water level 20 s after the instant load rejection. He suggested this was probably due to the pump-turbine entering the 'reverse pump' region; this was supported by other delegates. It was explained by M. Kashiwayanagi that the measurements had been made by civil engineers, without corresponding measurements of the pump-turbine conditions; this brought out a general recommendation that, for such a test to produce overall benefits, complementary measurements by civil and plant engineers should ideally be carried out.

The session then considered the comparison of site measurements with the results of computer analysis of hydraulic transients. Dr A.P. Boldy, of the University of Warwick, displayed such a comparison made for Steenbras pumped storage scheme in South Africa, with four 45 MW reversible sets under a head of 275 m. He explained that the results were very sensitive to the form of the pump-turbine characteristics in the turbine brake and reverse pump regions and how these were represented in the computer analysis.

Dr Grein referred to prototype tests undertaken in

Italy to check the so-called 'hysteresis' region of a pump-turbine head-discharge characteristic: no untoward effects had been found during tests made over the unstable region identified during model tests.

A lively discussion then took place over the validity of model test results, derived under steady conditions but then applied for the study of transient behaviour of the plant and its hydraulic system. E. Goldwag, of GEC Alstom, was of the opinion that one should not expect too much correlation to apply, while L.E. Lindestrom, of Kvaerner Turbin, in support of this view explained their practice of carrying out additional quasi-static model tests, obtaining oscillograph traces.

The session was closed by Dr Grein who pointed to the desirability of site tests and analytical work being developed further whenever opportunities became available.

Use of submersible ROVs for the inspection and repair of hydroelectric station tunnels

R.E. HEFFRON, PE, Subsea Engineering International Ltd,
Richmond, VA, USA

SYNOPSIS

Advanced underwater robotic technology is providing, for the first time, an alternative to the dewatering process for the inspection, evaluation, and repair of hydroelectric station water conveyance tunnel systems. The traditional process of dewatering has many drawbacks in the eyes of station management including the extended disruption of station operations. In addition, tunnel inspections performed under maximum head, full pressure conditions yield data which is much more representative of the normal operating water pressure conditions.

At Virginia Power's Bath County Pumped Storage Station, the largest station of its kind in the world, this robotic technology was pioneered through the development of a purpose-built Remotely Operated Vehicle (ROV) to gain access to the tunnels. Unmanned ROV's have been used by the offshore oilfield community and the military since the early 1970's and have seen significant improvements over the years.

The "Hydrover I" system, developed by Virginia Power and their partner, Allegheny Power Systems, was successful in performing a record tunnel excursion of 1850m. One important element in this

----

1 Director, Tunnel-Vision(sm) Inspection Program, Subsea Engineering International Ltd, 2803 Sunrise Blvd, Richmond, Virginia 23233

success was the development of a 2120m neutrally buoyant umbilical which limits frictional drag on the tunnel liner.

Building upon this success, the "Super Hydrover II" system is currently under development which will be capable of excursions as long as 10km. The real key to the success of this technology however, lies not simply in getting to the end of the tunnel, but rather on the techniques used and experience of the Civil Engineers operating the system.

The "Hydrover" systems incorporate an array of video, stereoscopic, and photogammetric cameras arranged to complement the unique requirements of tunnel inspections. An on-board dye release system has proven to be very useful in detecting the movement of water through cracks in the tunnel liner.

The "Hydrover I" system was initially deployed in October, 1988 and has been used extensively for the inspection of Bath County's three tunnels. This very successful application of this advanced technology has proven to be a safe, practical and very economical alternative for the inspection of hydroelectric station tunnels.

INTRODUCTION

The inspection of hydroelectric station water conveyance tunnels has been a problem for electric utilities worldwide. The tunnels are generally too deep and long for divers and dewatering can be very expensive and disruptive. Such was the problem faced by the Bath County Pumped Storage Station in rural western Virginia.

The Bath County Pumped Storage Station, with a combined generating capacity of 2100 megawatts, is the largest station of its kind in the world. Owned by Virginia Power and their partner, Allegheny Power System, the station was placed into operation in 1985.

The station consists of the upper and lower reservoirs, the powerhouse, and a series of three tunnels linking them. The major portion of the 20 story powerhouse is submerged beneath the lower reservoir. The powerhouse contains six turbine pump generators capable of producing 350 megawatts each. The three tunnels were excavated through the mountain rock to provide a link between the reservoirs and the powerhouse and are lined with concrete. The penstocks are 5.5m in diameter, tapering to 2.75m in diameter at the spherical valves.

As indicated in Figure 1, each of the three tunnels is approximately 3.2 km in length and is made up of the upper low pressure tunnel, the vertical power shaft, the lower high pressure tunnel, and two steel lined penstocks. The tunnels are each 8.6 m in diameter and are lined with concrete. The penstocks are 5.5m in diameter, tapering to 2.75m in diameter at the spherical valves.

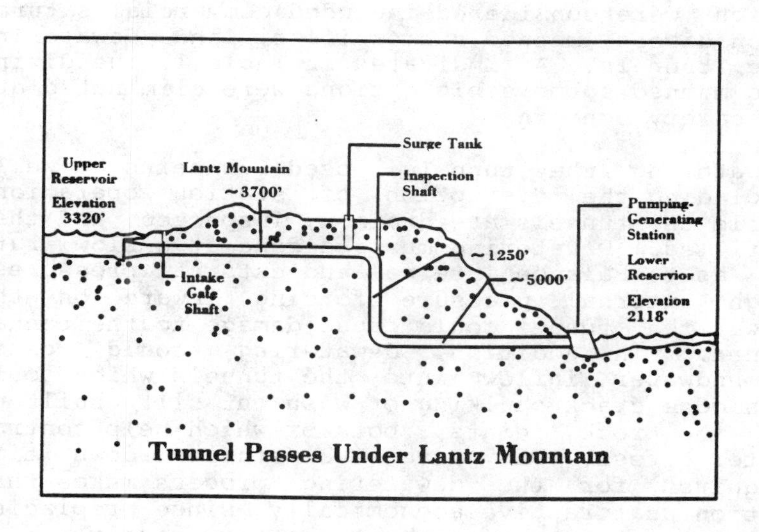

Figure 1. Elevation View of Typical Tunnel

With a maximum operating head of 410m, the water pressure and nearly continuous water flow subject the concrete liner to significant forces. Discovery of water egress from the tunnels into the highly jointed mountain rock led to an extensive remedial grouting program, delaying plant startup for several months. Since the grouting program, leakage from the tunnels has been reduced considerably. The leakage presently observed has displayed seasonal characteristics. Water flow from drainage tunnels, provided in the mountain to relieve pressures, is much greater in the winter months than in the warmer seasons. As the rock mass and concrete liner contract, the cracks open wider to allow more flow.

Since the station has been operational, the condition of the tunnels has been assumed to be satisfactory based on performance, though the actual condition was largely unknown. When it was determined that the tunnels should be inspected, it became readily apparent that the use of an ROV would provide the safest and most efficient means of inspection.

INSPECTION ALTERNATIVES

Alternative means of conducting the inspection which were considered included commercial saturation divers, manned submersibles, and dewatering the tunnels. As indicated in Table 1, the diving and manned submersible options were eliminated due to safety concerns.

Dewatering the tunnels posed concerns as well, including the disruption of station operations while the tunnels are drained, inspected, and then refilled. Dewatering must be done at a slow rate so as to balance internal and external pressures. High external pressure forcing inward on the dewatered tunnel could cause damage to the tunnel liner. In addition, dewatering could cause groundwater inflow into the tunnels which could dislodge crack caulking or wash out silt built-up in the rock joints, both of which help control water egress. The extensive station down time required for the dewatering process makes this option unattractive economically since replacing the lost generating capacity is very costly.

Dewatering the tunnels also relaxes the outward pressure on the tunnel liner, allowing cracks, if present, to close. The resulting inspection would not be representative of operating water pressures.

The use of an ROV offers the advantages of a safe and economical inspection with the least amount of disruption to station operations. The ROV also offers the advantage of being able to view and plot water egress from the tunnels under normal operating water pressures.

TABLE 1
COMPARISON OF INSPECTION ALTERNATIVES

METHOD	ADVANTAGES	DISADVANTAGES
Commercial Saturation Diving	Allows u/w inspection Allows first-hand observation	Unsafe Limited by divers inspection qualifications
Manned Submersibles	Allows u/w inspection Allows first-hand observation	Unsafe compared to other methods Limited down time due to batteries and air supply Limited by pilots inspection qualifications
Remotely Operated Vehicle	Allows u/w inspection Allows Professional Engineers direct control over inspection Safe and economical	Requires development of new equipment
Dewatering	Allows first-hand observation Allows Professional Engineers direct access to tunnels for inspection	Requires additional time for draining and refilling Requires setting up lighting and scaffolding for access to crown Possible damage to liner from high external pressures Possible loss of crack caulking and silt build-up in cracks due to ground-water inflow Not representative of operating conditions due to relaxed pressures.

PROJECT BACKGROUND

The original intention of the owners was to contract the inspection work out to a commercial ROV service contractor. However, disappointing proposals received from several of the leading contractors underscored the differences between this unique tunnel inspection task and the offshore oilfield related work normally serviced by these contractors.

The offshore oilfield market has, in the past, been the largest commercial employer of ROV services. Typical offshore tasks consist of deep dives, to as much as 3,000m or more, and relatively short horizontal excursions. This requires, among other things, a cage or garage which houses the vehicle during deployment.

Fig. 2 2120 meter neutral umbilical

The garage is lowered from the deck of a pitching and rolling launch vessel using an armored electro- mechanical lift wire/umbilical. Generally, expensive heave compensation parameters must be designed into the handling system to accommodate rough sea conditions. Thus, to deploy such a system designed for use offshore on an in-shore tunnel inspection project, the contractor must leave these expensive components of his system idle. Instead, a new type of umbilical is required which is neutrally buoyant over its entire length.

In contrast to the offshore oilfield related work which requires deep vertical and relatively short horizontal excursions, tunnel penetrations require just the opposite. The Bath County tunnels require a maximum depth of only 410m, but after turning around a 90 degree elbow, must proceed nearly 1500m horizontally.

The differences that tunnel inspections pose to the application of ROVs do not stop with the umbilical and handling requirements. The purpose of deploying an ROV inside a tunnel is not simply to see if it can reach the end. Rather, it is to perform a specific mission upon deployment. In tunnel inspections, the mission is to visually inspect surfaces, at any orientation, in search of concrete cracking patterns, steel corrosion, liner erosion or deterioration, and water leakage. Too often in the past, ROVs with standard offshore camera configurations and equipment have been deployed in tunnels and penstocks only to find that they can see in front of the vehicle just fine. However, the vehicle must be turned ninety degrees to face the wall to facilitate a visual inspection, requiring considerable maneuvering of the ROV. The crown and invert of the tunnels were never properly inspected due to the limitations of the camera configurations.

THE 'HYDROVER' SYSTEM

The goal in developing the system for the inspection of the Bath County tunnels was to overcome each of these past deficiencies. The owners decided to develop an ROV system for the expressed purpose of inspecting tunnels. A leading submarine manufacturer located in Vancouver, B.C. was selected to design and manufacture a submersible tunnel inspection ROV system to the owners' specifications. The result of this development effort is the 'Hydrover' ROV shown in Figure 2.

The potential for limited visibility water conditions and the requirement for close-up detailed inspection of concrete crack and steel corrosion characteristics dictated that the inspection camera be no more than twelve inches, at maximum, from any portion of the tunnel wall. This was accomplished by mounting the primary video inspection camera on a swing arm that rotates 360 degrees around a horizontal axis through the longitudinal center of the vehicle. This swing arm is positioned at the bow of the vehicle inside a protective frame. The unique arrangement of this color video camera facilitates the close-up inspection of the crown, invert, and side walls of the tunnel while maintaining the vehicle in a heading parallel with the tunnel direction. This may seem to be a minor factor at first, but by eliminating the need to turn the vehicle 90 degrees in each direction to face the walls, the speed of the inspection as well as the accuracy of ROV positioning is greatly increased.

Fig. 3 The 'Hydrover' Submersible ROV

In addition to the rotating camera, two other video cameras are included. The bow-mounted color camera points forward and has an internal pan, tilt, and rotate mechanism for the lens. This camera is used for navigation and for inspecting vertical surfaces. The third camera is black and white and is mounted aft. It is used primarily for tracking the position of the umbilical behind the vehicle.

A stereo photo camera is used for photographing cracks, spalls, or other features in the tunnels. The resulting slide pairs are mounted to produce a 3-dimensional effect for depth perception.

Among Hydrover's other unique capabilities, a dye release system is used to release a small quantity of highly concentrated red dye near a crack to observe potential water egress through the tunnel liner. Two dye release ports are provided, one in view of the rotating camera and the second on the end of the manipulator arm. The five function manipulation is used for accurately positioning the dye port. The manipulator, equipped with a stiff bristle brush, is also used for cleaning surfaces to allow closer inspection. This was especially useful in evaluating corrosion in the steel lined penstocks.

Of particular importance to navigating and positioning in the tunnels are the dual head sonar system and the gyrocompass. The 675kHz sonar head, with a fan transducer, is mounted vertically on the upper left front corner of the vehicle's buoyance module. This head is useful for long range navigation and general obstacle avoidance. In addition, the 675 kHz head was very useful in tracking the vehicle position in the vertical tunnel sections relative to the tunnel circumference. The second sonar head is a high resolution 2 Mhz head with a profiling transducer. This unit proved to be invaluable in tracking the vehicle position in horizontal tunnel sections relative to the tunnel circumference. The sonar proved to be so valuable that, without it, the documentation of observed cracks and features could not have been performed to an acceptable degree of accuracy. The gyrocompass, used for navigation, avoids magnetic interference from the steel in the penstocks and reinforcing steel in the concrete.

The dimensions of the vehicle were driven by the need to access the power tunnels through the 2m diameter inspection shaft. Final dimensions are approximately .9m wide x 1.2m high x 1.5m long. Four thrusters facilitate movement in the forward, reverse, lateral, and vertical directions.

The most significant difference between the 'Hydrover' system and conventional ROVs is the 2120 umbilical which is neutrally buoyant over the entire length. Manufactured in the U.K., this is the longest neutrally buoyant umbilical known to exist. At a final diameter of 44mm, the umbilical is jacketed with a pressure resistant material of low specific gravity to counter balance the weight of the conductors such that the umbilical neither floats nor sinks under pressure. Neutral buoyancy was critical in order to keep the umbilical from dragging against the concrete liner of the tunnel.

Frictional resistance due to dragging was kept to a minimum with a neutral umbilical. Low frictional resistance thus reduced the forward thrust requirements and hence the physical size of the vehicle.

The umbilical provides power to the electrical and hydraulic systems on the vehicle and transmits telemetry commands and feedback between the control console and the ROV. Research and testing performed on the coaxial cable over which the video signal is transmitted ensured that attenuation was sufficiently low to achieve exceptional quality video over the 2120m length. Extensive testing was also performed in a pressure vessel to verify neutrality characteristics of the umbilical in accordance with the specification.

The handling system consists of a winch and drum for storage of the umbilical and a jib crane for the deployment and retrieval of the vehicle. The large winch is sized to provide optimum cooling of the umbilical by having only three wraps on the drum when on the surface. This eliminates the need for supplemental cooling such as a water bath. The swinging jib crane positioned next to the tunnel access shaft is used to lower the vehicle through the air interface to the waterline. A latching mechanism on a steel cable, with a remote release, is used for launch and recovery to avoid strain on the umbilical.

The horizontal position of the ROV within the tunnel is very accurately determined by measuring the amount of umbilical deployed. Slight forward thrust on the vehicle is used to maintain a tight umbilical, eliminating slack which could lead to measurement errors. An electronic counter on the winch transmits the payout to the pilots heads-up video display. A back-up mechanical counter is used for redundancy. In addition, the umbilical is marked in 1.5m increments to provide further assurance of positioning accuracy.

SYSTEM OPERATION

Operation of the vehicle is provided from a control house which contains the control console, sonar processor, power distribution unit, annotation keyboard, and the video tape recorders and monitors. As shown in Figure 3, the pilot controls the vehicle functions while the navigator/documentor keeps track of vehicle position, maps the inspection results, operates the video equipment and narrates the video.

The inspection results were documented on a scrolling map of the tunnel, providing a continuous presentation of the tunnel layout. Crack widths and other feature dimensions were determined from the monitors which were equipped with graduated engineering scales, corrected for the effects of camera position and underwater magnification.

Fig. 4 Engineers Operating 'Hydrover' System

Consistent with an international trend among highway authorities, military branches, and utilities with underwater structural/geotechnical facilities, the owners elected to have licensed professional engineers perform the underwater inspection work. This trend reflects the increased emphasis on the importance of civil works inspections in light of dam, bridge, and powerhouse collapses in recent years. Experienced engineers, trained in diving skills, are taking their expertise below the water line where they are in a better position than the typical contract diver to evaluate cracks, spalls, scour, etc. and recommend appropriate repairs if necessary. Virginia Power Civil

Engineers, experienced in diving inspection work, were trained in the operation and maintenance of the ROV system and produced a detailed report of their findings.

The tunnels, while displaying some leakage from the vertical shaft areas and the area where the main tunnel branches into two penstocks, were in very good condition overall with no present concerns regarding structural stability. Virginia Power is currently investigating the feasibility of slowing the rate of water egress through artificial siltation or grout injection. Hydrover will also be used in support of this effort.

The performance of the ROV system exceeded all expectations while providing the owners with assurance of the integrity of the tunnels. Even at a cost of about $1 million (USD) to manufacture, test, and deploy the system, the ROV option has proven to be a bargain compared to the past accepted practice of dewatering.

Near term plans include the development of a family of such vehicles designed to optimize the size, power and on-board equipment to a variety of work environments. These vehicles will range from small but powerful systems capable of penetrating pipelines, to larger systems capable of chemical grouting to seal tunnel leaks, to tracked vehicles capable of debris removal. Development is also proceeding on a first-of-its-kind fiber optic, neutrally buoyant umbilical which will allow penetrations to 10km or more.

UK experience of inspecting hydraulic systems using remotely operated submersible equipment

T.H. DOUGLAS, FICE, FIStructE, FIWEM, MConsE, James Williamson & Partners, C.F. LADD, formerly Subsea Offshore Limited, and C.K. JOHNSTON, MICE, Hydro-Electric

ABSTRACT

In the past, hydraulic tunnels and shafts have been physically inspected from time to time, wherever possible during station outages. These operations require careful planning and control to ensure that the survey and remedial work, if any, is carried out so as to reduce the effect of the lack of availability on the electrical system.

At Dinorwig, in North Wales, the first survey was carried out by ROV 5 years after initial filling of the system and gradually increased operation. The vehicle used in the high pressure system was able to traverse the whole system including the steel lined tunnels a short distance upstream of the underground power station. The survey was carried out in 9 days and the results showed that all the hydraulic passages were in an excellent condition with no major defects.

At Cruachan and Foyers, in Scotland, ROVs have been used to back up periodic dewatered inspections and have shown the economy of a flexible but systematic approach to the maintenance requirements of the hydraulic systems of pumped storage power stations.

INTRODUCTION

The hydraulic systems of the 3 earlier pumped storage power stations in the UK (Ffestiniog - 360 MW, N. Wales; Cruachan - 400 MW, and Foyers - 300 MW, Scotland) have been dewatered and inspected periodically.

The Dinorwig Power Station (1800 MW) was planned to be dewatered and inspected periodically but the successful operation of the Station with its key role in assisting in the control of the National Grid of England and Wales precluded earlier or unnecessary outages which might have allowed such inspections to take place.

Pumped storage. Thomas Telford, London, 1990.

DINORWIG - 1986 INSPECTION

Background

During the design and construction of Dinorwig, considerable effort was expended by the Consulting Engineers, James Williamson & Partners, and the Contractors, MBZ, to ensure that the hydraulic system was completed to the highest practical standards. Special attention was paid to the layout and design of the largely unreinforced concrete tunnel linings, and steel linings were coated with epoxy, wherever possible in controlled conditions, outside the underground environment. Great care was taken with tunnel shutter design, placing methods - especially in the high pressure shaft which was slipformed.

The hydraulic system is shown on Figure 1 and was filled during the period 29 September to 9 November 1981. Pumping operations were initiated on 1 December 1981 while first generation was undertaken on 8 December 1981. The reliability runs on all six machines were completed by August 1984.

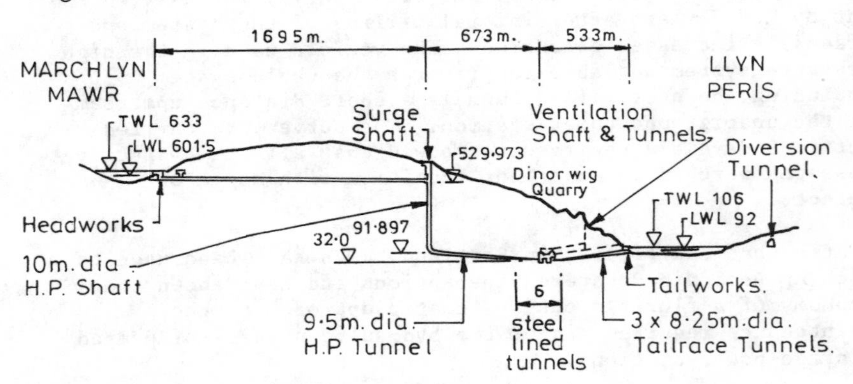

Fig. 1 Dinorwig hydraulic system

It was originally suggested that a first inspection should be carried out within a year or two of the start of full commercial operation of the Station. When the time came to put this work in hand, the Board faced the option of trying relatively new technology (ROVs had been contemplated at the design stage but experience was limited in the mid 70's in the use of this equipment) or, alternatively, mounting a fairly major exercise of dewatering coupled with the hire or purchase of inspection gear.

By 1986 when 129,000 mode changes of the plant had taken place, the Station Manager, Central Electricity Generating Board, decided to implement an ROV inspection of the principal components of the hydraulic system.

Survey

The survey was implemented in the period 18 to 27 September 1986 during which time the hydraulic system was shut down but remained full. Two ROVs were used, the Scorpee 8 (designed and built by Ametek Straza) and the Pioneer XVII (designed, built and operated exclusively by Subsea Offshore Ltd). Scorpee 8 was used to survey the headworks, tailworks and the upstream end of the low pressure tunnel. Pioneer XVII was used for the downstream section of low pressure tunnel and for the main deepwater survey which included the high pressure shaft and tunnel and the manifold (Figure 2).

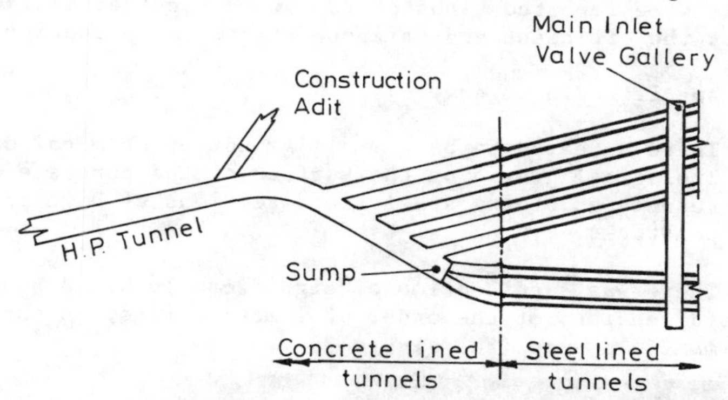

Fig. 2 Dinorwig manifold

The position of each ROV relative to the launch point was determined by measuring the amount of umbilical deployed. This was accurate to within a few metres except when the umbilical was not tight. This occurred occasionally when the ROV was required to back track for a short distance. The deployed umbilical length was input manually into a computer at 6 metre intervals. The computer calculated the position in the tunnel, in the case of vertical shafts above AOD, in the case of the low pressure tunnels as a distance from the headgates, and in the high pressure tunnel as a distance from the HP shaft/tunnel intersection. This information was overlaid on the video display.

Two methods were used to determine the vehicle orientation within the vertical shaft. A nylon rope with a weight attached was lowered down the shaft and the pilot used this as a guide. After each pass, the rope was removed around the shaft a further 15° (in case of HP shaft). In areas where the shaft changed in diameter (e.g. between the orifice and HP shafts) the vehicle's gyro compass was used.

In the near horizontal tunnels, the sonar was used to give an indication of the vehicle's circumferential position within the tunnel.

The overall time from arrival on site to departure from site was 13 days, the first 4 days of which were spent setting up the vehicles and awaiting station shutdown.

Results

All the areas surveyed were found to be in good condition and there were no significant defects. Several items of debris were reported although none were considered likely to affect the efficient and safe operation of the Station.

Some specific findings were:

1. There appeared to be a small amount of chemical or biological growth on the surface of the concrete tunnels adjoining steel lined sections of high pressure penstocks.

2. There was a collection of significantly sized boulders (dimensions of the order of 1 metre or so) in the manifold sump (see Figure 3).

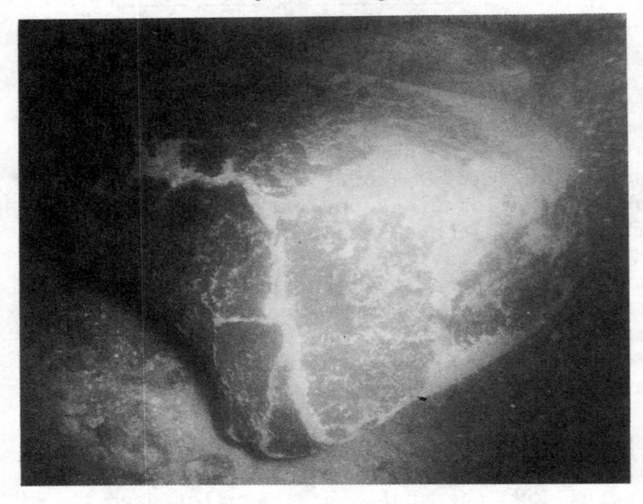

Figure 3 : Dinorwig-Boulder in Manifold Sump

The sump is formed at the end of the High Pressure Tunnel where the tunnels servicing machines 1 and 2 continue on towards the Machine Hall. The chord across the inverts of the smaller tunnels (width of sump) is 6.90 m while the depth to the invert of the High Pressure Tunnel is 1.30 m. The tunnels all grade at 10% so that the total length of the sump amounts to 13 m.

3. Very small amounts of erosion or degradation of concrete were noted in a few places.

4. A bent screen bar was noted at the tailworks.

5. A small amount of debris/rocks was noted in the lower velocity areas at screens.

It was significant that although some limited cracking was evident no special movements were apparent in the unreinforced concrete tunnel linings. This was of some concern since the amount of high pressure grouting had been severely restricted from that adopted normally and had been completely omitted as a systematic operation.

It was intriguing to note that a small boulder had reached the 'wrong' side of the tailworks screen.

Future Inspections

The cost of the ROV inspection amounted to £65,000.

The cost of a dewatered inspection would carry a significant additional penalty due to Station outage being extended for both emptying and filling the hydraulic system. An estimate of the cost of the inspection including the purchase or hire of equipment, cost of personnel time and the electrical units to empty and recharge the system would depend on the length of time required but could vary between £75,000 and £100,000.

Two special points should be addressed:

1. The overriding commercial pressure on the Station Manager to maintain the plant in operation.

This pressure is tempered with the need for overall economy of the Station and the knowledge that if repairs are needed in the hydraulic system limited work may prevent continued or extended degradation or deterioration.

2. The effect on the concrete linings of dewatering the system.

   While the unreinforced linings have been designed to sustain external water pressures, they have never been subjected to these loads. If there are any flaws at joints or cracks, dewatering would become the stage at which these would show up.

CRUACHAN AND FOYERS

Hydro-Electric who operate Cruachan and Foyers have been using ROV technology since 1981 to complement the engineering inspection of tunnels and other underwater structures. Based on their experience, in order to gain the full advantages of an ROV inspection certain key steps should be taken in planning the operation.

It is important that there should be a close liaison between the Engneer and the ROV operator on the selection of appropriate equipment which may vary depending on the location and geometry of the installation and water conditions likely to be encountered.

The Engineer directing the work should have a good knowledge of the structure which is to be inspected and should be aware of the type of defects likely to be encountered and thus be able to interpret the visual record as the inspection proceeds. The Engineer should also have the experience and background to allow him to interpret the significance of any defects which are found and direct any follow up action.

Hydro-Electric's view is that ROV inspections should not be considered as a substitute for a direct (dry) inspection and that, if at all possible comprehensive direct (dry) inspections should be carried out wherever possible prior to an underwater inspection to establish appropriate reference conditions against which the underwater results can be compared. In this regard it is particularly useful to establish markers or reference points in order to guide the ROV during a direct inspection to appropriate locations.

Particular surveys based on these guidelines have been made by Hydro-Electric at Cruachan Intake, Tailrace Tunnel and associated structures and at Foyers Intake with satisfactory results.

ACKNOWLEDGEMENTS

The survey at Dinorwig was commissioned by Mr. M. Hancock, formerly Station Manager, Dinorwig Power Station, on behalf of the Central Electricity Generating Board, while the work at Cruachan and Foyers was directed by Mr. F.G. Johnson, formerly Head of Engineering, North of Scotland Hydro-Electric Board, on behalf of the Board.

Design of high pressure concrete linings for the Drakensberg pumped storage scheme

P.A.A. BACK, BSc, DPhil, FEng, FICE, Sir Alexander Gibb & Partners Ltd

SYNOPSIS The Drakensberg Pumped Storage Scheme in the Republic of South Africa is a 1000 MW Station with an operating head of around 500m which is used both as a conventional pumped storage facility and as a high head pumping station to transfer water over the Drakensberg escarpment from the Tugela Catchment into the Vaal Catchment. Final feasibility studies began in September 1974 with a programme of geotechnical investigations and the first of the four 250 MW units was commissioned early in 1981 with the remaining machines following shortly thereafter. The concrete lined waterways for the project total 8.5 km and for the high pressure sections have to withstand internal pressures up to 720m water head. Extensive studies were made to develop a concrete lining capable of withstanding such heads - thus greatly reducing the use of heavy steel linings with consequent significant savings in cost. The paper describes the technique which was used to develop high prestress in the concrete linings and their subsequent performance in operation.

INTRODUCTION

1. The four 250 MW pump turbines at Drakensberg are fed by two 6m dia, 1450m long low pressure tunnels, and two 5.5m dia vertical shafts dropping 260m from their intersections with the low pressure tunnels.

2. From the base of the vertical shafts two concrete lined pressure tunnels each 5.5m diameter and approximately 850m long, sloping at a grade of 1:10 connect into two steel lined penstocks, each 4.8m diameter, which bifurcate to 3.4m diameter approximately 100m upstream of the machines. (See Figure 1)

3. The operating head on the machines varies between 424m to 467m. In the generating mode the flow per machine is 75 m^3/s and in the pumping mode 65 m^3/s. Transient

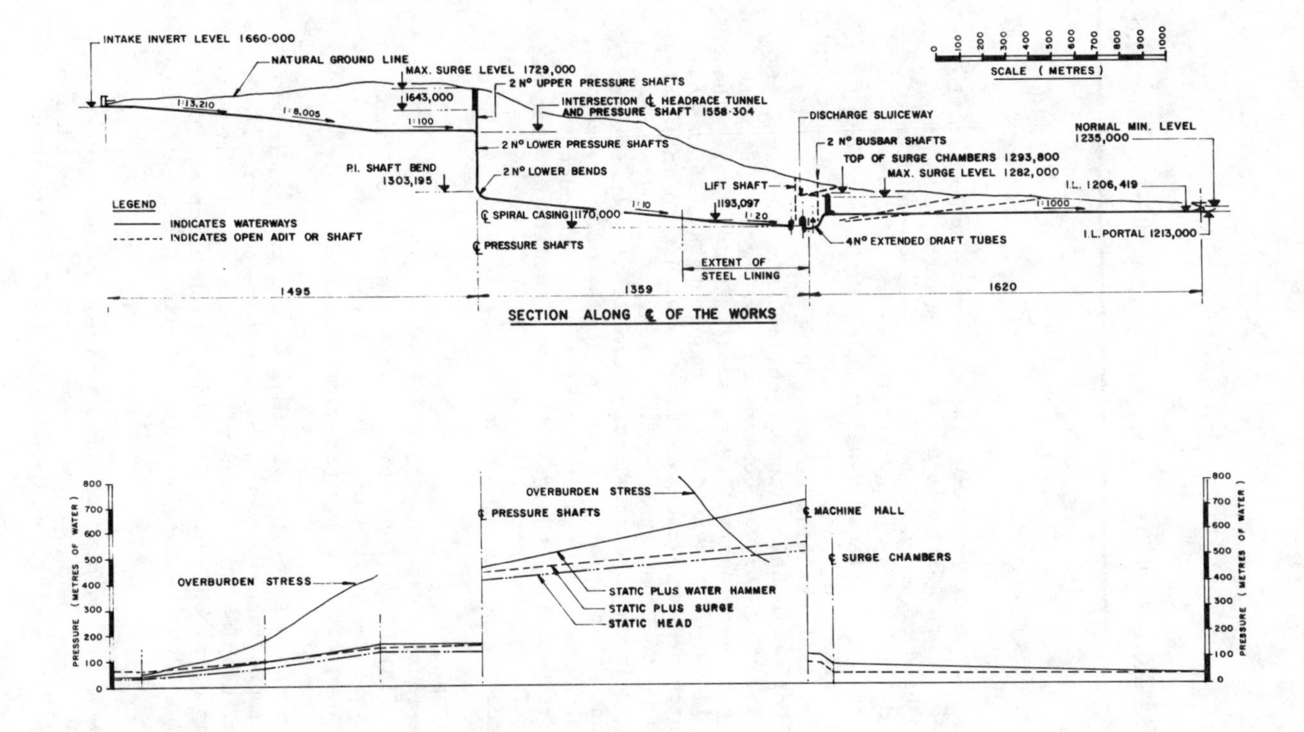

Fig. 1 — PRESSURE PROFILE ALONG WATER TUNNELS

pressures up to 720m water head have to be resisted by the high pressure tunnels during rapid shut down of the machines.

Geology

4. The geology of the site consists mainly of flat lying alternating beds of sandstones, mudstones and siltstones with cross bedding and discontinuities.

General Considerations

5. The rock formations through which the waterways pass are generally impermeable but weak and erodible with the mudstones particularly susceptible to disintegration from loss of confinement and change of moisture. The need for lining the high pressure shafts and tunnels was thus clear, but its cost likely to be high in view of the high transients to be accommodated and the high deformability of the rock which would give minimal support for the lining.

Lining Design

6. A number of alternative designs for the high pressure tunnels and shafts were examined. These included steel linings, watertight plastic membrane linings, cable stressed concrete linings and concrete linings prestressed by interface grouting.

7. The prestress by interface grouting was the solution finally adopted after an exhaustive study of its use in somewhat less demanding applications elsewhere (Refs. 1 & 2) Professor G. Seeber of Innsbruck University who had been involved in the development of the methods (known as TIWAG method after Tiroler Wasserkraft AG of Austria) was retained to provide expert advice on the development of the system for the Drakensberg application.

8. The principle behind the design is to induce a prestress into the lining of sufficient magnitude that after allowing for all the processes of destressing (creep of supporting rock, creep of concrete lining, drying shrinkage of concrete, thermal shrinkage, etc.) there will remain a stress within the concrete so that even under the conditions of highest internal pressure the concrete lining will not be cracked.

REFERENCES

1. SEEBER, G. Moeglichkeiten und Grenzen im Drukstulienbau Schweizer Ingenieur und Architekt Vol.29 1981
2. OLIVE, R T. The prestressing of Gordon Power Tunnel lining by interface grouting. Civ. Eng. Trans. The Institution of Engineers, Australia, 1976.

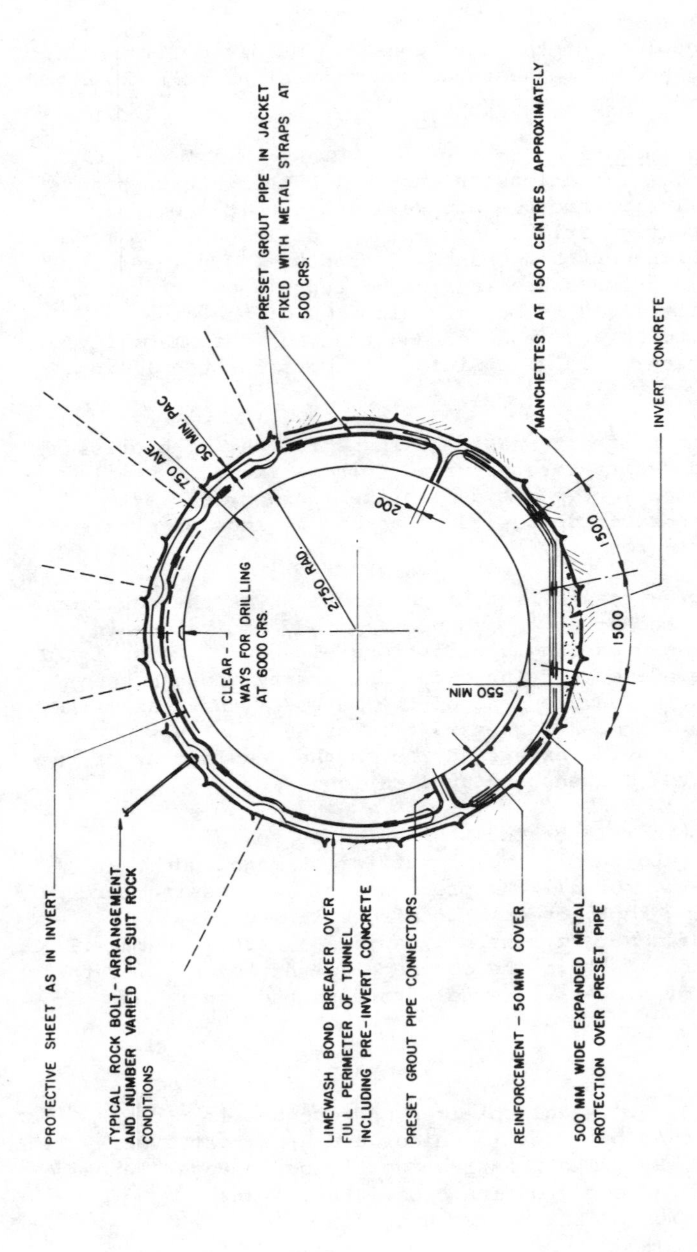

Fig. 2 — TYPICAL CROSS SECTION THROUGH HIGH PRESSURE TUNNEL

9. Analysis of the predicted behaviour of the concrete lining indicated that approximately 55% of the initial prestress would be lost through the above effects. The combination of relatively high rock deformability and high internal pressure required that the pressure tunnels should be prestressed to a much higher level than previously attempted. A strain of 1000×10^{-6} m/m was required at the downstream ends of the pressure tunnels reducing to 500×10^{-6} m/m at the top of the pressure shafts. The maximum theoretical required diametrical contraction on the 5.5 m diameter pressure tunnels amounted to 5.5. mm during the prestress grouting operation. In practice due to the non uniformity of applied load and lining thickness variations it was necessary to increase the diametrical contraction to around 8mm to ensure sufficient compression stress was retained over the circumference of the lining.

Construction Method (See Figure 2)

10. The method of achieving prestress is to inject grout through preset pipes fitted with non return valves and placed in the interface between the concrete lining and the rock. The pipes are covered with a thin plastic sheet to facilitate the initial penetration of the grout from the pipe into the interface by providing a surface on which the grout pressure can act. Prior to concreting the lining, the rock is profiled to ensure that the lining thickness does not vary unacceptably. (At Drakensberg, the profiling using shotcrete to infill depressions required a final lining thickness within the range 750 mm ± 100 mm). Limewash is then applied to the tunnel surfaces to ensure that the concrete lining can detach from the rock or shotcrete surface during the prestress operation.

11. The preset grout injection pipes comprising separate invert and crown loops were fixed at 2.4 m centres along the tunnels with grout arrestors every 36m to prevent uncontrolled longitudinal travel of grout along the tunnel. The arrestors took the form of folded fleece (non woven polyester material) nailed to the surface of the excavation around the circumference prior to concreting. The arrestors allowed bleed water from the grout mix to pass but choked the passage of the cement particles.

Prestressing Operation

12. The prestressing operation commenced at the downstream end of the pressure tunnel and continued round the clock towards the upstream. Three rigs were used which leap-frogged each other up the tunnel and which were operated to produce a progressive wave of compression of the tunnel lining. The whole operation was carefully monitored by means of convergence measurements on four diameters at

arrays every 6m along the tunnel. These measurements allowed axial and bending strains to be evaluated as the work proceeded and thus to avoid local overstressing and failure of the lining. Secondary control was maintained by monitoring the grout injection pressure although it was sometimes necessary to adopt very high pressures (± 100 bars) to initiate grout travel behind the lining. The level of prestress to satisfy the final operating conditions required that the initial level of prestress was close to the crushing strength of the concrete and therefore monitoring of the operation was strictly enforced and in practice worked very satisfactorily.

13. However, in view of the very close tolerances which had to be worked to and the relatively small margin of safety against tension developing in the lining, it was decided to provide reinforcement of the concrete in the form of 25mm diameter high-yield bars circumferentially at 200mm centres. The steel was proportioned and placed to ensure even distribution of cracks in the concrete.

Operational Experience

14. To date all the evidence indicates that the linings have performed very satisfactorily. A visual inspection carried out some two years after commissioning revealed conditions almost exactly the same as at original completion. There was no evidence or cracking or distress in the lining, or that there was either inflow or outflow through the lining.

Influence of neotectonic activity on the pumped storage scheme tunnel lining behaviour and failure

J. OBRADOVIC, PhD, Enegoprojekt Consulting, Yugoslavia

Summary

This report deals with the problems of the Neotectonic activity and its influence on the safety of Civil Engineering structures.

The knowledge of the recent tectonic activity is equally important for the completion of input design data as well as construction and operation conditions.

Many, very important civil engineering structures are damaged or less safe, due to the immediate influence of neotectonic movements, or induced forces resulting from the new state of equilibrium generated by the creation of a new boundary condition in the area which is already neotectonically sensitive.

The Pumped Storage Tunnel failure which is in question was directly caused by the neotectonic sensitivity of the surrounding area.

Many proofs provided by the analyses and field evidence are elaborated in this Report.

The problem of the change of the main principal stresses arrangement, established in seismicity unstable Block II in relation to the quazi stable Block I, represents the fact of migration of the main principal stresses, due to neotectonically induced changes of state of stress.

The figures given in the Report illustrate in the best way the above mentioned statement.

Introduction

In spite of the fact that Neotectonic activity of the earth crust is evident and described by many of the authors all over the world, no standard practice or guidance for the civil engineering works, especially for the hydropower structures deals with it.

In the last ten years of our practice, some problems have been recorded relating to the Neotectonic activity and its immediate influence on the design procedure, as well as the erection works keeping safety in regard. One of these cases is the Pumped Storage Scheme with the long Tunnel which shall be treated in this Paper. In that respect the Author will try to give an explanation in a shortest possible way of the method of investigation applied and to point out the significance and consequence of this phenomenon which influences many Pumped Storage structures.

After the evidence of neotectonic movement, as well as an immediate practical consequence of tunnel lining failure, a comprehensive programme comprising the detailed neotectonic study, structural analysis of rock masses, as well a stress distribution analysis was carried out. In that respect Neotectonic Maps, statistical analysis of rock fissuring with corresponding diagrams and stress distribution and its migration in time induced by recent neotectonic movements were performed.

This Paper deals with the experience of the Tunnel operation and evidences of Tunnel lining fissuring, as well as the records of numerous anomalies which have been established in an overall surrounding of Pumped Storage area.

The Pumped Storage Scheme is located in the middle part of Yugoslav territory inside the mountain region, belonging to the interior Dinarid belts, characterized in all aspects as a part of Alpine Orogeny. The corresponding geotechnical set up and geological composition, comprising the limestone rock masses have had a decisive influence on its response to the neotectonic movements, i.e. neotectonic feature frequency mode of occurrences and their effects on the tunnel structure itself.

In spite of the fact that the positive faulting structure was proven during the investigation campaign and tunnelling excavation, no special attention had been given at that time on the possible negative effect of these phenomena on the tunnel lining stability.

In the last period of eight years of tunnel operation, the normal and routine annual inspection of the tunnel and other appurtenant structures (surge tank, inlet, outlet, etc.) was carried out. No special and serious problems or changes have been observed in the first six years of operation. But in 1987 a visible and continuous crack across the tunnel alignment was recorded. The cracking of the concrete lining occurred at the tunnel section composed of limestone rocks, where prestressed grouting had been applied in order to prevent any leakage of the water from the tunnel, as well as to restore a state of stress around the concrete lining, being destroyed by the excavation works.

A minor part of the tunnel towards the exit is built up in Flysch series consisting of alteration of marls and thin bedded limestone rocks. In this section of the tunnel only the consolidation grouting was carried out, keeping in mind that the water losses are not in question due to the fact that the elevation of the tunnel is below that of the reservoir.

That means, that even theoretical chance for the water losses from this tunnel section, does not exist at all.

The method of investigation

The main goal of further investigation was to explore the global Neotectonic framework of the wider tunnel area and to clarify and define the presence of active or potentially active fault or neotectonic zone, which could be considered as a source of induced forces enable to provoke further fissuring of concrete lining.

In the scope of the Neotectonic Study the following works were carried out:

- A detailed photogeological analysis and remote sensing
- The first trend of relief energy analysis
- Preparation of theoretical Model of Relief development in time, without the influences of tectonic movement
- Definition of the anomalies of a real relief in accordance with constructed theoretical Model.

Namely, all anomalies as well as anomalies of the first trend of relief energy represent the main proof of the youngest tectonic movements.

In order to increase the objectivity of the neotectonic exploration, both these parameters are reduced to an unique summary parameter, given as a weighted average of the sum of the first trend of relief energy and first trend of anomaly of recent relief from theoretical model itself.

For the computation of this parameter an original computer programme was used. All measurements were done on the map on scale 1:25000. The results of systematic counting and computation of weighted average of the sum of the first trend of relief energy and first trend of anomaly difference of recent relief from the theoretical model are given on the Map on scale 1:25000. The positive values of parameters show the neotectonic active areas. Neotectonic active zone can be presented by gravitational faults and transcurrent faults. Reverse fault could not be excluded either.

The negative values of the parameters response to the central portion of the blocks with the vertical differential movements; that could be the upper part of relatively uplifted blocks but also the bottom of the blocks which are relatively lowered.

The middle part of the tunnel is located just in so called 'Crazy Plane'. The implication of an existence of such an active zone might be of far-reaching meaning. It is supposed that all Tunnel defects encountered at the chainage of 5+500 km could be the result of activity of this neotectonic upheaving process. On the other hand, relative upheaving movements of the 'Crazy Plane' imply the lowering of ground water, i.e. increasing of karstification process. The block of 'Crazy Plane' is bounded from the south-west direction with the greatest neotectonic faulty structure of NW-SE strike. This faulty structure is manifested as complex graben unit steplike sloping down. In that respect the Konjska River graben represents the lowest part of this very complex structure unit. All structures within this neotectonic zone of NW-SE direction are younger ones and very active.

The boundary structure which bounds the block of 'Crazy Plane' from the NW side crosses the tunnel at the chainage of 5+222 up to 5+233 km.

As a whole the block of so called 'Crazy Plane' relatively upheaves. But with unequal velocity its western part is rotated declining westwards in this part.

The above mentioned kinematic process of 'Crazy Plane' movements, with formation of typical graben unit, as well as the rotation of the block westward has resulted into damage of concrete lining of the PS tunnel at the chainage of 5 km. It is not excluded that the further neotectonic activity could produce a new concrete fracturing at the same location or in its vicinity.

In addition, using very rich data basis obtained by data processing, a new parameter of weighted average of local anomalies (differences) of the first trend of relief energy, i.e. the first trend of anomalies of recent relief from the theoretical model was obtained. Accordingly the local anomalies show how the neotectonic activity looks in separate regions giving a much more detailed picture of the younger tectonic movements. In order to provide as good as possible presentation of the relevant neotectonic structure, the levelling of all data by the computer ('running mean values' method, with the basis of 9 numbers) was carried out. The results of such an analysis are presented by the contour line map on scale 1:25000 (Fig. 1).

As the second step of neotectonic investigation, the following works was carried out:

1. Determination of properties of the rock fissuring and jointing;

2. Comparative analysis of joint distribution in different faulty blocks;

3. Comparative analysis of joint pattern inbetween rock masses and concrete lining;

4. Back analysis of the regional neotectonic set up based on detailed statistical consideration of rock fissuring and concrete lining failure results.

In the scope of investigated area the joints and cracks as well as faulty fractures represent prevailing penetrative element of the rock structure. The faults are present occasionally, but bedding planes occur in the form of thick bedded

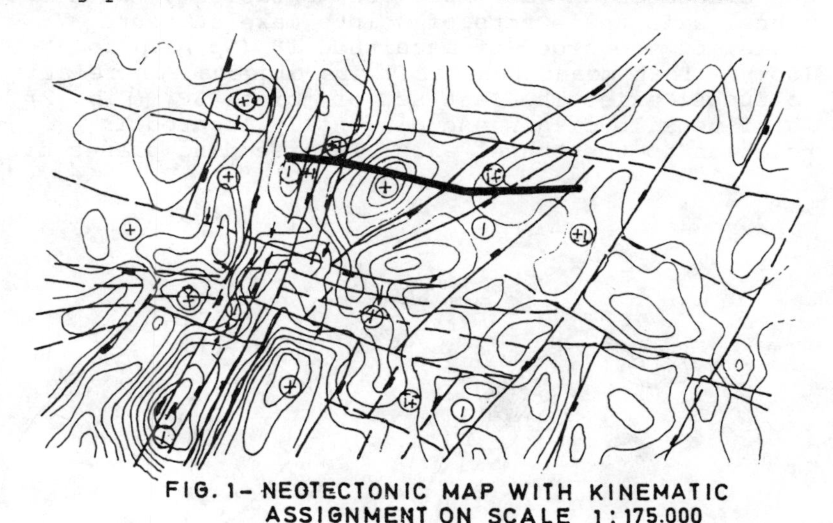

FIG. 1 - NEOTECTONIC MAP WITH KINEMATIC ASSIGNMENT ON SCALE 1 : 175.000

EXPLANATION:

CONTOUR LINES OF LOCAL ANOMALIES OF FIRST TREND OF RELIEF ENERGY AND ANOMALIES OF RECENT RELIEF FROM THEORETICAL MODEL OF RELIEF DEVELOPMENT

STATISTICAL AND SCHEMATIC POSITION NEOTECTONICALY ACTIVE FRACTURE

FAULT WITH HORIZONTAL MOVEMENT

TUNNEL AXIS

KINEMATIC ASSIGNMENT OF BLOCK MOVEMENT

DROPDOWN BLOCK

layers. Different genetic types and geometry of joint are developed.

By the statistical analysis of joint distribution the number and relationship of all sets of joints are defined. These results were used as a basis for the separation of relatively homogenous rocky blocks.

As far as the comparative analysis of two adjacent faulty blocks is concerned the following procedure was accepted:

Keeping in mind that recorded tunnel fracturing is located at the chainage of cca 5 km, a separate joint distribution analysis was done for each block, located southward and northward from the fault line.

In the rocky block located northward of fault structure, four sets of joint are distinguished (Fig. 2). At the same diagram and same statistical sample, the bedding plane is also defined, as a primary mechanical discontinuity, along which the differential movement has occurred. The sets of joints Sp_1-Sp_2 and probably Sp_3 represent the shearing planes proved by the field inspection and the evidences in 'as built documentation'. With the tunnel axis these sets of joints make an acute angle, of the order of magnitude 18 (Sp_1) up to 45 (Sp_3). That means that they are diagonal in relation to tunnel axis. The last set of joints SP_4 (Fig. 2) is of tensile origin and diagonally located in relation to the tunnel axis. Interbedding planes are

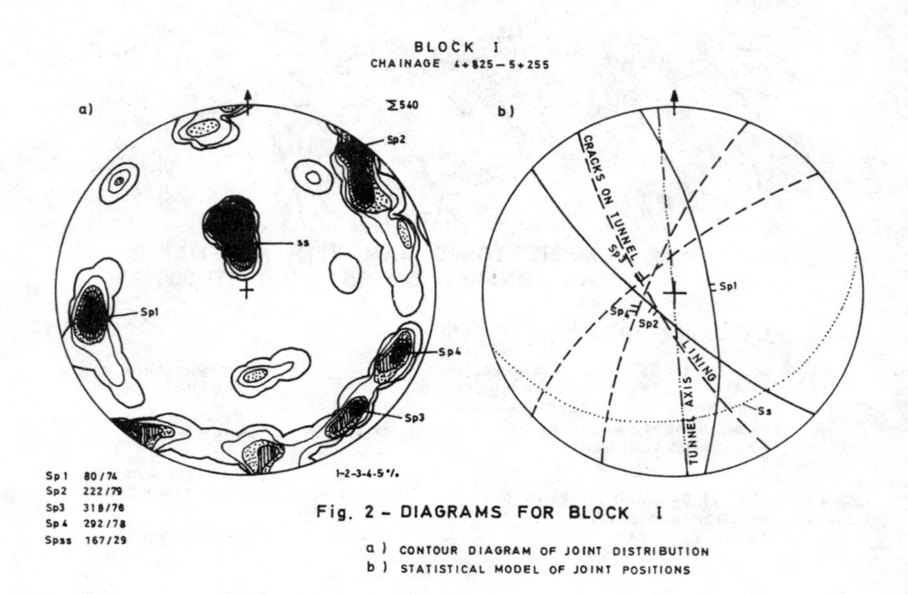

Fig. 2 – DIAGRAMS FOR BLOCK I

a) CONTOUR DIAGRAM OF JOINT DISTRIBUTION
b) STATISTICAL MODEL OF JOINT POSITIONS

spaced at 10-20 m′ and transversally located in relation to the tunnel axis.

In the second rocky block the same set of joints are present, but different much more in their spatial distribution and positions (Fig. 3).

The set of joints Sp_1 is the most relevant (Figs 3, D_4) and represented by the young and neotectonically most active rock separations. They are often with parallel arrangement. The set of joints Sp_5 and Sp_3 (Fig. 3) belongs to the shearing type of discontinuities in first block also (Figs 3, D_3). The set of joints Sp_5 (Fig. 3) occur only in this block. In relation to the tunnel axis this set of joints is perpendicular but with opposite dipping direction than the bedding planes. These are generally short distance cracks with gentle dipping angle (16) appearing probably as a consequence of block rotation and tension development in superficial sections.

Comparative analysis of rock and concrete lining fracturing

In order to get reliable data on the origin and nature of joints in concrete lining of the tunnel as well as in rock masses, an adequate structural analysis of concrete and rock fissuring was carried out.

Namely, at the chainage of 5+240-5+250 an extensive

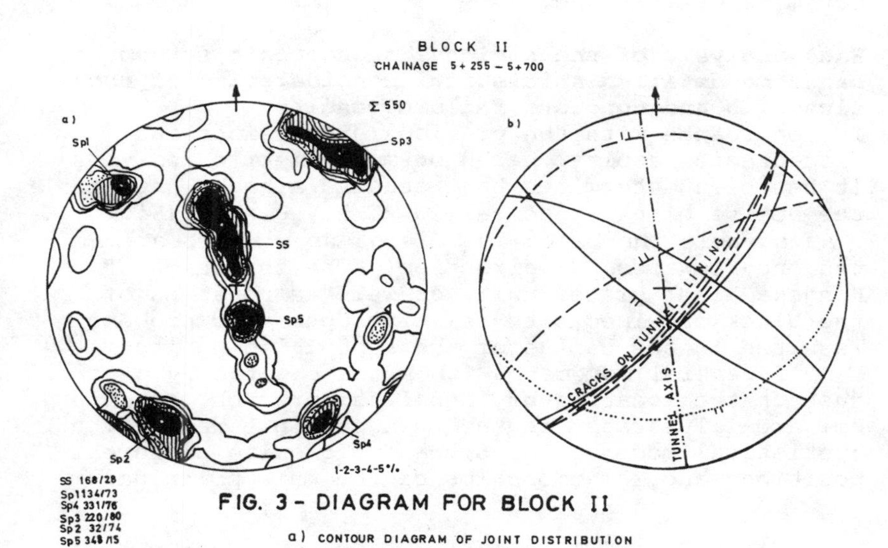

FIG. 3 - DIAGRAM FOR BLOCK II

a) CONTOUR DIAGRAM OF JOINT DISTRIBUTION
b) STATISTICAL MODEL OF JOINT POSITIONS

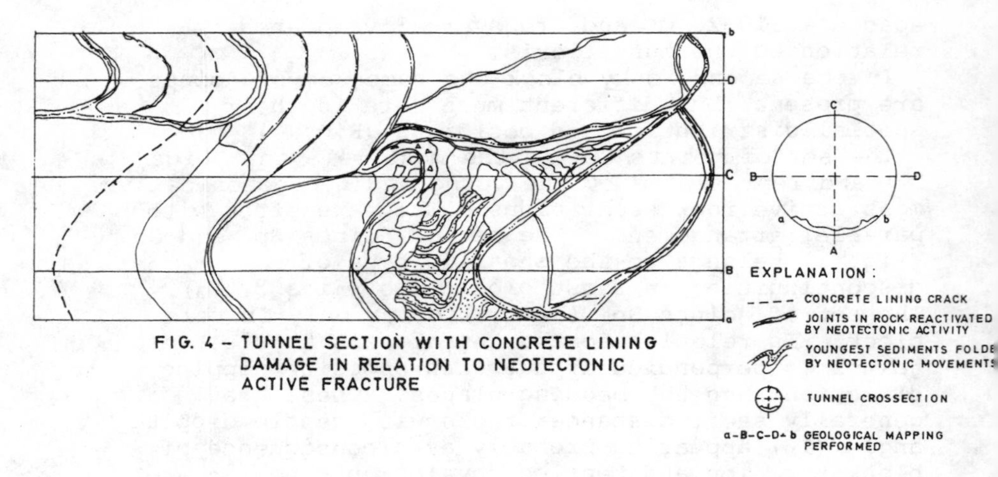

FIG. 4 - TUNNEL SECTION WITH CONCRETE LINING DAMAGE IN RELATION TO NEOTECTONIC ACTIVE FRACTURE

crack in concrete lining was observed with the dipping elements of 240/80-90. At the same chainage the prominent joint in rock masses was recorded too (Fig. 3). The same coincidence of an occurrence of jointing in rock and concrete lining was reported at the chainages of 5+321+-5+330 km. This fact certainly indicates neotectonic origin of the concrete fracturing. Accordingly, the direction and course of the striae (Fig. 4) as well as the position of reconstructed stress axis (Figs 4, D_4) on one hand and the direction and course of shearing displacement of tunnel concrete lining on the other hand, coinciding to each other prove gravitations movement of ? ranges.

Back analysis of the regional neotectonic set up based on detailed statistical consideration of rock fissuring and concrete failure results

In accordance with the previous discussions and neotectonic results, based on morphometric analysis it has been determined that studied area belongs to the active block of so called 'Crazy Plane' (Fig. 7).

As a whole the block is raising up with rotation tendency towards WSE direction. The formation of Konjska River graben unit, as well as rotation of the block combined with relative upheaving process, resulted in the damage of the tunnel lining itself.

Differential movements inbetween the faulty blocks, just on the location of tunnel failure (ch. cca 5.0 km) are illustrated in the best possible way by the statistical model (Fig. 5), showing quite different positions and relationships of the main principal stresses $_1$, $_2$ and $_3$ in block I and II. Namely, the part of the tunnel northward of the concrete failure (ch. 5+300), belongs to the relatively stable block in accordance to the southward one, that is proved by the position of the main principal stresses, especially the spatial position of the main principal stress $_1$, gently inclined towards the maximum stress action direction. That position could be related to the tectonic movement which had occurred much before the latest neotectonic activity. But, the position of the main principal stress $_1$ into statistical model relating to the south block II, where $_1$ is shifted into almost vertical position, shows that all principal stresses are transported into their new position being subjected by the latest neotectonic forces (Fig. 6).

Statistical cumulative model shows the type of possible migration of principal stresses in relation to the position of tunnel axis and cracks occurring in the concrete lining of the tunnel.

It would be of practical interest for the conclusive consideration to mention that beside all theoretical approaches provided in the neotectonic study, there exist also the field evidences which confirm the above mentioned statement. Namely, in the vicinity of tunnel failure during the tunnel excavation and on 'as build geological mapping',

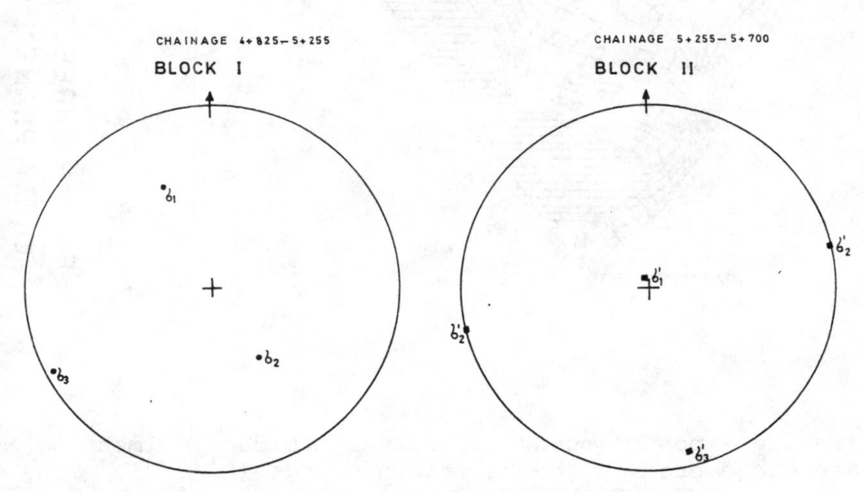

FIG. 5 – STATISTICAL MODEL SHOWING THE POSITION OF MAIN PRINCIPAL STRESSES IN BLOCK I & II

σ_1 AXIS OF MAX. STRESS
σ_2 AXIS OF MEDIUM STRESS
σ_3 AXIS OF MIN. STRESS

σ_1' AXIS OF MAX. STRESS
σ_2' AXIS OF MEDIUM STRESS
σ_3' AXIS OF MIN. STRESS

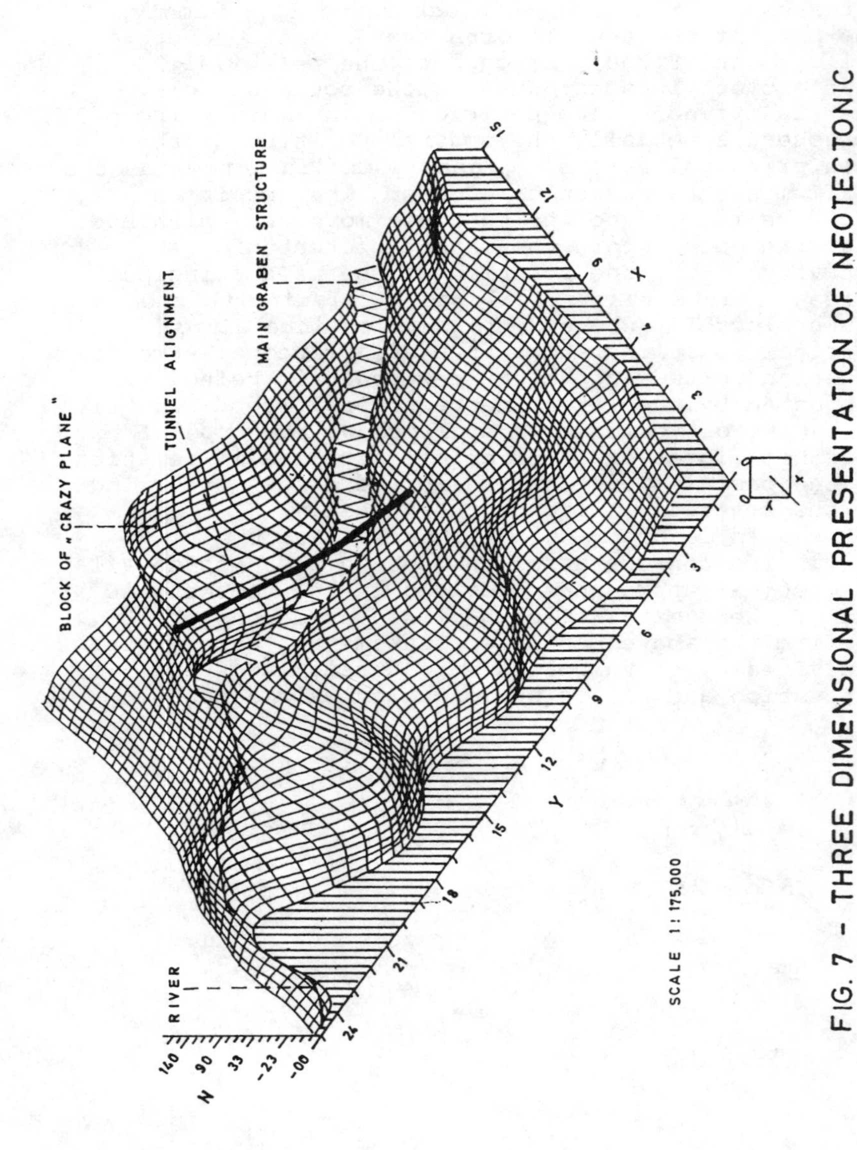

FIG. 7 - THREE DIMENSIONAL PRESENTATION OF NEOTECTONIC „CRAZY PLANE" GRABEN STRUCTURE ON TUNNEL AXIS DIRECTION

finding a postpliocene intensively folded sediments are discovered in the karstic openings (Fig. 7). It is supposed that this disconformity would have been reactivated in recent time, as well.

Conclusion

All evidences and results discussed in this paper are consistent, making a general picture of the neotectonic concept of the studied area, including the result of the geodetic monitoring works, as well as induced seismicity records which are not discussed this time.

Interaction of the field and structures defined during this investigation shows, on the best way, that it is necessary to provide neotectonic investigation on time avoiding any knowledge gap, which could make many difficulties in design, construction and operation procedure.

REFERENCES

(1) SHERARD J.L., CLUFF L.S. and ALLEN A.R. Potentially active faults in dam foundations, Geotechnics 1974, **24**, 3, 367-428.

(2) CARULLI C.B., Neotectonics and its implication on engineering geology rock mechanics Suppl. Springer Verlag 1980, **10** 3-45

(3) GUZINA B. and OBRADOVIC J., XVI Congress des grands barrages, San Francisco, 1988.

Workshop summary

F. G. JOHNSON, North of Scotland Hydro-Electric Board

F.G. JOHNSON, ***formerly Scottish Hydro-Electric***

Dr P. Back of Sir Alexander Gibb and Partners introduced his paper 'The design of high pressure linings for the Drakensburg Pumped Storage Scheme'. In particular, these include a prestressed concrete lining designed for internal pressures of up to 720 m water head, created by grout injection between the lining and the supporting rock. Steel linings have also been employed for sections where the safety factor of 1.2 could not be met in relation to overburden stress.

The audience questioned the Author on the performance of the concrete linings. Dr Back said that he had inspected the tunnel two years after it had come into service and it was in pristine condition.

Mr Graber of Eskon, RSA reported that the tunnels had been inspected after eight years operation and that they were in very good condition except for 2-3 mm of softening of the concrete over the inside of the tunnels.

Dr Back was questioned on the current evidence for a satisfactory level of prestress in the lining and whether it was being checked. Dr Back considered the margins included in the design would ensure the prestress until the end of the life of the tunnel, approximately 100 years hence. The prestress could be examined by dimensional checks on the reference points incorporated in the lining. Mr Graber said some dimensional checks had been made and found to be fully satisfactory.

Another speaker raised the alternative of prestressing the lining by prestressed wires incorporated in the lining as had been adopted in mainland Europe. Dr Back did not favour this solution on account of the vulnerability of the wires to corrosion.

Other subjects raised included relief drainage tunnels, cracking due to thermal effects, lengths of tunnelbays for concreting and the actual grouting sequence for prestressing.

Professor Cao commented that prestressing by grouting was employed in China for tunnel linings.

Mr Johnson questioned Dr Back on whether the strength of the concrete of the lining was the only criterion. He explained that in Hydro Electric's tunnels, erosion, abrasion and chemical attack were also important parameters and had to be accommodated. Hydro-Electric were interested in enhancing the integrity of concrete linings by impregnation of the surface layer by low viscosity, water curing epoxy resins. On the Rannoch Tunnel, constructed over 50 years ago, softening and disintegration of the cement fondu concrete lining had occurred.

Mr Douglas questioned the great disparity of costs of steel linings in the UK (£3200/t), South Africa (£1600/t) and China (£800/t). He suggested an alternative design of tunnel lining comprising a steel membrane sandwiched between concrete which might be an excellent compromise between a steel lining and a prestressed concrete lining.

Mr Douglas introduced the paper by Dr Obradovic. He said that it was of special interest in that, although other pumped storage power tunnels have been subjected to seismic or tectonic forces, this was, to his knowledge, the first recorded failure of such a tunnel. He questioned the term 'Crazy Plane', which is a local morphological name for a karstified plane with many cavities filled with clay or even open. He also questioned the reason for the 'visible and continuous crack' observed in 1987, which has been found to be a delayed response to a long-term movement

Mechanical seals for pump turbine duties

R.G. ALBERY, BTech, Bestobell Seals, Slough, UK

SYNOPSIS. Some of the special problems associated with sealing pump turbine main shafts, such as high pressures and changes of mode, are briefly described, including the short-comings of conventional sealing approaches. Concise details of the history and design of a unique family of split mechanical seals are given underlining their record of success and particular suitability for these applications.

MECHANICAL SEALS FOR PUMP TURBINE DUTIES

1. The traditional methods of sealing a water turbine main shaft all have severe limitations which may make them totally inappropriate for pump turbine applications. A problem for the seal is the bi-directional rotation of the shaft, which may be attended by a change in other operating conditions such as pressure and vibration. As the tail water head is normally higher than for conventional turbines, higher pressures have to be sealed. Also the machines are often used pressurized with air in the "blow-down" condition, sometimes for hours on end. Such factors often result in troublesome pressure surges.

2. The leakage of air through labyrinth or close tolerance bush devices in the "blow-down" condition necessitates economically unacceptable demands on the compressors, whilst soft packings require a clean water flush and frequent adjustment to prevent high leakage or severe wear on the shaft sleeve. The pressure capability of soft packings is necessarily somewhat limited by these shortcomings.

3. Multilayer carbon segment seals require great care in handling and assembly as well as being susceptible to vibration and abrasive damage. Leakage of air under "blow-down" conditions again can be very high. Specialist carbon faced hydrostatic seals used by some turbine manufacturers require a lot of clean injected water. They are as delicate as other carbon seals and have a very limited

life due to the restricted amount of wear permissible within the seal design.

4. It might at first sight seem desireable to scale up the ubiquitous mechanical seal found in many shaft sealing applications but the manifest need for clean conditions plus a very high cost of engineering tend to rule out such an approach. The prime difficulty in producing a large seal of this type is lapping the radial faces to the necessary flatness. If components are to be split, which is invariably necessary for assembly and maintenance, then the problems of manufacturing are multiplied.

5. The mechanical seal may be seen as a very attractive proposition however, if the right sealing face materials can be found. Ideally these would allow for split faces which are tolerant of abrasive particles in the water but will "bed-in" themselves during the first few hours of operation and preferably "bed-in" again after a period of abuse or overloaded running.

6. A large family of mechanical seals with the above characteristics have been developed by Bestobell Seals of Slough for use on water turbine main shafts. Over a period of two decades two basic designs; mechanically loaded conical face and hydraulically loaded radial face seals, have been designed and developed to suit the majority of water turbine applications.

7. These seals were initially conceived because of difficulties encountered on pump storage sites. Following laboratory testing of scaled down models the first full size seal was installed in the North of Scotland Hydro-Electric Board's station at Cruachan in the Summer of 1969 (ref. 1). Within two years successful operation of this prototype, all four machines were fitted with similar seals.

8. Since then more than four hundred seals have been supplied both as original equipment and as retrofit improvements to approximately two hundred water turbine sites worldwide. It would seem the users believe this design of seal has outperformed the expected capabilities of the alternative types. The versatility of the design in coping with a great number of speeds and pressures is evidenced by the range of installations from mini-turbines to some of the largest machines made. Whilst seals have been supplied for sealing diameters up to 1.8m this does not represent the upper limit of practical sizes.

9. The basic design of the seal was a development of the Company's proven C.W. (cooling water) design which in its turn had been a development of a smaller seal (the T-Y mechanical seal). This had originated in the early 1950's as a seal

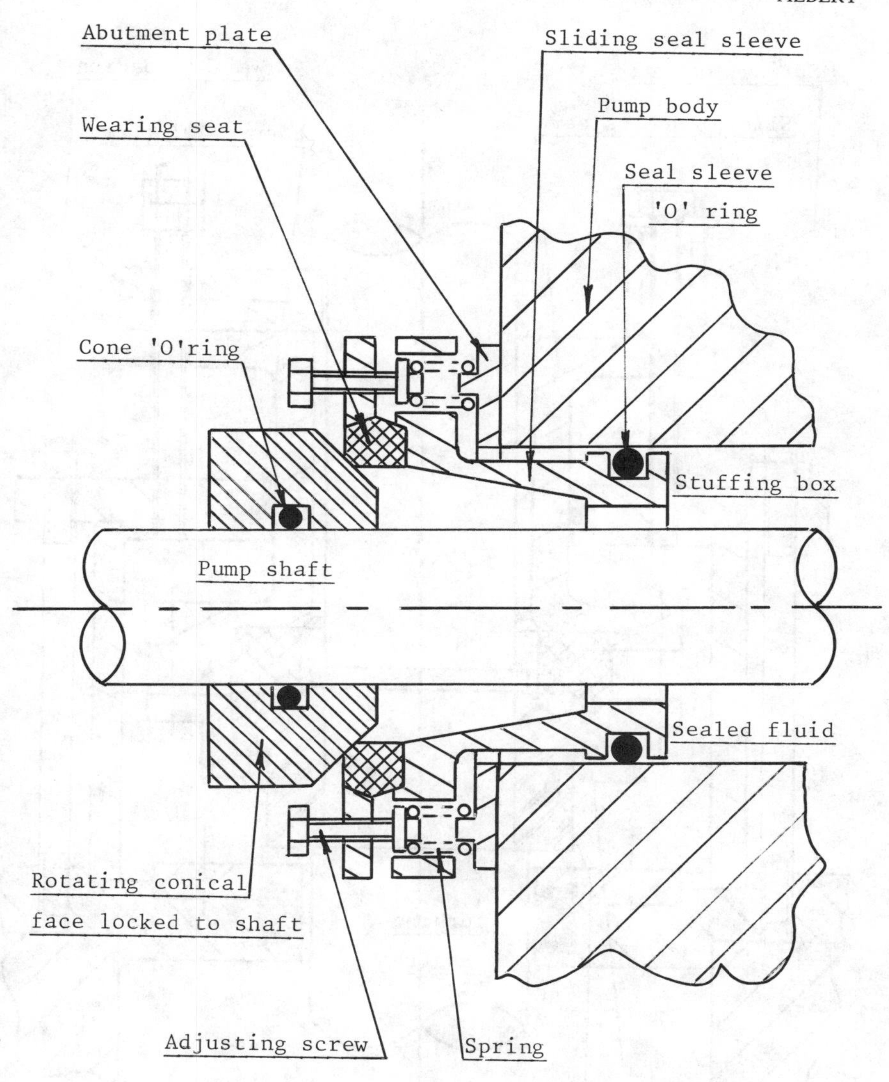

Fig. 1. Basic design layout of smaller conical type Bestobell mechanical seals.

for process pumps operating in the very dirty or abrasive conditions found in the sugar and sewage industries. The outline design of these smaller seals is shown in Fig.1. The C.W. seal enjoyed immediate popularity especially on large condenser cooling water pumps. Increasingly these tended to be situated in remote locations, typically estuaries, where water quality was not good. The tolerance of the basic design to this, with only a small penalty of reduced life between overhauls, combined with simple maintenance ensured

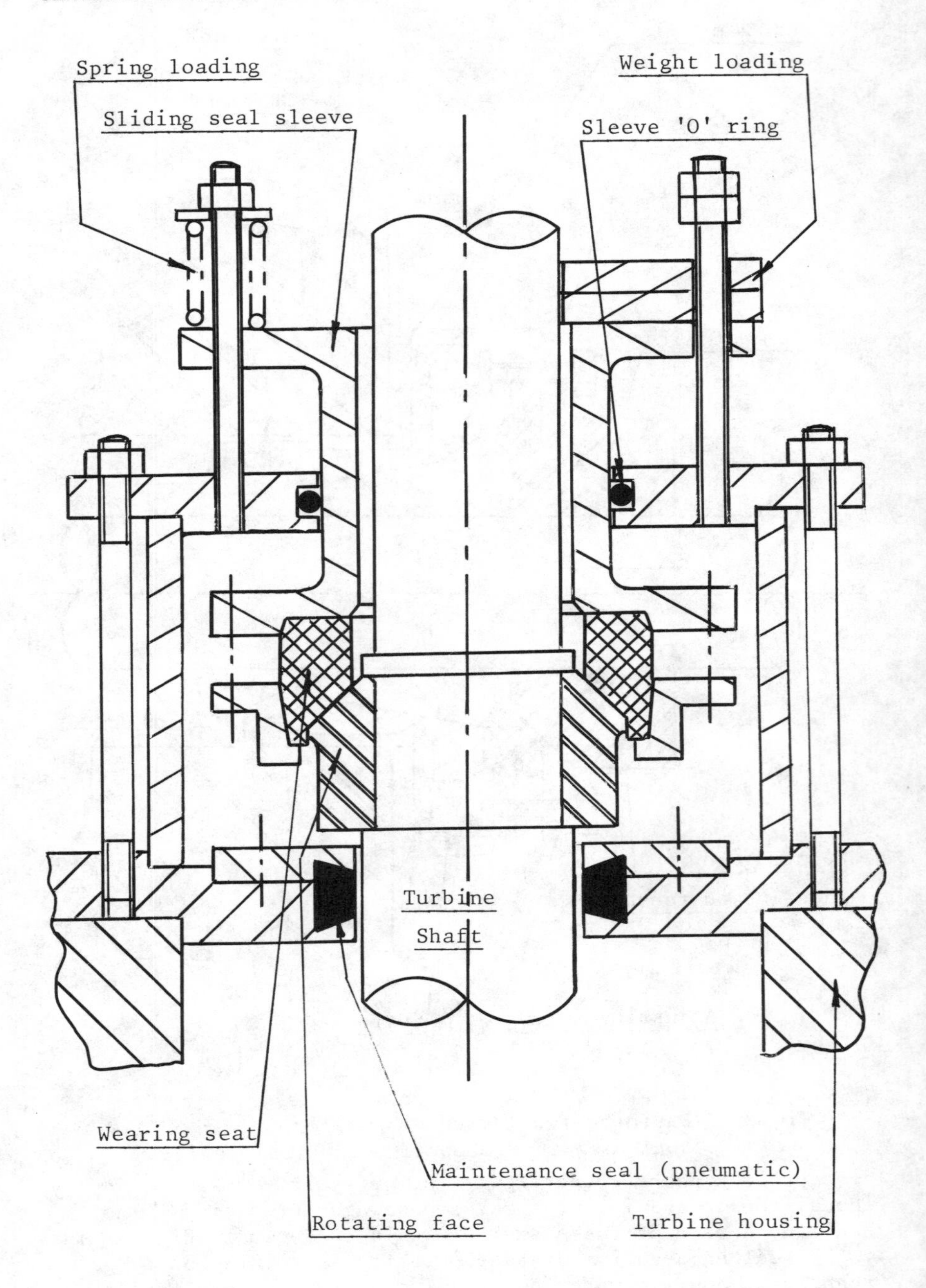

Fig. 2. Conical face water turbine seal

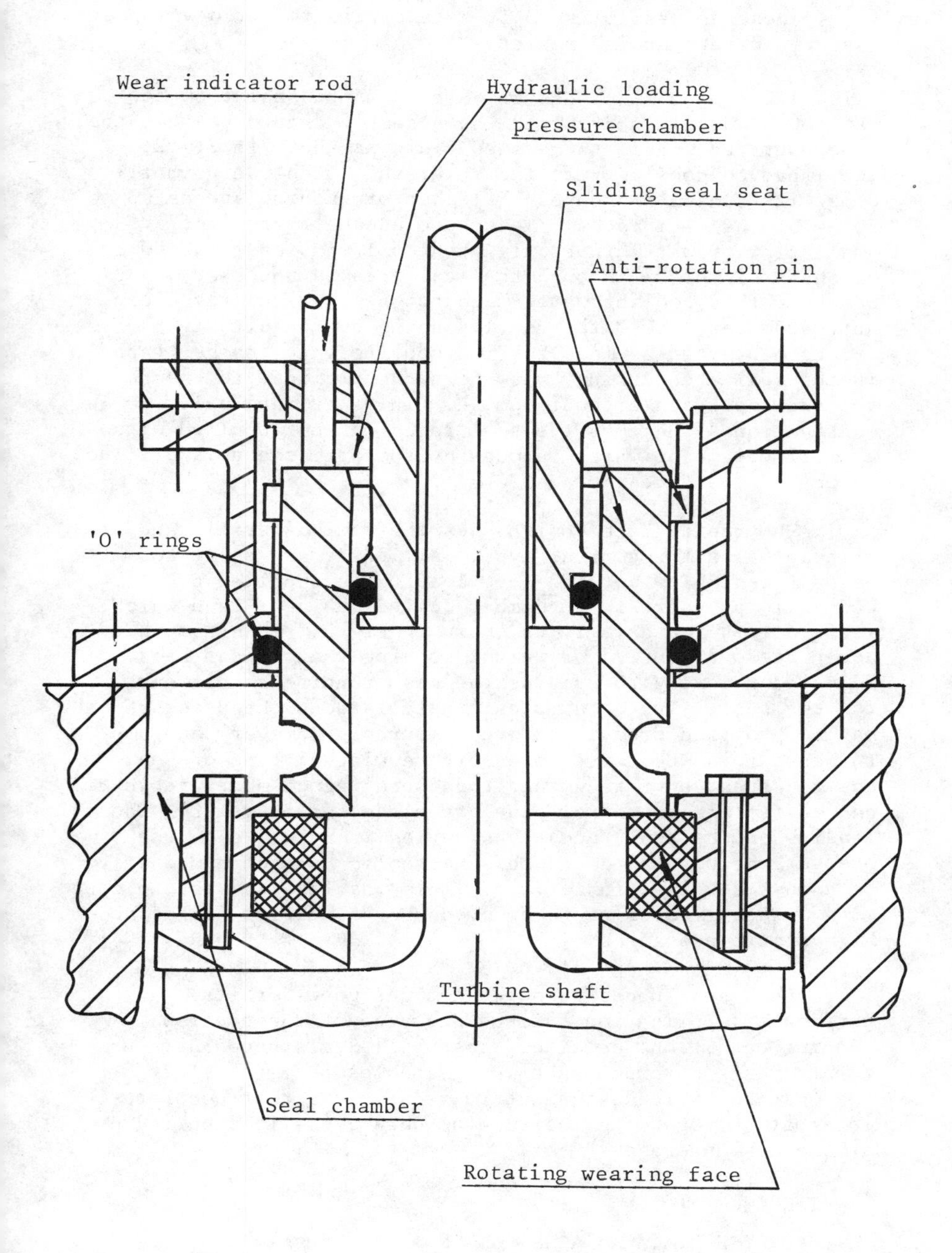

Fig. 3. W.P.R. water turbine seal

its success and established it as a candidate for development into the water turbine market.

10. The conical face design water turbine seal took the basic C.W. design and turned the assembly around so that the cone operated within the sealed fluid as shown in Fig.2. The improved cooling resulted in a seal which can normally cope with sealed pressures of 10 bar or rubbing speeds up to 30m/s at the seal faces. To ensure adequate cooling, a small flow of water (normally about 4 l/min for each 100mm of shaft diameter) through the seal area is necessary. However, this cooling water need not be particularly clean, untreated penstock water usually being quite suitable. Arrangements can be made for the cooling water to be trapped in the seal area during "blow-down" conditions, to ensure the seal operates without damage. In other marginal lubrication situations the seal faces can be supplied with a small volume of water through hydrostatic channels drilled in one of the faces.

11. The radial face design uses the same materials and principles as the conical type. All parts are split for ease of assembly, but the face loading is supplied by hydraulic pressure in a chamber behind one seal face which is free to slide axially as shown in Fig.3. This has the advantage over spring or weight loading that it can be adjusted remotely to provide the best running conditions for the seal. In the simplest installations the pressure may be provided from an external source. However, on pump turbines this WPR (water pressure regulated) type of seal has the great advantage that the operating chamber pressure can be fed via a control panel from the sealed water. The loading on the seal faces thus varies in relation to the pressure being sealed and this hydraulic balance can easily be adjusted whilst in service. This enables the seal to cope with a much greater range of pressure as well as surges and high static pressures.

12. Each seal needs to be custom designed for its particular application taking into account the precise running conditions on site. Bestobell Seals have built up twenty years of experience which enables them to select the most suitable design but it is critical that accurate information regarding the running parameters is supplied at the design stage.

REFERENCE

1. CRAVEN S.A. and ARMSTRONG N.A. The development of a mechanical seal for the main shafts of pump-turbines. Water Power, August 1973.

The effect on Dinorwig power station plant of providing the majority of the British power systems reserves of frequency control with the resulting plant development necessary to maintain that capability

W.S. WILLIAMS, National Grid Company plc

I PLANT DESCRIPTION

1. Hydraulic

Upper reservoirs with 634 meter a.o.d. maximum level and 600 meter a.o.d. minimum level and Lower reservoir 106.25m a.o.d. maximum level and 92m a.o.d. minimum level, each with 7,000,000 cubic metre capacity, and thus providing a differential head of 542 metres maximum and 494 metres minimum.

There is one concrete lined tunnel/shaft of mean diameter of 10 metres to the high pressure penstock manifold. The vertical shaft is 555m high including the 60 metre high by 30 metre diameter Surge shaft and 2500 square metre by 14 metre high Surge Pond. The H.P. tunnel from shaft to Penstock manifold is 9.5 metres in diameter, 700 metres long with 10 per cent gradient.
The penstock manifold barificated into six penstocks of concrete 3.8 metre diameter for the first 80 metres that interfaces into 110 metres of 3.5 to 2.5 metre diameter steel lined penstock in pressure grouted concrete that terminate with an exposed steel penstock at the Main Inlet Valve (MIV) in the Main Inlet Valve Gallery (MIVG).

The 2.5 m diameter intermediate penstock connects the MIVs to the spiral of the pump-turbine in the Machine Hall, with the draft tube leading to the Draft Tube Valve (DTV) in the Draft Tube Valve Gallery. The 3.75 metre Draft Tube Valve is an offset centreline butterfly valve with bypass valves, used for maintenance and dewatering of turbine and tailrace and capable of closing in an emergency and of opening to relieve high pressure in an emergency.

Pumped storage. Thomas Telford, London, 1990.

There are three 8.5m diameter concrete lined tailraces, each connected to two draft tube valves and the lower lake via tailgates and screens.

2. Main Inlet Valve

The six M.I.Vs are spherical rotary valves, each of nominal 2.5 metre diameter, opening in 5 seconds by two hydraulic oil cylinders and closing in 20 seconds by two closing weights of 16 tonne each. The clearance between the moving valve rotor and the valve body is closed whenever the valve is closed by a steel ring that slides axially in the body under hydraulic control and is referred to as the service seal and located downstream. There is a similar maintenance seal upstream that is only used for providing double isolation for maintenance. The rotor trunnion and valve body attached to the penstock taking an axial thrust of approx 2,800 tons when the valve is closed and the service seal is applied, and may close under full flow discharge in an emergency.

3. Pump Turbines

The six vertical pump turbines are Francis, reversible, fixed speed 500 rpm, with 3.76 metre diameter stainless steel runner coupled through an intermediate shaft to the generator-motor. The turbine has 24 guide vanes and an overall diameter of 7 metres, with an efficiency of 92.5% turbining and 91.7% pumping and demanding 290 MW pumping and providing 317 MW output-turbining at maximum differential head. Although, during system emergencies, they have provided 330 MW for short periods.

A station compressed air system provides air to blowdown the water in the turbine when the MIV is closed thus allowing the runner to rotate in air when machine is synchronised and rotating in either direction, thus allowing fast transition into pumping or generation.

4. Generator-Motors

The six reversible, vertical shaft, salient pole, air cooled generator-motors provide 313.5 MW, 330 MVA when generating and demand 292 MW, 312 MVA when pumping.

The rotor has 12 laminated poles, thus 500 rpm synchronous speed and can withstand transient overspeed of 763 rpm and 650 rpm steady state overspeed following load rejection. The total unit rotating weight of 510 tonne is supported at the top of the generator-motor via a thrust bearing with 10 water cooled thrust pads each on 59 springs supported by the top bracket of the generator motor. The bearing being designed originally as a self-lubricating type with jacking oil for starting and stopping.

The 22 metre length of the unit shaft has three guide bearings, one at the top of the turbine, another just below the rotor and the other above the rotor but below the thrust bearing, each being self lubricating with tilting pads. The shaft braking is by electrical dynamic braking although an emergency mechanical braking system is available.

The excitation system is supplied by static rectifier, controlled by fast acting automatic voltage regulators.

The machine is started and synchronised in the pump direction by static variable frequency equipment - two sets available, with the alternative of back to back starting.

5. High Voltage Electrical System

Each 18kV generator-motor is connected to its association 18/420 kV 330 MVA transformers and auxiliary transformers via phase segregated busbars and switchgear. The main generator-motor circuit breaker is an air blast circuit breaker and the circuit breaker connecting the three phase short circuit bar to the generator terminals for dynamic braking is an air operated circuit breaker.

The isolators for reversing the rotation of the machine, for connecting the starting equipment, for back to back operation are all motorised and electrically interlocked to allow safe operation during sequence control of any transition from one mode to another.

The six generator-motor transformers are connected via 400kV load braking isolators in pairs to a single 400 kV busbar divided into three sections (one section for each pair) by two bus section circuit breakers, with two feeder circuit breakers to connect the two outer sections to two 400 kV oil filled cable circuits to the National Grid.

The 400Kv substation is located in the Transformer Hall above the transformers and equipped with 420 kV metal clad SF6 gas filled switchgear.

6. Control System

The transition of each unit from one mode to another (including shutdown mode) is controlled by automatic sequence control equipment based on a micro-processor which is programmed to monitor and control the state of plant and associated auxiliaries, and to issue step by step commands as appropriate during progression of sequence. All mode changes are fully automatic requiring only an initiation by the operator, a number of mode changes can also be initiated automatically by the operation of frequency sensitive relays.

Alarm and data logging systems continually monitor state of plant, with a separate relay system for plant protection.

7. Modes of Operation

1. Standstill
2. Spin Pump
3. Pump
4. Spin Generation
5. Generation
6. Emergency Generation from Pump

The first five modes of operation are automatically available from any other mode with no manual involvement in the transition. The emergency generation from pump is also fully automatic with the water column braking the machine and accelerating it in the opposite direction to be resynchronised for generation.

When any unit is at standstill, pumping, spin generation or generation mode, any power system low frequency level can be selected, and then should the frequency drop to the selected level the unit will automatically provide a pre selected amount of power to the Power System in the following times e.g.:-

from Pump - 200MW in 7 seconds
285MW in 100 milleseconds
from Spin Generating - 200MW in 10 seconds
from 150MW Generating - 150MW in 4.5 seconds
from Standstill - 300MW in 90 seconds

In the generating mode the unit is operated with a governor droop characteristic of 1% i.e. as the system frequency varies by ±0.5% of 50 cycles the unit output varies by ±150 MW to reduce or increase the system frequency accordingly.

II PLANT OPERATION AND AVAILABILITY

Since the station was fully commissioned in late 1984 the plant has been continuously used by the UK Grid system, mainly for peak lopping, frequency control and emergency reserve. A typical daily station operating cycle consists of:-

Pumping 00.00 to 01.00 hrs	08.00 to 01.00 hrs
Generation/Spin Generation	01.00 to 03.00 hrs
Pumping/Spin Pump	03.00 to 07.00 hrs
Generation/Spin Generation	07.00 to 24.00 hrs

This results in onerous operating cycles of an average of 37 mode changes per day, peaking occassionally at 80. Up to December 1989 the total number of mode changes carried out by the station was 270,000.

Within the Generation/Spin periods the units can be called in by frequency relays to Generate up to 313MW each within periods of 10 to 20 seconds, the duration of the generation can vary from 1 minute to 5 hours and will differ on each set. Similar cycling occurs in the pumping/spin pump mode.

The transition of plant status instructed by these mode changes result in onerous duties on the hydraulic, mechanical, electrical and control equipment. This result in an abnormal maintenance cycle with respect to other power stations because in a 2 year period the generator main and excitation circuit breaker maintenance requirement is the same as is required every 6 years in a coal fired power station, and once in the lifetime of a nuclear power station. Similarily the maintenance of the main inlet is 2 years, which is equivalent to 15 years in a normal hydro station, and the varied stresses imposed on the stator and rotor of the generator is of the order of 10 times that of a normal hydro machine. Thus, to provide a similar plant life, the necessary maintenance and development work is proportionally greater.

The operational reliability has been maintained at the design intent of 99% while the average target availability of 93% has been 88% up to April 1989, mainly because of the installation time required to carry out development work - see Fig.1; and the average winter time availability target of 97% has been 96.5%. Since April 1989, the availability was as shown in Fig. 2 because it was decided to modify and install a new thrust bearing in all the units to bring availability back to target, this was achieved over the winter of 1989/90.

III MAINTENANCE AND ASSOCIATED DEVELOPMENT

1. General

Normal maintenance is based on two yearly statutory inspections and the guaranteed safe number of operations of the main circuit breaker in between major overhauls thus resulting initially in three outages of 5 weeks duration per annum. All other major maintenance being carried out in the same periods. These periods have now been reduced to 3.5 weeks through purchasing of a rolling spare circuit breaker so that the overhaul is carried out in between outages, and this principal has been applied to other plants to reduce the peaking of manpower during overhauls.

Single unit overhaul including turbine overhaul is carried out with the high pressure hydraulic isolation given by application of the service and maintenance seal of the associated MIV, thus avoiding a station outage and the resulting loss of revenue.

2. Hydraulic System

Because of the cost penalty, through loss of revenue, dewatering of the high pressure tunnels and shaft is avoided by the adoption of a remote operated submersible vehicle (R.O.V.) for inspection. The first inspection was carried out in 1986 with no distress found, while some axial fine cracks were noted and some concrete pour interface joint dressing had been removed.

From first inpounding in 1981 the overall leakage of water into the power station of 350 litres per minute had reduced to 50 by the end of 1988 due to siltation of the leakage paths. Continuous monitoring of the leakage identified a gradual increase in station leakage during 1989, up to 120 litres per minute. The 70 litre increase identified to be coming from the concrete adjacent to the No. 1 steel penstock and proved to be from the HP system. After development of procedures and equipment to search for the source within the H.P. system, using an R.O.V., the station was shut down for two successive week-ends at the end of March 1990 for tunnel inspections and to locate the source of the leak. The source was located and filmed so that grouting procedures can now be developed and the inspection found no other abnormalities. The source of the leak is via a concrete joint, 120 metres from the MIV Gallery, through a grouted construction drainage plastic pipe left embedded in the mass concrete surrounding the steel penstock - a lesson to all. The joint crack seems of the order of 4mm, with spoiling of joint edge up to 50mm, for a length of approximately 100mm in one location but the detailed analysis is not complete at present.

3. <u>High Pressure Steel Penstock structural integrity</u>

This consists of the High Pressure penstock, intermediate penstock and spiral casing. During manufacture and installation all welds were fingerprinted using radiographic, magnetic particle and ultrasonic examination. Then a pessimistic fracture mechanics assessment of known defects and of those that may be there but were below detection level was carried out based on intended operational cycles. This determined the inspection frequency in service. After in service inspections during outages, based on the actual operational cycles, it is now considered safe to extend all frequencies (years), e.g.

H.P. (Section A) penstock	from	20	to	135 years
Intermediate penstock	from	6	to	57 years
Spiral	from	2	to	19 years

Should the operating cycle increase then the revised frequency (years) would have to be reduced. However, to date the selection of steel adopted in the design and the rigorous quality control of manufacturing and welding, both at works and on site, has proved to be a success.

4. Main Inlet Valve

The number of operations of the six valves to December 1989 was 252,000 with one having operated 49,000 times. To date there has only been one mal-operation where the service seal was damaged by the rotor; within 5 days a redesign of the hydraulic valve that caused the mal-operation had eliminated the possibility of a similar incident.

During outages a modification to the trunnion sleeve retention arrangement has been installed, some trunnion seals have had to be replaced while the 'D' ring seals of the service seals have a mean life of approximately 3 years, i.e. 21,000 operations.

At present design studies are being carried out with a view to developing a system for replacing the trunnion bearings without having to dewater the penstock when their wear goes outside tolerances.

5. Pump Turbine

Due to the frequency correction activity of the turbine, the guide vane sleeve wear results in the top sleeves having to be replaced every 3 to 4 years during outages. The turbine runners have small cavitation damage that are welded and dressed at each outage. The spiral stay vanes had substantial cavitation damage while pumping that none of the propriety resin based fillers could protect for more than a few days. Site tests were then initiated with a company, to develop a resin and a graded ceramic filler, this has resolved the problem resulting in only a small amount of maintenance of the filler on one or two vanes after two years in operation.

6. Thrust Bearing See Fig.3

The thrust bearing supports the 510 tonne of rotating parts and during operating mode transitions, i.e. approximately 10,000 per annum per unit, the dynamic load oscilates between 100 and 850 tonne and the bearing was at its deisgn limit, resulting in frequent fatigue failures of the thrust pad white metal. Mean failure rate being twice per annum up to the end of 1988 and deteriorating rapidly.

Reasons for the failures were

(a) the collar is dowelled to the polished runner disc while the disc is insulated from the collar.

(b) the bearing has a high specific loading and operated at high temperatures - 96^{0}C -just below the white metal of the pads when operating hydrodynamically.

(c) the oil was severely airated in the bath and in the pad to collar oil wedge.

(d) for these conditions the runner disc was too thin and overall there was insufficient cooling.

(e) there is no method available to clean oil in service.

Because of the heat generated in the thin oil wedge, heat could not transfer from the disc to the collar via the insulation, a differential expansion occurred between the collar and disc resulting in stresses in both but accentuated in the disc and causing a permanent deformation in the form of a circumferential ripple of up to 0.008 inch peak to peak inbetween each of the 12 dowels. This caused 6000 pressure pulses per minute on the white metal surface of the pad and thus fatigued the white metal at these high temperatures.

The resolution to the problem was:

(a) provide a disc three times the thickness also increase the bulk of the collar.

(b) remove the differential stresses caused by the dowels by using radial keys between collar and disc and removing the dowels.

(c) change the oil system to provide a continuous oil feed to centre of pad but ensuring the oil is de-airated to avoid pump damage and to improve oil content of the wedge.

(d) provide a continuous oil filtration/cooling loop for the oil bath.

A prototype bearing was installed in 1988 under a joint development agreement with the manufacturers and has been running since then with maximum mean pad temperatures of 78^{0}C. To date this unit has only had one outage to inspect the pads and to prove that no deformation of disc face has occurred. This modification has, therefore, eliminated the cause of failures. The new bearing was installed during 1989 in five units and they have operated with no problem to date, however, further optimisation development work is still being carried out to further improve the bearing.

7. Rotor Pole to Pole Connections

Fatigue failure of the top of pole laminated copper connections described in another paper could have resulted in very long outages for repair and installation of new connections, since the original design of the generator required the rotor to be removed in order to remove and replace the rotor poles. This was not acceptable, thus the following equipment was developed so that the poles could be removed quickly, without rotor removal, in order to forward the poles to the manufacturer for attaching new laminated connections:-

1. A hydraulic operated baring gear that can rotate the rotor at speed and accurately stop it at any location.

2. Ultrasonic heating elements that could be fitted and operated to accurately and safely desolder and resolder the copper laminates in the interpole space.

3. Hydraulic tools, to unbolt and bolt the rotor pole vee clamps, operating within the air gap; plus numerous tools to ensure the integrity of bending the laminates to the correct dimensions without introducing stress raisers during assembly. The top bracket of the unit was also modified to allow a pole to be removed and repositioned through a section of it.

A revised design of the arc in the pole to pole connection was installed and thermocouples and strain gauges fitted to it and monitored in operation, the arc being similar to the existing bottom of pole connection. The operational test results, together with finite element analysis of the strain levels anticipated from connector movement calculated due to variation in centrefugal forces and rotor copper temperature, confirms with confidence that the connections will have a life of at least 20 years.

Because a minimum life of 20 years is required the pole to slipring connection was also modified to provide the same life expectancy.

Due to the development of the fast removal of poles and the benefit of being able to quickly visually inspect each stator slot by standing or sitting on a cradle in the pole space and rotating the rotor slowly, it was decided to develop a bolted pole to pole connection to avoid the risk of initiating laminate failures through soldering and desoldering the connections for regular stator inspections.

8. Stator

Due to the onerous operating cycle, relaxation of the stator slot contents is anticipated in future years, and to assist with attempting to identify the initiation of relaxation, capacitor couplers have been fitted to monitor in operation at all modes the initiation of discharge in each of the parallel paths of each phase of the six units.

During outages since 1984 each unit has been tested at full phase voltage for its total discharge level, and for individual slot discharge levels, although the levels have varied slightly over the years it is concluded that to date the stator slot conductor insulation has not suffered any measurable degradation.

Now that poles can be removed and replaced easily a programme of slot wedge relaxation inspection during outages has been initiated over the last two years to attempt to forestall the initiation of slot discharge. Because relaxation of slot content is certain to occur eventually with modern conductor insulation systems, a design study is being carried out to attempt to develop a slot wedge that enables the re-establishment of slot forces during outage inspections.

Should the monitoring systems fail to avoid the need to rewind a machine at a future date the other items of the generator that have been under review for improvements, i.e.

(1) Top ring bar layout redesign to eliminate high temperatures measured on the copper of these bars and caused by eddy currents resulting from proximity effects in adjacent non-transposed conductor. This design is now available after investigations and analysis by the CERL with the co-operation of the manufacturers and ex-CEGB staff.

(2) Monitoring of the core laminates identified ratcheting of the bottom two laminates during the two years of operation and thus a ceramic dowelling system was developed to inhibit this if necessary. Since then there has been no further movement and the equipment is in store ready to be used should this movement re-initiate in future and thus threaten the integrity of the slot conductor insulation.

IV CONCLUSION

Although the items referred to in this paper are the major plant developments that have been necessary to maintain availability when the plant is subject to such onerous duties, numerous other small developments have been necessary, e.g. installation of p.l.cs on auxiliary plant control systems, installation of power system stabilisers with turbine output compensation and installation of cavern smoke control system to ensure safety of personnel underground should a major fire occur.

The lessons learnt to date are:

1. Attempt to learn of problems encountered by others with similar plant since what occurs to their plant over a period of 30 years of normal operational activity could well occur to Dinorwig plant in the first 5 - 10 years.

2. Set up monitoring and inspection systems to anticipate failures.

3. Initiate design studies well in advance based on probability of occurrence, since a design failure will occur on all units within a period of 12 months of each other.

4. Close co-operation is essential between manufacturers and the owners in order to ensure success and longer life of the modifications.

5. Minimise outages by having rolling spares that can be refurbished and tested inbetween unit outages.

The benefit of joint assessments and redevelopment are substantial to both parties since the manufacturers obtain information from plant that is far better than any test equipment he can establish at his works, and the owner has the benefit of all the original design data and pre-history being availabe for his modifications. However, the greatest beneficiaries are the future purchasers of similar plant from these manufacturers who will have implanted Dinorwig's experience in their design.

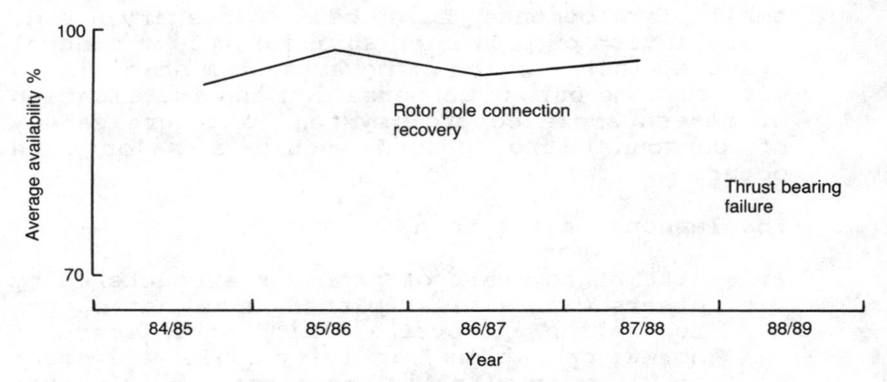

Fig. 1 Dinorwig power station average station availability (winter)

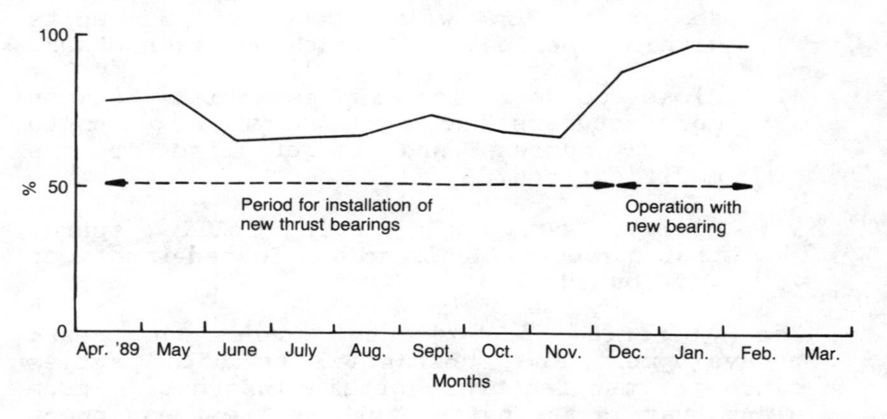

Fig. 2 Dinorwig power station average capability available 1989/1990

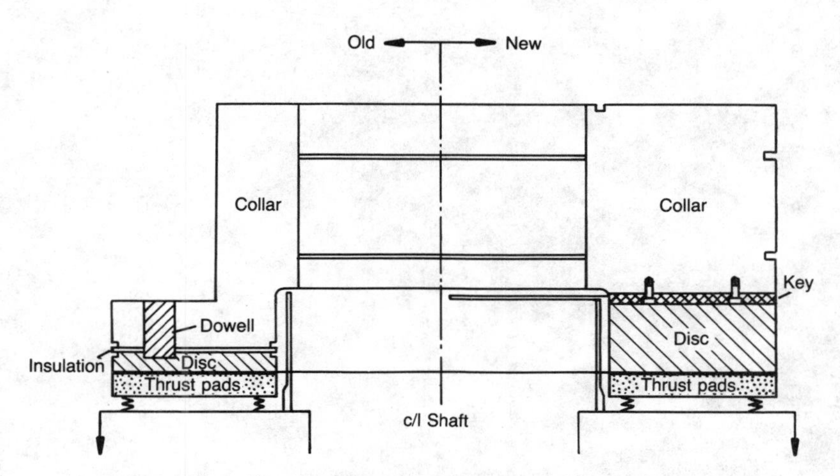

Fig. 3 Thrust bearing modification

Workshop summary

A. SIDEBOTHAM, North of Scotland Hydro-Electric Board

A. SIDEBOTHAM, *Scottish Hydro-Electric*
Two papers were presented, one on seals by R. Albery of Bestobell Seals, UK and one on the Dinorwig experience by W. Williams.

There was discussion of the methods of loading conical seals, with springs, weights and water pressure regulated systems. Some of the problems which had been encountered, particularly those of air entrainment in the water pressure regulation system during blown down operation were considered.

Another form of shaft seal was shown by a Norwegian delegate. This consisted of a laminated seal with a small detail acting as a pumping element which has been fitted to 200 machines in Norway. This seal requires an inflatable element to limit leakage under standstill conditions.

There was some discussion of the experience with Ni-resist iron for the seal cone rings of composition seals. This can lead to higher wear rates due to the tendency of Ni-resist iron to flake crazing. It is fortunate that the flake crazing areas tend to fill with the friction material after wear and this reduces the rate of further wear. Stainless steel rings are now used in order to avoid the problem.

A further problem had been experienced at Steenbras in Capetown with the cooling water holes in Ni-resist iron clogging with corrosion.

The problem of the wave around the Dinorwig thrust collar face and the manner in which this had been modified to increase the life of trust pad metal was also discussed.

Pumped storage: the environmentally acceptable solution
Introduction to the Paper by R. WATTS
This Paper accepts the need for pump storage as being proved. It is intended to illustrate a way in which this need may be satisfied in south-east

England by the use of the sea for the upper reservoir and a series of caverns for the lower reservoir. The resulting layout is very similar to that described in Mr Willett's paper on the Summit pumped storage project. The big difference, however, is that he has found an abandoned limestone mine at a depth of 670 m, whereas this Paper proposed the excavation and sale of limestone at a depth of 900 to 950 m, currently worth some £8 to £9 per tonne. The second purpose of the Paper is to locate the most useful mechanical aids to the construction. All but the first shaft may take advantage of the use of raise boring techniques. Two lifts would currently be required to reach the 950 m depth. Secondly the shaft muck removal can be carried out with a series of three vertical conveyors.

For the cavern construction the use of large diameter TBMs is envisaged in the region of 10 to 12 m dia. These will produce material that is readily handled by a conveyor system and will only need secondary crushing to produce a saleable product. The second attraction of a TBM excavation is the small amount of damage inflicted on the remaining rock cavern walls.

Lastly the muck removal must be arranged on a continuous and reliable system. There is no better system on offer than a conveyor. The problem in the past has been arranging for the extension of a conveyor behind a TBM. This has now been solved by overlap and magazines systems developed in the Mining Industry. Boretec have recently produced a system that is extending to a length of 3.7 km and includes a 500 m ramp at 13% down grade with production at 300 m/wk.

With a ready source of power from a UK nuclear station or from the French Nuclear station via the cross Channel link nearby, with the ready sale of Carboniferous limestone in the south-east of the country and with the use of all the most modern and well proved mechanical developments there seems to be little doubt that the proposed scheme and methods could produce a useful contribution to the power system of the UK with very little damage to the environment.

Summary of the proceedings of the conference

I. W. HANNAH, formerly National Power Division CEGB

It is obviously not practicable to attempt to paraphrase the papers presented at this conference. I am confining my task to presenting my personal impressions of the event, to the identification of a few of the common threads noted in the paper and discussions and finally to listing the principle lessons which should be recalled if there is ever the glorious opportunity to plan, design or build a pumped storage plant in the future.

Two immediate impressions made this conference rather special for me. As a civil engineer working across the full range of the power industry, whose connections with hydro-electric projects have been confined largely to Dinorwig pump storage station, I have been greatly impressed by the worldwide family atmosphere which so clearly exists in this specialist area. This shows itself not only in the friendly professional atmosphere that has pervaded this gathering but also, more importantly, in the openness and frankness of most of the contributions.

The second general impression that I shall take away is the width and complexity that pumped storage has achieved worldwide. Not too many years ago the whole idea of pumping water uphill only to allow it to fall back through turbines, incurring efficiency losses in both directions, seemed to many senior engineers nonsensical. Yet we have learned this week that the apparently flawed principle has already been adopted in some 300 schemes, representing some 22 000 GW of plant, and that plans exist for increasing that remarkable total to 50 000 GW. The spread of this large tranche of plant is not confined to the developed world and I am sure that Mr Gudge's account of his efforts to take the technology to Western Samoa rated one of the highlights of the meeting.

In considering these two general impressions, may I suggest that the openness of the hydro engineering community and the wide and rapid spread of the pumped storage principle are linked to some extent. Four possible reasons for this helpful openness may have some validity:

(a) All pump storage schemes are individually dictated by local topography and hence the chance of direct engineering plagiarism is limited.
(b) Hydro perhaps offers the engineer one of the last obvious opportunities to direct the great forces of nature for the use and convenience of man - a challenge well worth sharing.
(c) Hydro projects are almost always of a long term nature and hence inimical to the short term commercial concerns that have dominated UK government and financial thinking over recent decades.
(d) Schemes of such size and importance arise so rarely and take so long that, as Mr Sadden pointed out, very few engineers can expect to undertake more than one in a working lifetime.

In preparing this summary I have been helped on the mechanical and electrical (M&E) aspects by Mr Strongman. He, too, emphasized the broad co-operation that abounds in hydro work and attributed it to the need, in any successful project, for the freest possible interchange of information, experience and knowledge between the civil, hydraulic, power system, pump turbine and motor generator interests. Mr Strongman also highlighted the unusually close relationships that are normally forged between the client, designers, manufacturers and construction parties on hydro work, due in large measure to the remoteness of most sites. That degree of co-operation was a major feature of the conference.

Several Papers emphasize the inherent flexibility of pumped storage concepts. Modern schemes offer: energy reserve, peak lopping, outage cover for conventional plant, energy transfer capability in time, load following, frequency control and synchronous condenser. Beyond these standard duties there are additional functions such as the water supply pumping linkage at Drakensburg and Palmiet, wherein the combined effectiveness of the dual duties has been demonstrated to be significantly greater than simple addition would suggest. Mr Haws' Paper describes the adaption of pumped storage principles to the Mersey tidal barrage proposal, while a remarkable example of lateral thinking at Sellain Silz in Austria uses pumped storage in an exchange deal for firm power base load to the Tyrol.

In summary the papers clearly show the extensive flexibility of the pump storage principles. It is perhaps legitimate to question whether so valuable a system, well able to undertake so much of the tricky, difficult aspects of power production and control, is not being undervalued by utilities. More financially rewarding assessments of the pump storage system may go some way towards countering the short term thinking that dominates many commercial value judgements.

In relation to the systems that offer similar attributes to pumped storage, some comfort could be derived from the minimal expansion of the underground compressed air schemes beyond Huntdorf (290 MW) and Alabama (110 MW). However the principal challenge in Europe is likely to come from the combined cycle gas turbine, which has sprung into great prominence following the discovery of large supplies of natural gas, together with EC and governmental recognition that its value as a chemical feedstock is less than hitherto supposed. Although not able to offer equivalent rapid response characteristics to pump storage and unable to offer power storage the CCGT's low first cost, rapid construction and efficiency of operation make it a formidable rival to pumped storage.

Several papers deal with the problems encountered in commissioning and early operation of pumped storage plants. Details are better obtained directly from the papers themselves, but the general picture is encouraging. Despite the readiness of authors and contributors to reveal details of problems encountered, the great majority were apparently minor M&E matters for which solutions had been readily found. Several papers rightly refer to the potentially serious consequences of not paying due regard to such problems, but comfort and encouragement can be drawn from their solvability. Mr Sadden's paper gives a particularly valuable listing and analysis of worldwide experience in this area, conducted on a commission from the Electrical Power Research Institute (EPRI), which will be of prime use to commissioning engineering and operating staff.

Several useful papers were presented on pumped storage maintenance. Of particular interest were those giving precise detail of experience in civil inspection findings, notably on the enhanced concrete erosion damage experienced in Scottish pumped storage plants compared with typical erosion damage in conventional hydro stations. This significant additional maintenance requirement was attributed to the two-way flow and stop-start regimes imposed in pumped storage operations. The further problems inherent in repairing erosion damage in inclined tunnels compared with vertical shafts, is clearly portrayed. Proper corrosion protection of steel tunnel linings and of small bore pipework are obvious areas for initial concern during construction of any pumped storage station, as confirmed in many of the papers.

The use of Remote Operated Vehicles (ROVs), for tunnel inspection without dewatering, has moved forward technically in recent years, offering adequate video photographic coverage of surface deterioration and water-borne obstructions for engineering evaluation, together with substantial reductions in the necessary system outage times.

Operational safety is mentioned by several authors. The reports are generally favourable in dealing with a broad span of hazards, such as fire, smoke, gas, chemicals, seismic and security in underground stations. The EPRI analysis presented by Mr Sadden is a prime data source, wherein consideration of flooding in underground plants is perhaps the principal area requiring special care. Dam safety in UK stations is legally covered by separate legislative requirements, but the rapid water level changes in pumped storage reservoirs, compared with water supply technology, calls for enhanced dam inspection and surveillance, particularly during the early years of scheme operation.

As with all major engineering artefacts, the influence of environmental issues has massively increased in importance worldwide in recent years. Several authors refer to the detailed requirements imposed on their scheme to satisfy environmental demands and it is certain that this trend will become progressively more dominant. These added impositions may well have very significant cost consequences, but the underlying merit of ready compliance is often undeniable. Environmental issues arising both during construction and operation are referred to by authors. In short the pumped storage engineer in schemes proposed throughout the developed world would be ill advised to contest any reasonable demand, using his engineering skill largely to determine cost effective solutions. The acceptance of later schemes will increasingly depend on attitudes encountered by the planning professionals and environmental activists on earlier projects.

In this regard the importance of first class visitor facilities on existing stations is mentioned by some contributors, perhaps most notably by Mr Ferreira of the Northfield Mountain Station.

What are the future prospects for pumped storage? In the longer term several speakers gave optimistic prognoses. Many admirable sites have been identified in several countries, including Great Britain, and conceptual design studies for many of them are already available.

In the short and medium term, however, it is not so easy to substantiate a similarly bullish view. Perhaps all too typical of the probable future was the saga of engineering dedication related by Mr Willet, who has already devoted some 15 years to the planning stages of the Summit Project in Ohio. This scheme, if approved by the Federal Energy Regulatory Commission, would literally break new ground as the first major application of the underground pumped storage principle, although the lower reservoir would utilize an old mine rather than being specially excavated. To be able to show that such schemes are viable and economic would undoubtedly open up a vast area of potential for pumped storage, freeing the siting locations from the stranglehold of having to find two reservoir sites in close horizontal proximity but of greatly differing elevations.

The use of the sea as one of the two reservoirs has long appealed to pumped storage engineers, several schemes having proposed it as the lower lake. Mr Watts neatly inverted this tradition by his concept of using it as the upper reservoir, also introducing the underground reservoir technology for his lower one. Although Mr Watts' scheme has been seen as somewhat simplistic the lateral thinking behind it should not be ignored as it too has a potential for relieving the site location stranglehold. The underlying need to sidestep the use of mountainous terrain could hardly be better illustrated than by considering the use of pumped storage in the Low Countries as described by Mr Van Tongeren et al. It is not surprising that economics has forced the shelving of the several ingenious designs to overcome their lack of high hills, but it is probable that their problem is far more widespread than might initially be suspected, as few power utilities would wish to transmit via a major dogleg to their mountains and back if their pumped storage could be located on the direct path between supply and demand.

Mr Cowie et al. present a sound review of the current prospects for pumped storage in the UK. They too emphasize the inherent merits of the underground reservoir concept despite the almost plethora of suitable topography for conventional schemes in Scotland. There is left little doubt that the main challenge facing those involved with pumped storage is to produce economic designs of the underground reservoir principle as soon as possible!

The largest pumped storage project in China at Panjiakou, described by Professor Cao, offered yet another example of the extraordinary potential flexibility of the pumped principle. There, a single conventional hydro unit of 150 MW will shortly be joined by three pumped storage units, giving a total installed capacity of 420 MW, all operating off the same upper reservoir in order to provide a facility for multi-year water regulation.

In wet years the pumped storage units will offer a degree of extra hydro capacity, some flood control and will reduce the overall water wastage. In the drier years water economy and a more conventional pumped storage role will predominate. Again the pumped storage capacity will hopefully add very effectively to the total hydro output of the station in terms both of water control and power generation in a way better than the simple summation of their individual characteristics would suggest.

Perhaps the acid test of the value of conferences where a wealth of the world's talent in a particular technology is assembled in so free and open a forum, is to list the lessons made available for the benefit of the fortunate engineer who is invited to plan, design and construct a future pumped storage plant. My own slight regret is that such an array of expertise could not have been gathered together before the CEGB launched the Dinorwig Scheme, as hindsight and related experiences are such powerful teachers! Accepting that my own listing is unlikely fully to match that of others, I would suggest the following:

(a) Planning points.

(i) Refer to the accumulated experience of the great majority of pumped storage operators as listed in Mr Sadden's schedules prepared for the Electric Power Research Institute, EPRI. Items and areas meriting particular care at the planning stage, and later, can be readily abstracted from his accounts of earlier debatable and erroneous decisions.

(ii) Treat the operational duties and parameters given for the project with heavy scepticism if not outright disbelief. Universal experience suggests that the plant's functions will be very different from those envisaged at early planning stages and it would be folly to impose unnecessary limitations into the planning or design too early. (This injunction must however not be used as an excuse for gross indecision.) Mr Lowen's account of the Dinorwig role changes is instructive on this point.

(iii) Undertake the most thorough geotechnical investigations that can be afforded at the planning stage. Good examples abound but perhaps Mr Kohli's account of the rebuilding of Koepchenwerk pumped storage plant is worth special attention in its search for a reasonably conservative design, rather than merely protecting the designer's reputation with an oversafe, uneconomic approach.

(iv) Do not underestimate the pressures that the environmental and ecological interests will bring to bear on the scheme. A sensible budgetary appreciation of such influences at the planning stage affords the luxury of being able to debate each demand logically and professionally when it arises, which will almost certainly be inconveniently late in the design programme.

(b) Design and construction points.

(i) Ensure that all the possible basic forms of station are considered and logically eliminated down to the optimal solution. It is surprisingly easy to plump for a particular form of station by precedent, prejudice or even whim and to carry that superficial judgement beyond the point of no practicable return. Options include the machines in shafts, in surface stations, in underground stations or in underground stations with an underground lower reservoir.

(ii) Be generous with space, especially in the design of underground station areas. More contributors from the operations side complained of inadequate lay down and movement area than anything else. It is a particularly galling mistake to skimp on space since the massive developments in excavation and mining equipment since the 1950s have effectively cheapened the provision of man-made space to a unique degree.

(iii) Recognize the vital importance and consequential cheapness of thorough, detailed design prior to construction. Examples given in the confessional part of papers and contributions fully justify the inclusion of this restatement of a fundamental formula for the success of almost every engineering enterprise in any listing of lessons learned.

(iv) Study the twin papers and contributions given by the North of Scotland Hydro Board engineers, Mr Sidebotham and Mr Kennedy on the M&E aspects and the Johnson and Johnston civil contributions. In a conference remarkable for its frankness, the detail given therein is exceptional.

(v) On plant design aspects a study of the logical review of the salient parameters given by Dr Vögele of ABB should be compulsory reading for the engineers of all disciplines involved. Rarely can the underlying principles have been more succinctly restated.

(vi) On specific machine design beware of the dangers of thermal stress damage to the generators due to operational complexities and high numbers of mode changes not envisaged, or even imagined, at the design stage. (This warning, of course, relates to the general uncertainties on plant roles and functions referred to above.)

(vii) As a corollary to point (vi) rejection of water cooling of the generators on grounds of prime cost may prove to be ill advised.

(viii) In formulating the design specification for dams it is important to recognize that the extremely high rates of water flow needed for operation may result in water level and pressure changes at the dam that are well above any equivalent figures arising in water supply engineering. Hence additional conservatism in the design of the dam to cover both stability and leakage requirements should be accepted as fully justifiable.

(ix) In the design of hydraulic tunnels and shafts the adoption of inclined shafts to minimize the total length of water conduits may well prove to be a very false economy, not only due to the innate difficulties of driving and lining such shafts, but again during service where concrete erosion is probably much increased in magnitude and in the complexity of its repair.

(x) At the seemingly trivial edge of the design process our conference highlighted the dramatic consequential difficulties that can result from the thoughtless embedment of small bore pipework unnecessarily in concrete. Corrosion or blockage then present repair and access costs which are wholly disproportionate to the apparent scale of the problem. Again detailed design should recognize the potential need to inspect and/or replace every embedded item.

It would be easy to extend any such list of lessons almost indefinitely. In so doing the number of references to individual papers would become satisfyingly more comprehensive, but I fear too long, and I must there apologize to those whose work I have not alluded to specifically, and trust that their contributions to this conference will in no way be seen to have been diminished in value.

In my opinion this has been an outstandingly successful event in attracting so many of the true exponents of the relatively new power technology of pumped storage and finding them all so willing to share the wealth of their expertise and experience. I devoutly hope that by the time we hold the next such comprehensive conference on pumped storage our commercial and financial colleagues will have had the grace to recognize the folly of the short-term commercial assessment of capitally intensive projects that so effectively lines a few pockets today at the expense of so much of the world's prosperity tomorrow. Given that, I believe pumped storage has an important part to play in that world prosperity.